PURE AND APPLIED MATHEMATICS

NAYFEH—Perturbation Methods
NAYFEH and MOOK—Nonlinear Oscillations
PETRICH—Inverse Semigroups
PIER—Amenable Locally Compact Groups
PRENTER—Splines and Variational Methods
RAO—Measure Theory and Integration
RICHTMYER and MORTON—Difference Methods for Initial-Value Problems, 2nd Edition
ROCKAFELLAR—Network Flows and Monotropic Optimization
SCHUMAKER—Spline Functions: Basic Theory
SHAPIRO—Introduction to the Theory of Numbers
SIEGEL—Topics in Complex Function Theory
Volume 1—Elliptic Functions and Uniformization Theory
Volume 2—Automorphic Functions and Abelian Integrals
Volume 3—Abelian Functions and Modular Functions of Several Variables
STAKGOLD—Green's Functions and Boundary Value Problems
STOKER—Differential Geometry
STOKER—Nonlinear Vibrations in Mechanical and Electrical Systems
TURÁN—On A New Method of Analysis and Its Applications
WHITHAM—Linear and Nonlinear Waves
ZAUDERER—Partial Differential Equations of Applied Mathematics

LINEAR SPACES AND DIFFERENTIATION THEORY

LINEAR SPACES AND DIFFERENTIATION THEORY

ALFRED FRÖLICHER
University of Geneva, Switzerland

ANDREAS KRIEGL
University of Vienna, Austria

A Wiley–Interscience Publication

JOHN WILEY & SONS
Chichester · New York · Brisbane · Toronto · Singapore

Library of Congress Cataloging-in-Publication Data:

Frölicher, Alfred.
Linear spaces and differentiation theory.

(Pure and applied mathematics)
"A Wiley-Interscience publication."
Bibliography: p.
1. Manifolds (Mathematics) 2. Calculus.
3. Vector spaces. I. Kriegl, Andreas. II. Title.
III. Series: Pure and applied mathematics (John Wiley & Sons)
QA614.5.F75 1988 516.3′6 87-29659
ISBN 0 471 91786 9

British Library Cataloguing in Publication Data

Frölicher, Alfred
Linear spaces and differentiation theory.
(Wiley-Interscience series in pure and applied mathematics).
1. Vector spaces 2. Calculus, Differential
I. Title II. Kriegl, Andreas
515.7′3 QA186

ISBN 0 471 91786 9

Typeset by Macmillan India Ltd, Bangalore 25.
Printed and bound in Great Britain by Anchor Brendon Ltd, Tiptree, Essex

To

Brigitte and Maria

CONTENTS

PREFACE

The main subject of this book is differentiation in linear spaces of arbitrary dimension. Infinite-dimensional spaces appear in particular as function spaces, and for many purposes an appropriate calculus for such spaces would be useful. We give two examples here.

The first one concerns commutation of integration and differentiation. For a smooth (i.e. infinitely often differentiable) function $g\colon \mathbb{R}\times I\to\mathbb{R}$, where $I:=[0,1]$, we consider the function $f\colon \mathbb{R}\to\mathbb{R}$ defined by $f(t):=\int_0^1 g(t,s)\mathrm{d}s$. Thus $f(t)$ is obtained by first associating to t the function $g(t,\_)\colon I\to\mathbb{R}$ and then integrating this function over I. This means that f is actually a composite, $f=\mathscr{m}\circ g^{\vee}$, where $g^{\vee}\colon \mathbb{R}\to C^{\infty}(I,\mathbb{R})$ is defined by $g^{\vee}(t):=g(t,\_)$ and $\mathscr{m}\colon C^{\infty}(I,\mathbb{R})\to\mathbb{R}$ by $\mathscr{m}(h):=\int_0^1 h(s)\mathrm{d}s$. We remark that the natural topology of uniform convergence of the derivatives turns $C^{\infty}(I,\mathbb{R})$ into a non-normable (nuclear) Fréchet space. Nevertheless one should have a natural notion of smoothness such that both maps $g^{\vee}$ and $\mathscr{m}$ are smooth. And since $\mathscr{m}$ is linear, differentiation of $f=\mathscr{m}\circ g^{\vee}$ should give $f^{\cdot}=\mathscr{m}\circ(g^{\vee})^{\cdot}$. Since the derivative of $g^{\vee}$ corresponds to the partial derivative of g with respect to the first variable, i.e. $(g^{\vee})^{\cdot}(t)(s)=\partial_1 g(t,s)$, one would have an elegant proof that f is smooth and $f^{\cdot}(t)=\int_0^1\partial_1 g(t,s)\mathrm{d}s$.

As the second example we consider flows on a compact smooth manifold M. They can be considered as the 1-parameter subgroups of the group Diff(M) of smooth diffeomorphisms of M. The smoothness of a flow should be expressible by means of a differentiable structure of some kind on Diff(M), and Diff(M) should behave similarly as a classical Lie group. For a finite-dimensional Lie group G, the tangents at 0 of the 1-parameter subgroups form the domain for a natural chart at the neutral element. Since the tangent at 0 of a flow on M is just the corresponding vector field on M, one concludes that the infinite-dimensional Lie algebra of vector fields on M should be the tangent space at the identity of Diff(M). The space of vector fields which is thus the natural candidate for a modelling vector space of Diff(M) is again in a canonical way a non-normable Fréchet space.

Classical differentiation in linear spaces of arbitrary dimension uses Banach spaces. Its main deficiency is the fact, illustrated by the two given examples, that most function spaces are not Banach spaces. Owing to the need for a differentiation theory also involving non-normable linear spaces, many generalizations of Banach space calculus have been proposed during the last few decades. Compared with these, the theory presented in this book has several advantages. We first emphasize its naturalness and conceptual simplicity.

Former approaches usually involved several somehow arbitrary choices. The first was the choice of some class of linear spaces. Should one replace the norm by a topology, or a bornology, or a convergence structure, or by some other type of structure? Then one used to choose a differentiability condition in order to replace the classical condition of Frechet which involves the norms. And finally, in order to consider higher derivatives, one needed a structure on the function spaces $L(E, F)$ of linear morphisms. We shall proceed differently, defining k-fold as well as infinite differentiability classes not inductively but directly by reduction to the respective differentiability of functions $\mathbb{R}\to\mathbb{R}$. It then turns out that starting with locally convex or convex bornological or convergence vector spaces (or even others, such as smooth vector spaces) actually leads to the same category of linear spaces, provided one looks for greatest possible generality and identifies structures yielding the same differentiability.

In order to explain the fundamental idea of reducing differentiability of general maps to that of functions $\mathbb{R}\to\mathbb{R}$, we mention that one of the important theorems which will be proved gives in particular the following new aspect for Banach space calculus. For maps between Banach spaces one has:

(1) $c: \mathbb{R}\to E$ is a smooth curve if and only if for all $\ell \in E'$ (E' the usual dual of E) the composite $\ell \circ c: \mathbb{R}\to\mathbb{R}$ is smooth;
(2) $g: E\to F$ is a smooth map if and only if for each smooth curve $c: \mathbb{R}\to E$ the composite $g \circ c: \mathbb{R}\to F$ is a smooth curve of F.

These two results show that in order to characterize smooth maps between Banach spaces one only has to know the duals of E and F; one can forget the norms, the topologies and the bornologies.

One hopes of course that analogous results hold for finite order differentiability. Simple examples show that (1) and (2) both fail if 'smooth' is replaced by 'k-times differentiable' or by 'k-times continuously differentiable', (2) fails even in the finite-dimensional case. However, (1) and (2) both hold if one replaces 'smooth' by 'k-times Lipschitz differentiable', where Lipschitz differentiable means Gâteaux differentiable with locally Lipschitzian derivative (Frechet differentiability then follows).

We mention two further reasons why 'Lipschitz differentiable' behaves better than 'continuously differentiable'. Already for functions $f: \mathbb{R}\to\mathbb{R}$, differentiability and continuity are of rather different nature; however, one simple condition expresses simultaneously k-fold differentiability and Lipschitz condition: f is k-times Lipschitz differentiable if and only if its difference quotient of order $k+1$ is bounded on bounded sets. And for a Gâteaux differentiable function $g: E\to F$ between Banach spaces, the derivative $f': E\to L(E, F)$ is locally Lipschitzian if and only if the differential $\mathrm{d}f: E\times E\to F$ is; however, continuity of $\mathrm{d}f$ does not imply continuity of f'.

The given reasons justify why we shall generalize the notion 'k-times Lipschitz differentiable' rather than the notion 'k-times continuously differentiable'. The way to do this is absolutely natural. One needs vector spaces structured by a given dual space E', where E' may at first be any subspace of the algebraic dual (not necessarily separating points of E). We call such spaces dualized vector spaces and define for $0 \leq k \leq \infty$ (a function $f\colon \mathbb{R} \to \mathbb{R}$ is called ∞-times Lipschitz differentiable iff it is smooth):

(1′) $c\colon \mathbb{R} \to E$ is a $\mathscr{Lip}^k$-curve of E if and only if for all $\ell \in E'$ the composite $\ell \circ c\colon \mathbb{R} \to \mathbb{R}$ is k-times Lipschitz differentiable;

(2′) $g\colon E \to F$ is a $\mathscr{Lip}^k$-map if and only if for each $\mathscr{Lip}^k$-curve $c\colon \mathbb{R} \to E$ the composite $g \circ c\colon \mathbb{R} \to F$ is a $\mathscr{Lip}^k$-curve of F.

For a fixed vector space E, the set of all possible duals (i.e. all dualized vector space structures) can be decomposed in equivalence classes as follows: Two duals are called equivalent if the identity is a $\mathscr{Lip}^k$-diffeomorphism, or, equivalently, if they yield the same $\mathscr{Lip}^k$-curves. Luckily enough it can be proved that these classes do not depend on k. Each class contains a canonical representative, namely the maximal one, and a given dual is such a maximal representative exactly if it satisfies the saturation condition that it is not only included in, but equal to, the set of all linear $\mathscr{Lip}^k$-functions. Since replacing a given dual for E by an equivalent one (in the above sense) does not change the sets of $\mathscr{Lip}^k$-maps with range or domain E, it is reasonable and no loss of generality to consider only these representatives. This means that we will work with structures which not only determine the $\mathscr{Lip}^k$-maps, but are conversely determined by the $\mathscr{Lip}^k$-maps.

Of course, one would like $\mathscr{Lip}^k$-maps to have unique derivatives up to order k. It turns out that this holds, provided that at least smooth curves (i.e. $\mathscr{Lip}^\infty$-curves) have unique first derivatives. By a derivative of $c\colon \mathbb{R} \to E$ we understand a curve $c^\cdot\colon \mathbb{R} \to E$, such that $(\ell \circ c)^\cdot = \ell \circ c^\cdot$ for all $\ell \in E'$, or, equivalently, such that $c^\cdot(t) = \lim_{s \to 0} (c(t+s) - c(t))/s$, the limit being taken with respect to the weak topology. Therefore one adds two restrictions to the dualized vector spaces. The first one is necessary and sufficient for the uniqueness of $c^\cdot$ and is of course the condition that E' separates points of E. Existence of $c^\cdot$ is implied by various possible completeness conditions. Among these we choose the weakest one, which is not only sufficient but also necessary. By imposing only these two additional conditions, we obtain the most general spaces for our approach. They will be called convenient vector spaces.

We have so far indicated why for a differentiation theory as outlined the convenient vector spaces form the natural frame and are as general as possible. But much more important than great generality is the fact that they form a class with excellent properties, which means that it is closed under important constructions. So one has arbitrary products and direct sums, as well as a tensor product with the property that $L(E_1 \otimes E_2, F)$ not only corresponds to the space of bilinear differentiable maps $E_1 \times E_2 \to F$, but also to $L(E_1, L(E_2, F))$. Moreover, closedness with respect to the formation of function spaces is achieved in great generality. For very general domains X (including any subset of a convenient vector space or any classical manifold), the k-times Lipschitz differentiable maps $X \to E$ form a convenient vector space $\mathscr{Lip}^k(X, E)$ provided E is one. Similar results hold also for spaces of sections of vector bundles with convenient vector spaces as fibres. In the case of infinite differentiability, we get an 'exponential law' expressing the fact that a map $X \to \mathscr{Lip}^\infty(Y, E)$ is smooth provided the associated map $X \times Y \to E$ is smooth. This is important not only for the examples mentioned at the beginning, but also for far-reaching applications that will be discussed.

Convenient vector spaces were defined as certain dualized vector spaces, because this is the simplest description and allows an easy access to $\mathscr{Lip}^k$-maps. But there are several other descriptions within classes of linear spaces traditionally used in analysis. We mention the following aspects.

- On any convenient vector space E there exists a canonical locally convex topology (the finest one) such that E' becomes the topological dual. This identifies the convenient vector spaces with the separated bornological locally convex spaces that are locally complete (terminology of [Jarchow, 1981]). Since metrizable implies bornological and complete implies locally complete, all Fréchet spaces and in particular all Banach spaces are convenient.
- On any convenient vector space E there exists a canonical convex bornology (the coarsest one) such that E' becomes the bornological dual. This identifies convenient vector spaces with the separated topological complete convex bornological spaces (terminology of [Hogbe-Nlend, 1977]).
- On any convenient vector space E there exists a canonical convergence structure (called Mackey convergence) such that E' becomes the continuous dual. This identifies convenient vector spaces with certain convergence vector spaces. Completeness amounts exactly to completeness of the Mackey convergence.
- On any convenient vector space E there exists a canonical smooth structure in the sense of [Frölicher, 1982] (the coarsest one) such that E' becomes the space of smooth linear functionals. This identifies the convenient vector spaces with certain smooth vector spaces. Similarly for $\mathscr{Lip}^k$-structures.
- Convenient vector spaces can also be identified with certain compactly generated vector spaces and with certain nuclear spaces.

Though the convenient vector spaces have exactly the properties which ensure that smooth curves have a unique derivative, one gets, nevertheless, those theorems of calculus one can reasonably hope for. In particular one obtains for $0 \leq k \leq \infty$ and any $\mathscr{Lip}^{k+1}$-map $g\colon E \to F$ the following: the differential $dg\colon E \sqcap E \to F$ exists ($E \sqcap E$ denotes the product of E with E), is unique, is a $\mathscr{Lip}^k$-map and is linear in the second variable; and the derivative $g'\colon E \to L(E, F)$ defined by $g'(x)(y) := dg(x, y)$ is a $\mathscr{Lip}^k$-map. Conversely, if dg exists and is $\mathscr{Lip}^k$, then g is a $\mathscr{Lip}^{k+1}$-map. From this one deduces that $\mathscr{Lip}^k$-maps which are defined directly can also be characterized recursively: g is $\mathscr{Lip}^k$ if and only if it is k-times differentiable and its derivatives are $\mathscr{Lip}^0$. In the case of finite-dimensional vector spaces this is due to [Boman, 1967]. In case of Banach spaces one shows that $\mathscr{Lip}^0$ is equivalent to being locally Lipschitzian and then easily deduces for k-times Lipschitz differentiable maps the statement (2) mentioned earlier. Furthermore, in our setting the chain rule not only holds, but is even easily proved.

Of course we will not only consider globally defined functions, but also functions $g: U \to F$, where $U \subseteq E$ is open and both E and F are convenient vector spaces. 'Open' is meant with respect to the so-called Mackey closure topology, and this is a very weak condition since it means that $c^{-1}(U)$ is open in $\mathbb{R}$ for every smooth curve $c: \mathbb{R} \to E$.

Let us mention a possible generalization of $\mathscr{Lip}^k$-maps. J. Boman established his result not only for Lipschitz differentiable maps but also for the case where the Lipschitz condition is replaced by the more general Hölder condition. A forthcoming paper of C. A. Faure will show that the important theorems of our differentiation theory hold also for the respective differentiability classes between convenient vector spaces.

In this book only real differentiability and therefore only real vector spaces are considered. It is, however, possible to develop a corresponding complex differentiation theory in infinite dimensions, cf. [Kriegl, Nel, 1985].

Now we give a short overview of the contents of the different chapters.

Chapter 1 contains some basic concepts and material which will be used later. It starts with a section on categories generated by a set of maps. In order to explain and motivate this notion we remark that for any classical smooth manifold or any Banach space (cf. (1) and (2) mentioned above) the family of smooth curves and the family of smooth functions determine each other in the following way: a map belongs to one of these families if and only if its composites with all the maps in the other family are in $C^\infty(\mathbb{R}, \mathbb{R})$. This suggests that one considers sets X structured by a family of curves $\mathscr{C} \subseteq X^{\mathbb{R}}$ and a family of functions $\mathscr{F} \subseteq \mathbb{R}^X$ that determine each other in the same way as above and to call them smooth spaces. Replacing $C^\infty(\mathbb{R}, \mathbb{R})$ by the set $\mathscr{Lip}^k(\mathbb{R}, \mathbb{R})$ of k-times Lipschitz differentiable functions, one obtains analogously $\mathscr{Lip}^k$-spaces and in particular, for $k=0$, Lipschitz spaces; all these spaces are introduced in section 1.4. It is advantageous, to consider similar structures also in the case where not a set of maps from $\mathbb{R}$ to $\mathbb{R}$ (such as $\mathscr{Lip}^k(\mathbb{R}, \mathbb{R})$) is given, but more generally any set $\mathscr{M}$ of maps from S to R, where S and R are fixed sets. One thus obtains the general notion of $\mathscr{M}$-spaces studied in section 1.1. The case $\mathscr{M} = \ell^\infty \subseteq \mathbb{R}^{\mathbb{N}}$ is investigated in section 1.2 and it is shown that the ℓ^∞-spaces form a useful class of bornological spaces. The result that they satisfy an exponential law (cartesian closedness of the category) becomes fundamental for showing in section 1.4 that the category of smooth spaces is also cartesian closed. The relation used between ℓ^∞-maps and smooth maps is provided by means of difference quotients. These allow, as shown in section 1.3, a simple characterization of smooth, and, more generally, of k-times Lipschitz differentiable functions $\mathbb{R}^m \to \mathbb{R}$.

Chapter 2 presents the many aspects of convenient vector spaces. It starts with preliminary considerations of linear spaces with additional structures: locally convex topologies and convex vector bornologies are discussed in section 2.1, convergence structures in section 2.2, ℓ^∞- and $\mathscr{Lip}^k$-structures in section 2.3. In section 2.4 it is proved that all these categories have subcategories that can be identified with each other; its objects are the so-called preconvenient vector spaces. By imposing an additional separation and a completeness condition one arrives at the convenient vector spaces. The various aspects of these two conditions are examined in sections 2.5 and 2.6, where it is also proved that with every preconvenient vector space there can be associated a convenient one (its completion) having the usual universal property.

Chapter 3 deals mainly with those constructions of new convenient vector spaces from given ones which involve only linear and multilinear maps. The initial and final structures presented in section 3.1 constitute a useful tool for the general results. Special cases are examined in the following sections: subspaces and quotients in section 3.2, products in section 3.3, direct sums in section 3.4, inductive limits in section 3.5, and linear function spaces in section 3.6. The appropriate multilinear maps and the respective function spaces are examined in section 3.7 and the corresponding tensor product in section 3.8. That every convenient vector space embeds into its bidual is shown in section 3.9, where the duals of products and direct sums are also determined. It is important that a uniform boundedness principle holds for various function spaces.

Chapter 4 contains the main core of the theory. Calculus for convenient vector spaces is developed there. According to the basic idea of characterizing differentiability of a map by means of its composites with curves, it starts in section 4.1 with differentiable curves, and in section 4.2 it is shown that the respective curves form convenient vector spaces. Section 4.3 gives the main properties of differentiable maps, in particular the equivalence of the $\mathscr{Lip}^k$-condition with k-fold differentiability. Of course the section includes standard results such as the chain rule and the symmetry properties of higher derivatives. In sections 4.4 and 4.5 it is shown that in great generality function spaces are again convenient, and that natural maps such as the composition maps have the differentiability properties one hopes for. Taylor developments are used for direct sum decompositions of the function spaces. The natural structure of the space of functions on a manifold modelled on convenient vector spaces is compared with classically considered topologies. In order to give in section 4.7 a new proof for the fact that the group of diffeomorphisms of a compact smooth manifold is a manifold we show in section 4.6 that spaces of sections of quite general vector bundles are also convenient. In section 4.8 we study k-fold Lipschitz differentiability of implicitly determined functions. This opens the way for an implicit function theorem for $\mathscr{Lip}^k$-maps under additional restrictions, cf. [Hamilton, 1982].

In Chapter 5 we associate with any $\mathscr{Lip}^k$-space X a convenient vector space λX (called free over X) which has the universal property that for any convenient vector space E, the $\mathscr{Lip}^k$-maps $X \to E$ are in bijection with the linear $\mathscr{Lip}^k$-

maps $\lambda X \to E$. If, in particular, X is a finite-dimensional smooth manifold, then λX is shown to be the space of distributions on X with compact support. Similarly, we associate to any ℓ^∞-space X a convenient vector space $\ell^1 X$ with the property that one has isomorphisms between $\ell^\infty(X, E)$ and $L(\ell^1 X, E)$. The classical Banach spaces formed by the absolutely summable functions on a set are recovered. All these spaces $\ell^1 X$ as well as the free spaces λX over smooth spaces are actually convenient co-algebras, i.e. convenient vector spaces with a compatible co-algebra structure, as shown in section 5.2. It is shown in section 5.3 that it is impossible to get certain closedness properties simultaneously with certain additional properties of the spaces for a class of convenient vector spaces. The chapter ends with investigations of the reflexivity of convenient vector spaces.

Chapter 6 deals mainly with the Mackey closure topology. This is the appropriate topology on convenient vector spaces for studying $\mathscr{Lip}^k$-maps. We compare it with other natural topologies and examine its compatibility with the vector space operations. The chapter concludes with a section on continuity and differentiability properties of convex functions.

In Chapter 7 we analyze how the important functors involving convenient vector spaces behave with respect to initiality, finality, limits and colimits. Various examples and counter-examples are also given. In particular, it is shown that a quotient of a convenient vector space by a closed subspace may have smooth curves which do not admit even a local smooth lifting. Similarly, smooth functions on a closed subspace do not always admit smooth extensions.

Let us finally mention the essential prerequisites. Since the differentiation theory presented here is self-contained, knowledge of classical Banach space calculus is required only where comparisons with it are investigated. We use some well known theorems of functional analysis such as the Hahn–Banach theorem and the Banach–Steinhaus theorem, which can be found in most books on functional analysis, and we will explicitly refer to [Jarchow, 1981]. Definitions and elementary facts from the theory of bornologies will also be used; for those we will refer to [Hogbe-Nlend, 1977]. Finally, we should mention that we use categorical language and basic facts on adjoint functors whenever this seems adequate. Chapter 8 is conceived as an appendix summarizing these notions and results for those who are not familiar with category theory.

It is a great pleasure to express our gratitude to several colleagues and friends.

We mention in particular Peter Michor. His influence during the preparation of this book was substantial. Stimulating ideas as well as suggestions for including certain topics are due to him. We have, likewise, obtained vital advice from F. William Lawvere. For several sections we had the opportunity to benefit from his profound mathematical insight. Our thanks also go to Bernard Gisin for his collaboration concerning the generalization of Boman's theorem.

Among those who encouraged us by their permanent interest and who offered us valuable discussions we mention Ernst Binz, Max Kelly, Anders Kock, Saunders Mac Lane, Louis Nel, Dieter Pumplün and Gonzalo Reyes. Claude

Alain Faure and Klaus Wegenkittl rendered helpful service by reading parts of the manuscript.

The authors are grateful for having had several periods of joint work at the University of Geneva. We owe a great debt to Gerhard Wanner for making possible and facilitating the respective visits as well as for the permanent interest in our work.

One of the authors is very much indebted to the other (A.K.) for having typed, with great skill and assiduity, the whole manuscript on his personal computer. The program used for processing the text is originally due to Louis Nel and we are grateful to him that besides his valuable help mentioned above he provided us with the source code.

Last but not least we want to thank Peter Hilton for having introduced the manuscript into this prestigious series of which he is editor, and to thank John Wiley & Sons for their efficiency and expertise during the production of the book.

A. FRÖLICHER, *Geneva*
A. KRIEGL, *Vienna*

1 FOUNDATIONAL MATERIAL

This chapter mainly provides the setting for the various structures which will be used later. Since our approach to differentiation theory uses structures which are not traditional, we give some motivations.

The Lipschitz condition certainly plays a fundamental role in many theorems of analysis. It is usually considered for maps between normed spaces, since its classical definition uses norms. However, since a map between normed spaces turns out to be locally Lipschitzian iff its composites with the locally Lipschitzian curves of the source and the locally Lipschitzian real valued functions of the range are locally Lipschitzian functions from $\mathbb{R}$ to $\mathbb{R}$, one can generalize the notion 'locally Lipschitzian' to maps between much more general spaces, namely to sets with a so-called $\mathscr{Lip}$-structure which consists of a given family of curves and a given family of real valued functions such that these two families determine each other by the condition that their composites are locally Lipschitzian functions from $\mathbb{R}$ to $\mathbb{R}$.

Among differentiable maps those which are smooth certainly form an extremely important class. Smoothness is a classical notion for maps between Banach spaces and for maps between smooth manifolds. But again one can prove that such a map is smooth iff its composites with the smooth curves of the source and the smooth real valued functions of the range are smooth (i.e. belong to $C^\infty(\mathbb{R}, \mathbb{R})$), and therefore one obtains a natural generalization of the notion 'smooth' to maps between much more general spaces, namely to sets with a so-called smooth structure which consists of a given family of curves and a given family of real valued functions such that these two families determine each other by the condition that their composites are smooth functions from $\mathbb{R}$ to $\mathbb{R}$.

A third situation where this kind of structure will be used is different in so far as a classical notion, namely that of bornology, is not generalized but restricted. A bornology on a set is a family of so-called bounded subsets satisfying simple axioms, and a map is called bornological if the image of every bounded set is bounded. Only special bornologies are determined by their bornological sequences (or curves) and their bornological real valued functions. But it seems that (almost) all bornologies which are important for analysis are of that restricted type. And those of this type allow interesting additional results which fail for general bornological spaces, cf. section 5.2.

The mentioned structures all fit in the same scheme which is developed in section 1.1. The special case of the specified bornologies is treated in section 1.2, that of Lipschitz structures and of smooth structures in section 1.4. Particular attention is given to the question of cartesian closedness. This is roughly speaking the problem whether the respective function spaces admit a structure of the same kind in such a way that morphisms into a function space correspond exactly to morphisms on the respective product space.

Section 1.3 is more classical. There it is shown that difference quotients provide a link between bornologies on the one hand and Lipschitz- and smooth structures on the other: functions of finitely many variables have certain differentiability properties if and only if certain of their difference quotients are bornological. Some technical complications are due to the fact that we consider functions on arbitrary open subsets of $\mathbb{R}^m$. The reader can avoid these difficulties by sticking to functions which are defined on the whole $\mathbb{R}^m$ or on a product of open intervals.

1.1 Categories generated by a set of maps

The examples mentioned in the introduction to the chapter motivate the consideration of sets X structured by a set $\mathscr{C}_X$ of 'curves' and a set $\mathscr{F}_X$ of 'functions', with the property that $\mathscr{C}_X$ and $\mathscr{F}_X$ determine each other by the condition that the composites $f \circ c$ have to belong to a given set of maps such as the set $\mathscr{Lip}(\mathbb{R}, \mathbb{R})$ of locally Lipschitzian functions or the set $C^\infty(\mathbb{R}, \mathbb{R})$ of smooth functions. The third example shows that it is useful to take as source for the 'curves' not only $\mathbb{R}$, but any fixed set S; and as range for the 'functions' any fixed set R (in [Frölicher, 1980] only the case $S = R$ was considered).

1.1.1 Definition. We suppose in the following, that a set $\mathscr{M}$ of maps from S to R is given, S and R being any fixed sets. For an arbitrary set A, maps $c: S \to A$ shall be called curves in A and maps $f: A \to R$ functions on A.

(i) Any set $\mathscr{C}$ of curves in A determines a set $\Phi\mathscr{C}$ of functions on A as follows: $\Phi\mathscr{C} := \{f: A \to R; f_*(\mathscr{C}) \subseteq \mathscr{M}\}$. Similarly any set $\mathscr{F}$ of functions on A determines a set $\Gamma\mathscr{F}$ of curves in A as follows: $\Gamma\mathscr{F} := \{c: S \to A; c^*(\mathscr{F}) \subseteq \mathscr{M}\}$. Obviously one has $\mathscr{C} \subseteq \Gamma\Phi\mathscr{C}$ and $\mathscr{F} \subseteq \Phi\Gamma\mathscr{F}$, $\mathscr{C}_1 \subseteq \mathscr{C}_2 \Rightarrow \Phi\mathscr{C}_1 \supseteq \Phi\mathscr{C}_2$ and $\mathscr{F}_1 \subseteq \mathscr{F}_2 \Rightarrow \Gamma\mathscr{F}_1 \supseteq \Gamma\mathscr{F}_2$; hence $\Phi\Gamma\Phi = \Phi$ and $\Gamma\Phi\Gamma = \Gamma$.

(ii) An *$\mathscr{M}$-structure* $(\mathscr{C}, \mathscr{F})$ on a set A consists of a set $\mathscr{C}$ of curves in A and a set $\mathscr{F}$ of functions on A such that these determine each other according to $\mathscr{F} = \Phi\mathscr{C}$ and $\mathscr{C} = \Gamma\mathscr{F}$. The elements of $\mathscr{C}$ are called the *structure curves*, those of $\mathscr{F}$ the *structure functions*.

(iii) An *$\mathscr{M}$-space* is a triple $X = (A_X; \mathscr{C}_X, \mathscr{F}_X)$, where A_X is a set (called underlying set of X) and $(\mathscr{C}_X, \mathscr{F}_X)$ is an $\mathscr{M}$-structure on A_X. An *$\mathscr{M}$-map* between

$\mathcal{M}$-spaces X and Y is a map $g: A_X \to A_Y$ which satisfies the equivalent conditions:

(a) $g_*(\mathscr{C}_X) \subseteq \mathscr{C}_Y$;
(b) $g^*(\mathscr{F}_Y) \subseteq \mathscr{F}_X$;
(c) $\mathscr{F}_Y \circ g \circ \mathscr{C}_X \subseteq \mathcal{M}$.

(iv) The category $\underline{\mathcal{M}}$ generated by $\mathcal{M}$ has as objects the $\mathcal{M}$-spaces and as morphisms the $\mathcal{M}$-maps.

(v) An $\mathcal{M}$-structure $(\mathscr{C}_1, \mathscr{F}_1)$ on A is called *finer* (we also say smaller) than a second one $(\mathscr{C}_2, \mathscr{F}_2)$ if the identity id_A is an $\underline{\mathcal{M}}$-morphism from $(A; \mathscr{C}_1, \mathscr{F}_1)$ to $(A; \mathscr{C}_2, \mathscr{F}_2)$, i.e. if $\mathscr{C}_1 \subseteq \mathscr{C}_2$ or equivalently $\mathscr{F}_1 \supseteq \mathscr{F}_2$.

1.1.2 Lemma. (i) *If $\mathscr{F}_0$ is any set of functions on A, then there exists a coarsest $\mathcal{M}$-structure $(\mathscr{C}, \mathscr{F})$ on A with $\mathscr{F}_0 \subseteq \mathscr{F}$. Furthermore, for any $\mathcal{M}$-space $X = (A_X; \mathscr{C}_X, \mathscr{F}_X)$ a map $g: A_X \to A$ is an $\underline{\mathcal{M}}$-morphism from X to $(A; \mathscr{C}, \mathscr{F})$ iff $g^*(\mathscr{F}_0) \subseteq \mathscr{F}_X$.*

(ii) *If $\mathscr{C}_0$ is any set of curves on A, then there exists a finest $\mathcal{M}$-structure $(\mathscr{C}, \mathscr{F})$ on A with $\mathscr{C}_0 \subseteq \mathscr{C}$. Furthermore, for any $\mathcal{M}$-space $X = (A_X; \mathscr{C}_X, \mathscr{F}_X)$ a map g: $A \to A_X$ is an $\underline{\mathcal{M}}$-morphism from $(A; \mathscr{C}, \mathscr{F})$ to X iff $g_*(\mathscr{C}_0) \subseteq \mathscr{C}_X$.*

Proof. Obviously the structure in (i) is given by $\mathscr{C} := \Gamma \mathscr{F}_0$ and $\mathscr{F} := \Phi \mathscr{C}$. In the dual way the structure in (ii) is given by $\mathscr{F} := \Phi \mathscr{C}_0$ and $\mathscr{C} := \Gamma \mathscr{F}$. □

1.1.3 Definition. (i) In the situation of (1.1.2) we say that $\mathscr{F}_0$ (respectively $\mathscr{C}_0$) *generates the structure* $(\mathscr{C}, \mathscr{F})$. An $\mathcal{M}$-structure generated by an empty set of curves is called *discrete*.

(ii) In the case where $R = \mathbb{R}$, an $\mathcal{M}$-structure on a vector space E admitting a generating set $\mathscr{F}_0$ that contains only linear functions is called *linearly generated*.

(iii) By *initial* (resp. *final*) *$\mathcal{M}$-structures* we mean initial (resp. final) $\mathcal{M}$-structures with respect to the forgetful functor to the category $\underline{\mathrm{Set}}$ of sets, cf. (8.7.1).

1.1.4 Proposition. *The category $\underline{\mathcal{M}}$ of $\mathcal{M}$-spaces has initial and final structures. Explicitly they are described as follows: Let X_j $(j \in J)$ be any family of $\mathcal{M}$-spaces. The initial structure on A induced by a given family of maps g_j: $A \to A_{X_j}$ is generated by the set of functions $\{f \circ g_j$: $A \to R$; $j \in J$ and $f \in \mathscr{S}_j\}$ provided $\mathscr{S}_j \subseteq \mathscr{F}_j$ generates the structure of X_j for all $j \in J$. It has as structure curves the set $\{c$: $S \to A$; $g_j \circ c \in \mathscr{C}_{X_j}$ for all $j \in J\}$. The final structure on A induced by a given family of maps g_j: $A_{X_j} \to A$ is generated by the set of curves $\{g_j \circ c$: $S \to A$; $j \in J$ and $c \in \mathscr{S}_j\}$ provided $\mathscr{S}_j \subseteq \mathscr{C}_j$ generates the structure of X_j for all $j \in J$. It has as structure functions the set $\{f$: $A \to R$; $f \circ g_j \in \mathscr{F}_{X_j}$ for all $j \in J\}$.*

Proof. Easy verification. □

Remark. In general one has no explicit description of the structure curves of a final structure and the structure functions of an initial structure. So the structure curves of a final structure may fail to lift to structure curves of the given spaces and the structure functions of an initial structure may fail to extend to the given spaces. For an example see (7.1.8). In (7.3.1) and (7.3.2) an explicit description of final morphisms will be given.

1.1.5 Corollary. *The category $\underline{\mathcal{M}}$ is complete and cocomplete. The forgetful functor from $\underline{\mathcal{M}}$ to $\underline{\text{Set}}$ has a left and a right adjoint. Limits (resp. colimits) in $\underline{\mathcal{M}}$ are obtained by forming them in $\underline{\text{Set}}$ and putting the initial (resp. final) $\mathcal{M}$-structure on them.*

Proof. (8.7.3) applies; cf. also the first remark after (8.7.1).

We mention in particular products of $\mathcal{M}$-spaces: The underlying set is the cartesian product of the underlying sets, and the structure curves of the product are those whose coordinates are structure curves of the factors. The product with factors X and Y will be denoted by $X \sqcap Y$.

1.1.6 Remarks. (i) A one-point set has two $\mathcal{M}$-structures if $\mathcal{M}$ does not contain all constant maps $S \to R$; cf. [Frölicher, 1979]. For our purpose we can restrict to the case that all constant maps $S \to R$ belong to $\mathcal{M}$. Then for any $\mathcal{M}$-space $(A; \mathcal{C}, \mathcal{F})$ the constant maps $S \to A$ belong to $\mathcal{C}$ and hence the one-point set has exactly one $\mathcal{M}$-structure. We denote the $\mathcal{M}$-space so obtained by $\{*\}$ and remark that $\{*\}$ is a terminal object of $\underline{\mathcal{M}}$ and yields a representation of the forgetful functor $\underline{\mathcal{M}} \to \underline{\text{Set}}$.

(ii) The set S has a natural $\mathcal{M}$-structure, namely $(\Gamma\mathcal{M}, \mathcal{M})$; we shall from now on also denote by S the $\mathcal{M}$-space so obtained. Similarly R shall also denote the $\mathcal{M}$-space having R as underlying set and $(\mathcal{M}, \Phi\mathcal{M})$ as $\mathcal{M}$-structure. This notation may look dangerous, because the $\mathcal{M}$-spaces S and R can be different even if the sets S and R are the same. But for the general considerations one uses anyhow two symbols S and R; and in all our examples where the sets S and R are equal, the $\mathcal{M}$-spaces S and R also coincide (this in fact holds iff $\mathcal{M}$ is a monoid).

By means of the special objects described above we get the following representations for any $\mathcal{M}$-space $X = (A_X; \mathcal{C}_X, \mathcal{F}_X)$:

$$A_X \cong \underline{\mathcal{M}}(\{*\}, X); \quad \mathcal{C}_X = \underline{\mathcal{M}}(S, X); \quad \mathcal{F}_X = \underline{\mathcal{M}}(X, R).$$

In particular one has $\mathcal{M} = \underline{\mathcal{M}}(S, R)$.

Using the canonical $\mathcal{M}$-structure on R one deduces that an $\mathcal{M}$-structure $(\mathcal{C}, \mathcal{F})$ on a set X is generated by $\mathcal{F}_0 \subseteq \mathcal{F}$ iff the $\mathcal{M}$-structure $(\mathcal{C}, \mathcal{F})$ is the initial one induced by the maps $f: X \to R$ with $f \in \mathcal{F}_0$.

1.1.7 Theorem. *Suppose that $\mathcal{M}$ is a set of maps $S \to R$ containing all constant maps and such that, with $\hat{c}(s, t) := c(s)(t)$,*

$$\mathcal{C}_{\mathcal{M}} := \{c: S \to \mathcal{M};\ \hat{c}\colon S \sqcap S \to R \text{ is an } \mathcal{M}\text{-map}\}$$

is the set of curves of an $\mathcal{M}$-structure on $\mathcal{M}$ (i.e. $\Gamma\Phi\mathcal{C}_{\mathcal{M}} \subseteq \mathcal{C}_{\mathcal{M}}$). Then one obtains for *any $\mathcal{M}$-spaces Y and Z an $\mathcal{M}$-space denoted $\mathcal{M}(Y, Z)$ having $\underline{\mathcal{M}}(Y, Z)$ as underlying set and the following structure curves:*

$$\mathscr{C}_{\mathcal{M}(Y,Z)} := \{e: S \to \underline{\mathcal{M}}(Y, Z);\ \hat{e}: S \sqcap Y \to Z \text{ is an } \mathcal{M}\text{-map}\}.$$

This function space structure behaves functorially and by the obtained functor $\mathcal{M}: \underline{\mathcal{M}}^{\mathrm{op}} \times \underline{\mathcal{M}} \to \underline{\mathcal{M}}$ the category of $\mathcal{M}$-spaces becomes cartesian closed. Thus a map $g: X \to \mathcal{M}(Y, Z)$ is an $\mathcal{M}$-map if and only if $\hat{g}$: $X \sqcap Y \to Z$ is an $\mathcal{M}$-map, and all $\mathcal{M}$-maps $X \sqcap Y \to Z$ are of this form.

Remark. The use of the same symbol $\mathcal{M}$ for the given set of maps and for the functor closing $\underline{\mathcal{M}}$ simplifies notation and will cause no confusion.

Proof. We first show that one has:

(a) $\mathscr{C}_{\mathcal{M}(Y,Z)} = \{e: S \to \underline{\mathcal{M}}(Y, Z);\ f_* \circ c^* \circ e \in \mathscr{C}_{\mathcal{M}}$ for all $f \in \mathscr{F}_Z,\ c \in \mathscr{C}_Y\}$.
So let $e: S \to \underline{\mathcal{M}}(Y, Z)$ be a map. Then $\hat{e}$ is an $\underline{\mathcal{M}}$-morphism $S \sqcap Y \to Z$ iff $f \circ \hat{e} \circ (\sigma, c) \in \mathcal{M}$ for all $\sigma \in \underline{\mathcal{M}}(S, S)$, $c \in \mathscr{C}_Y$ and $f \in \mathscr{F}_Z$. Since for $c \in \mathscr{C}_Y$ also $c \circ \tau \in \mathscr{C}_Y$ for any $\tau \in \underline{\mathcal{M}}(S, S)$ it follows that $\hat{e}$ is an $\mathcal{M}$-map iff $f \circ \hat{e} \circ (\sigma, c \circ \tau) \in \mathcal{M}$ for all $\sigma, \tau \in \underline{\mathcal{M}}(S, S)$, $c \in \mathscr{C}_Y$ and $f \in \mathscr{F}_Z$. Since $f \circ \hat{e} \circ (\sigma, c \circ \tau) = \hat{g} \circ (\sigma, \tau)$, where $g = f_* \circ c^* \circ e$, our first assertion follows according to the definition of $\mathscr{C}_{\mathcal{M}}$.

An immediate consequence of (a) is:

(b) For $\Psi \in \Phi\mathscr{C}_{\mathcal{M}}$, $c \in \mathscr{C}_Y$ and $f \in \mathscr{F}_Z$ one has: $\Psi \circ f_* \circ c^* \in \Phi\mathscr{C}_{\mathcal{M}(Y,Z)}$.

We now prove that $\mathscr{C}_{\mathcal{M}(Y,Z)}$ is the set of structure curves of an $\mathcal{M}$-structure on $\underline{\mathcal{M}}(Y, Z)$, i.e. that $\Gamma\Phi\mathscr{C}_{\mathcal{M}(Y,Z)} \subseteq \mathscr{C}_{\mathcal{M}(Y,Z)}$. So let $e \in \Gamma\Phi\mathscr{C}_{\mathcal{M}(Y,Z)}$, i.e. $e: S \to \underline{\mathcal{M}}(Y, Z)$ is such that $\varphi \circ e \in \mathcal{M}$ for all $\varphi \in \Phi\mathscr{C}_{\mathcal{M}(Y,Z)}$. Then, by (b) one deduces that $\Psi \circ f_* \circ c^* \circ e \in \mathcal{M}$ for $\Psi \in \Phi\mathscr{C}_{\mathcal{M}}$, $c \in \mathscr{C}_Y$, $f \in \mathscr{F}_Z$. According to the assumption this implies $f_* \circ c^* \circ e \in \mathscr{C}_{\mathcal{M}}$ for $c \in \mathscr{C}_Y$ and $f \in \mathscr{F}_Z$. Hence $e \in \mathscr{C}_{\mathcal{M}(Y,Z)}$ by (a).

Functoriality means that for $\mathcal{M}$-maps g: $Y_2 \to Y_1$ and h: $Z_1 \to Z_2$:

(c) $g^* \circ h_*: \mathcal{M}(Y_1, Z_1) \to \mathcal{M}(Y_2, Z_2)$ is an $\mathcal{M}$-map.
This in fact follows easily using (a). So let $e \in \mathscr{C}_{\mathcal{M}(Y_1, Z_1)}$. Then $f_* \circ c^* \circ (g^* \circ h_* \circ e) = (f \circ h)_* \circ (g \circ c)^* \circ e \in \mathscr{C}_{\mathcal{M}}$ for $f \in \mathscr{F}_{Z_2}$ and $c \in \mathscr{C}_{Y_2}$ since $f \circ h \in \mathscr{F}_{Z_1}$ and $g \circ c \in \mathscr{C}_{Y_1}$.

We next show that:

(d) The evaluations $\mathrm{ev}_Z^Y: \mathcal{M}(Y, Z) \sqcap Y \to Z$ are $\mathcal{M}$-maps.
For this we have to show that $\mathrm{ev}_Z^Y \circ (e, c): S \to Z$ is an $\mathcal{M}$-map for all structure curves $e: S \to \mathcal{M}(Y, Z)$ and $c: S \to Y$. This holds since $\mathrm{ev}_Z^Y \circ (e, c) = \hat{e} \circ (\mathrm{id}_S, c)$ and $\hat{e}$ is an $\mathcal{M}$-map.

(e) The insertions $\mathrm{ins}_X^Y: X \to \mathcal{M}(Y, X \sqcap Y)$ are $\mathcal{M}$-maps.
By definition one has $\mathrm{ins}_X^Y(x)(y) = (x, y)$. For any $x \in X$ the map $\mathrm{ins}_X^Y(x): X \to X \sqcap Y$ is an $\mathcal{M}$-map since the composites with the canonical projections of $X \sqcap Y$ both are $\mathcal{M}$-maps. For this one uses that any constant map $X \to Y$ is an $\mathcal{M}$-map, but we remark that this holds only because of the hypothesis that all constant maps $S \to R$ belong to $\mathcal{M}$ (without this hypothesis $\underline{\mathcal{M}}$ can still be cartesian closed, but not by a functor lifting the hom-functor; cf. [Frölicher, 1979]). For ins_X^Y to be an $\mathcal{M}$-map we have to show that for any $c \in \mathscr{C}_X$ the composite $\mathrm{ins}_X^Y \circ c$ is a structure curve of $\mathcal{M}(Y, X \sqcap Y)$, i.e. that $(\mathrm{ins}_X^Y \circ c)^\wedge: S \sqcap Y \to X \sqcap Y$ is an $\mathcal{M}$-map. This holds since $(\mathrm{ins}_X^Y \circ c)^\wedge = c \sqcap \mathrm{id}_Y$.

It remains to show the claimed bijection:

(f) For any $\mathcal{M}$-map $g: X \to \mathcal{M}(Y, Z)$ the map $\hat{g}: X \sqcap Y \to Z$ is an $\mathcal{M}$-map. In fact, $\hat{g} = \mathrm{ev}_Z^Y \circ (g \sqcap \mathrm{id}_Y)$.

(g) If $h: X \sqcap Y \to Z$ is an $\mathcal{M}$-map, then $h^\vee: X \to \mathcal{M}(Y, Z)$ has values in $\mathcal{M}(Y, Z)$ and is an $\mathcal{M}$-map. In fact, $h^\vee = h_* \circ \mathrm{ins}_X^Y$. □

Remark. It can be proved that for sets $\mathcal{M}$ that contain all constant maps the condition given above is in fact equivalent to the cartesian closedness of the category $\underline{\mathcal{M}}$; for the case where $\mathcal{M}$ is a monoid cf. [Frölicher, 1980].

The following result will be used later.

1.1.8 Proposition. *Suppose $\mathcal{M}$ is such that $\underline{\mathcal{M}}$ is cartesian closed, and let X, X_j, Y, Y_j be $\mathcal{M}$-spaces, $j \in J$.*

(i) *If $\{f_j: X_j \to X;\ j \in J\}$ is a final family such that $X = \bigcup_{j \in J} f_j(X_j)$, then $\{f_j^*: \mathcal{M}(X, Y) \to \mathcal{M}(X_j, Y); j \in J\}$ is an initial family.*

(ii) *If $\{g_j: Y \to Y_j;\ j \in J\}$ is an initial family, then so is*

$$\{(g_j)_*: \mathcal{M}(X, Y) \to \mathcal{M}(X, Y_j);\ j \in J\}.$$

Remark. The special case where $J = \emptyset$ will also be used later.

Proof. (i) Let $c: S \to \underline{\mathcal{M}}(X, Y)$ be such that $f_j^* \circ c$ is a structure-curve of $\mathcal{M}(X_j, Y)$ for all $j \in J$. Using the universal property of the function space structure twice one concludes that $(f_j^* \circ c)^\sim: X_j \to \mathcal{M}(S, Y)$ is a morphism, where $\tilde{h}$ is defined by $\tilde{h}(x)(y) := h(y)(x)$. By the same reasoning it is enough to show that $\tilde{c}: X \to \mathcal{M}(S, Y)$ is well-defined and a morphism. It is well-defined, since for every $x \in X$ there exists a $j \in J$ and an $x_j \in X_j$ with $f_j(x_j) = x$ and hence $\tilde{c}(x) = \tilde{c}(f_j(x_j)) = (f_j^* \circ c)^\sim(x_j) \in \mathcal{M}(S, Y)$. Since $(f_j^* \circ c)^\sim = c^\sim \circ f_j$, the map $\tilde{c}$ is a morphism. The assertion now follows, since for an arbitrary $\mathcal{M}$-space instead of S one can test by means of the structure curves.

(ii) Let $c: S \to \mathcal{M}(X, Y)$ be such that $(g_j)_* \circ c$ is a structure-curve for $\mathcal{M}(X, Y_j)$ for all j. Then $((g_j)_* \circ c)^\wedge: S \sqcap X \to Y_j$ is an $\underline{\mathcal{M}}$-morphism for all j (recall that $\hat{h}(x, y) = h(x)(y)$). Since $((g_j)_* \circ c)^\wedge = g_j \circ \hat{c}$, the hypothesis implies that $\hat{c}$ and hence c is a morphism. □

1.2 Bornologies and ℓ^∞-structures

We first introduce the category of bornological spaces and then study a certain subcategory which will play an important role. The bornological spaces usually used in analysis all belong to it. This subcategory admits a description in the sense of section 1.1 and hence is easy to handle and has excellent categorical properties.

1.2.1 Definition (cf. [Hogbe-Nlend, 1977, p. 18]). A *bornology* on a set X is a collection $\mathscr{B}$ of subsets of X such that

(i) $x \in X \Rightarrow \{x\} \in \mathscr{B}$;
(ii) $B_1 \subseteq B_2 \in \mathscr{B} \Rightarrow B_1 \in \mathscr{B}$;
(iii) $B_j \in \mathscr{B}$ for $j = 1, 2 \Rightarrow B_1 \cup B_2 \in \mathscr{B}$.

A *bornological space* X is a set (also denoted by X) together with a bornology $\mathscr{B}$ on it; the sets of $\mathscr{B}$ are called *bounded* subsets of X.

A *basis* for a bornology $\mathscr{B}$ is a subcollection $\mathscr{B}_0$ of $\mathscr{B}$ such that every $B \in \mathscr{B}$ is contained in some $B_0 \in \mathscr{B}_0$.

A *subbasis* for a bornology $\mathscr{B}$ is a subcollection $\mathscr{B}_0$ of $\mathscr{B}$ such that for every $B \in \mathscr{B}$ there are finitely many $B_i \in \mathscr{B}_0$ with $B \subseteq \bigcup_i B_i$ (any collection $\mathscr{B}_0$ of subsets of X which covers X is a subbasis of a unique bornology $\mathscr{B}$, obtained by forming all subsets of finite unions of sets belonging to $\mathscr{B}_0$).

The category $\underline{\text{Born}}$ has as objects the bornological spaces; $\underline{\text{Born}}(X, Y)$ is formed by the so-called *bornological* maps $g\colon X \to Y$, i.e. those maps g for which $B \subseteq X$ bounded implies $g(B) \subseteq Y$ bounded.

1.2.2 Remark. It is easy to verify that initial and final structures with respect to the forgetful functor $\underline{\text{Born}} \to \underline{\text{Set}}$ exist. Therefore all categorical limits and colimits in $\underline{\text{Born}}$ exist, cf. (8.7.3). We explicitly mention products which will be used later. The underlying set of a product $\Pi_{j \in J} X_j$ of bornological spaces X_j is the cartesian product of the underlying sets, and $B \subseteq \Pi_{j \in J} X_j$ is bounded iff $\text{pr}_j(B) \subseteq X_j$ is bounded for all $j \in J$.

1.2.3 Definition. With $\underline{\ell^\infty}$ we denote the category obtained according to (1.1.1) by choosing $S := \mathbb{N}$, $R := \mathbb{R}$ and $\mathscr{M} := \ell^\infty$, i.e. $\mathscr{M}$ is the set ℓ^∞ of bounded sequences of real numbers.

Thus notions like ℓ^∞-structure, ℓ^∞-space and ℓ^∞-map make sense, cf. (1.1.1).

1.2.4 Proposition. *The category $\underline{\ell^\infty}$ embeds into $\underline{\text{Born}}$ as follows: To an ℓ^∞-space X with ℓ^∞-structure $(\mathscr{C}, \mathscr{F})$ one associates a bornological space ιX by defining $B \subseteq \iota X$ to be bounded (shortly: $B \subseteq X$ bounded) iff B satisfies the equivalent conditions:*

(1) *every sequence $c\colon \mathbb{N} \to X$ with $c(\mathbb{N}) \subseteq B$ belongs to $\mathscr{C}$;*
(2) *$f(B)$ is bounded in $\mathbb{R}$ for all $f \in \mathscr{F}$.*

The embedding functor $\iota\colon \underline{\ell^\infty} \to \underline{\text{Born}}$ has a left-adjoint $\eta\colon \underline{\text{Born}} \to \underline{\ell^\infty}$ satisfying $\eta \circ \iota = \text{Id}$. Both functors preserve the underlying spaces and maps.

Proof. The equivalence of (1) and (2) is trivial. For any bornological space $(X, \mathscr{B})$, the set of bornological functions $X \to \mathbb{R}$ (where $\mathbb{R}$ is considered with the standard bornology) generates an ℓ^∞-structure $(\mathscr{C}, \mathscr{F})$ on X, for which in fact $\mathscr{F}$ is equal to the set of bornological functions, as one verifies easily. One puts $\eta(X, \mathscr{B}) := (X; \mathscr{C}, \mathscr{F})$. The verification that this yields a functor η with the stated properties is straightforward. □

Remark. We will show in (7.2.8) that η commutes with countable products. The analogue fails for products with at least $2^{\aleph_0}$-many factors, cf. (7.2.9).

By means of the embedding of (1.2.4) $\underline{\ell^\infty}$ can be identified with a full reflective subcategory of $\underline{\text{Born}}$. The following proposition gives equivalent descriptions of that subcategory.

1.2.5 Proposition. *Let X be a bornological space. Then the following statements are equivalent:*

(1) *The bornology of X comes from an ℓ^∞-structure;*
(2) *Any subset on which all bornological functions $f: X \to \mathbb{R}$ are bounded is bounded in X;*
(3) *Every unbounded subset of X contains an infinite countable subset whose only bounded subsets are the finite ones.*

Proof. $(1 \Rightarrow 2)$ follows from (1.2.4).

$(2 \Rightarrow 3)$ Let $B \subseteq X$ be unbounded. Then there exists a bornological function $f: X \to \mathbb{R}$ which is unbounded on B. Therefore one can choose for each $n \in \mathbb{N}$ a point $b_n \in B$ with $|f(b_n)| \geqslant n$. Then $\{b_n; n \in \mathbb{N}\}$ is a subset of B which has the stated properties.

$(3 \Rightarrow 1)$ We show that (3) implies the equation $X = \iota\eta X$. Because of the adjunction stated in (1.2.4) we have only to verify that the identity map from $\iota\eta X$ to X is bornological. So let $B \subseteq X$ be unbounded. By (3) there exists a sequence of points $b_n \in B$ with $b_n \neq b_m$ for $n \neq m$ such that every infinite subset of $\{b_n; n \in \mathbb{N}\}$ is unbounded. One defines a function $f: X \to \mathbb{R}$ by $f(b_n) := n$ and $f(x) := 0$ for $x \notin \{b_n; n \in \mathbb{N}\}$. Then f is a bornological function since on any bounded set it takes only finitely many values. Since $f(B) \subseteq \mathbb{R}$ is unbounded, the assertion $B \subseteq \iota\eta X$ unbounded thus follows. □

We remark that according to (3) the bornologies coming from ℓ^∞-structures are Kolmogorov-bornologies (cf. [Hogbe-Nlend, 1977, p. 118]), i.e. they have the property that each unbounded subset contains a denumerable unbounded set. However, condition (3) is slightly more restrictive as the following example shows: take $X = \mathbb{N}$ and define $B \in \mathscr{B}$ iff $\sum_{n \in B} 1/n < \infty$. Then $\mathscr{B}$ is obviously a Kolmogorov-bornology, but (3) fails for the unbounded subset $\mathbb{N}$. Furthermore the ℓ^∞-structure associated via η is the coarse one since all bornological functions are globally bounded.

1.2.6 Definition (cf. [Hogbe-Nlend, 1977, p. 21]). For any separated topological space X, the relatively compact subsets of X form a bornology, called the *compact bornology* on X.

1.2.7 Proposition. *Let Y be a metric space, $X \subseteq Y$ a subspace. The following two bornologies on X are identical:*

(1) *The bornology induced by the compact bornology on Y;*
(2) *The bornology associated to the ℓ^∞-structure on X generated by the set of those sequences in X that converge in Y.*

Proof. Call $\mathscr{B}_1$ and $\mathscr{B}_2$ the bornologies on X described in (1) and (2). Let $\mathscr{C}_0$ be the set of sequences mentioned in (2).

($\mathscr{B}_2 \subseteq \mathscr{B}_1$) Let $B \in \mathscr{B}_2$. We show first that every sequence in B has a subsequence belonging to $\mathscr{C}_0$: Suppose indirectly that there is a sequence $b: \mathbb{N} \to B$ without accumulation point in Y. Without loss of generality we may assume that $b_n \neq b_m$ for $n \neq m$. One considers the function $f: X \to \mathbb{R}$ defined by $f(b_n) := n$ and $f(x) = 0$ for $x \notin b(\mathbb{N})$. If $x: \mathbb{N} \to X$ is a sequence which converges in Y, then only finitely many b_n can belong to $x(\mathbb{N})$. Hence $f \circ x$ takes only a finite number of values and thus belongs to ℓ^∞. We conclude that $f \in \mathscr{F}$ and since $f(B)$ is unbounded we reach a contradiction to $B \in \mathscr{B}_2$.

Let now c be a sequence in the closure $\bar{B}$ of B in Y. We choose $b_n \in B$ with $\mathrm{d}(c_n, b_n) \leq 1/n$. Some subsequence of b converges to a point $y \in Y$. The corresponding subsequence of c has to converge to y also, and therefore $y \in \bar{B}$. Thus $\bar{B}$ is compact, being a sequentially compact subset of a metric space.

($\mathscr{B}_2 \supseteq \mathscr{B}_1$) Let $B \in \mathscr{B}_1$. So there has to exist a compact $K \subseteq Y$ with $B \subseteq K$. Let $f: X \to \mathbb{R}$ be a structure function of the ℓ^∞-structure on X generated by $\mathscr{C}_0$. Suppose it is unbounded on B. Then there are $b_n \in B$ with $|f(b_n)| \geq n$. Some subsequence of b has to converge in the compact set K and hence belongs to $\mathscr{C}_0$. But since f is unbounded on this subsequence we reach a contradiction. Thus we have shown that $B \in \mathscr{B}_2$. □

Remark. The bornology on X considered above is the compact bornology of X iff X is closed in Y.

1.2.8 Theorem. *The category $\underline{\ell^\infty}$ is cartesian closed (cf. section 8.6). The canonical ℓ^∞-structure on ℓ^∞ is linearly generated by the elements of ℓ^1 acting on ℓ^∞ (cf. (1.1.3)).*

Proof. We first remark that the bornology belonging to the canonical ℓ^∞-structure of $\mathbb{N}$ (cf. (ii) of (1.1.6)) is the coarse one ($\mathbb{N}$ itself is bounded), while for $\mathbb{R}$ it is the usual bornology. So we get for the set $\mathscr{C}_{\ell^\infty}$ of (1.1.7):

$$\mathscr{C}_{\ell^\infty} = \{c: \mathbb{N} \to \ell^\infty;\ \hat{c}(\mathbb{N} \times \mathbb{N}) \text{ bounded in } \mathbb{R}\}.$$

It is well known that the Banach space ℓ^∞ with the norm $\|x\|_\infty := \sup\{|x_n|; n \in \mathbb{N}\}$ is the dual of the space of absolutely summable sequences $\ell^1 := \{x : \mathbb{N} \to \mathbb{R}; \|x\|_1 := \sum_{n=1}^\infty |x_n| < \infty\}$; where the duality action is given by $\langle x|y\rangle := \sum_n x_n y_n$ for $x \in \ell^1$ and $y \in \ell^\infty$.

We show first that the elements of ℓ^1 act on ℓ^∞ as elements of $\Phi\mathscr{C}_{\ell^\infty} = \{\varphi : \ell^\infty \to \mathbb{R};\ \varphi \circ c \in \ell^\infty$ for all $\varphi \in \mathscr{C}_{\ell^\infty}\}$. So let $x \in \ell^1$, and $c \in \mathscr{C}_{\ell^\infty}$. Then $\hat{c} : \mathbb{N} \times \mathbb{N} \to \mathbb{R}$ is bounded, say by $A > 0$, and we obtain $|(x \circ c)(n)| = |\langle x|c(n)\rangle| = |\sum_{m=1}^\infty x_m \cdot \hat{c}(n, m)| \leq A \cdot \|x\|_1$, i.e. $x \circ c \in \ell^\infty$ and therefore $x \in \Phi\mathscr{C}_{\ell^\infty}$.

We show now that a sequence $c : \mathbb{N} \to \ell^\infty$ belongs to $\mathscr{C}_{\ell^\infty}$ provided $x \circ c \in \ell^\infty$ for all $x \in \ell^1$. For each $x \in \ell^1$ there exists a constant A_x such that $|x(c(n))| \leq A_x$ for all $n \in \mathbb{N}$, i.e. the $c(n)$ considered as elements of $(\ell^1)'$ form a pointwise bounded family. By the uniform boundedness theorem of Banach–Steinhaus, cf. [Jarchow, 1981, p. 220], we conclude that this family is bounded with respect to the norm, i.e. there is a constant A with $\|c(n)\|_\infty \leq A$ for all n. Hence $|\hat{c}(n, m)| \leq A$ for all $n, m \in \mathbb{N}$ and therefore $c \in \mathscr{C}_{\ell^\infty}$. This not only shows that $(\mathscr{C}_{\ell^\infty}, \Phi\mathscr{C}_{\ell^\infty})$ is an ℓ^∞-structure on ℓ^∞, but also that it is linearly generated by the elements of ℓ^1. □

Remark. If one is only interested in the cartesian closedness of $\underline{\ell^\infty}$, a much shorter proof is possible: one shows that the ℓ^∞-structure on ℓ^∞ which is generated by the function $\|\_\|_\infty : \ell^\infty \to \mathbb{R}$ has exactly $\mathscr{C}_{\ell^\infty}$ as set of structure curves. But since this function is not linear the obtained result would not be useful for investigating linear spaces and for showing the cartesian closedness of the category $\underline{\mathscr{C}^\infty}$ of smooth spaces; cf. (1.4.3).

The category $\underline{\ell^\infty}$ is even locally cartesian closed, cf. (8.6.5); since we shall not use this result and since the proof is not hard, we omit it. We shall, however, show in (7.1.6) that $\underline{C^\infty}$ is not locally cartesian closed.

1.2.9 Proposition. *Let X be any ℓ^∞-space and E a vector space with an ℓ^∞-structure that is generated by a set $\mathscr{S}$ of linear functions. Then the following families of linear morphisms are initial:*

(*i*) $c^* : \ell^\infty(X, E) \to \ell^\infty(\mathbb{N}, E) \quad (c \in \ell^\infty(\mathbb{N}, X))$;
(*ii*) $\ell_* : \ell^\infty(X, E) \to \ell^\infty(X, \mathbb{R}) \quad (\ell \in \mathscr{S})$;
(*iii*) $\ell^\infty(c, \ell) : \ell^\infty(X, E) \to \ell^\infty(\mathbb{N}, \mathbb{R}) = \ell^\infty \quad (c \in \ell^\infty(\mathbb{N}, X), \ell \in \mathscr{S})$.

Proof. This is an immediate consequence of (1.1.8).

1.2.10 Corollary. *Let X be an ℓ^∞-space and E a vector space with linearly generated ℓ^∞-structure. Then the structure of $\ell^\infty(X, E)$ is also linearly generated.*

Another way of obtaining the cartesian closedness of $\underline{\ell^\infty}$ is to restrict, according to the diagram (1.2.13), the functor Born: $\underline{\text{Born}}^{\text{op}} \times \underline{\text{Born}} \to \underline{\text{Born}}$ which describes the cartesian closedness of $\underline{\text{Born}}$. For bornological spaces X, Y the space Born(X, Y) is obtained by putting the following bornology on the set

$\underline{\mathrm{Born}}(X, Y)$: $B \subseteq \mathrm{Born}(X, Y)$ is bounded iff for any bounded $A \subseteq X$ the set $B(A)$ is bounded in Y. Cartesian closedness of $\underline{\mathrm{Born}}$ is then easily verified.

1.2.11 Lemma. *Let X and Y be bornological spaces, $A \subseteq X$ be bounded and $f: Y \to \mathbb{R}$ be bornological. Then the function $\varphi_{A,f}: \underline{\mathrm{Born}}(X, Y) \to \mathbb{R}$ defined by $\varphi_{A,f}(g) := \sup\{|f(g(x))|;\, x \in A\}$ is bornological.*

Proof. Let $B \subseteq \mathrm{Born}(X, Y)$ be bounded. Then $B(A) \subseteq Y$ is bounded and hence there exists an $M > 0$ with $f(B(A)) \subseteq [-M, M]$. Therefore $\varphi_{A,f}(B) \subseteq [-M, M]$. □

1.2.12 Lemma. *Let X and Y be bornological spaces. If the bornology of Y comes from an ℓ^∞-structure, then the same holds for the bornology of* $\mathrm{Born}(X, Y)$.

Proof. We use condition (2) of (1.2.5). Let $B \subseteq \mathrm{Born}(X, Y)$ be unbounded. Then we can choose a bounded $A \subseteq X$ with $B(A) \subseteq Y$ unbounded, and a bornological $f: Y \to \mathbb{R}$ with $f(B(A))$ unbounded. Hence, with $\varphi_{A,f}$ according to the lemma above, $\varphi_{A,f}(B)$ is unbounded, as to be shown. □

1.2.13 Corollary. *The following diagram commutes:*

$$\begin{array}{ccc}
(\underline{\ell^\infty})^{\mathrm{op}} \times \underline{\ell^\infty} & \xrightarrow{\ \ell^\infty\ } & \underline{\ell^\infty} \\
\Big\downarrow {\scriptstyle \iota^{\mathrm{op}} \times \iota} & & \Big\downarrow {\scriptstyle \iota} \\
\underline{\mathrm{Born}}^{\mathrm{op}} \times \underline{\mathrm{Born}} & \xrightarrow{\ \mathrm{Born}\ } & \underline{\mathrm{Born}}
\end{array}$$

In particular one has for ℓ^∞-spaces X and Y: $B \subseteq \ell^\infty(X, Y)$ is bounded iff $B(A) \subseteq Y$ is bounded for all bounded $A \subseteq X$.

1.2.14 Proposition. *The following diagram commutes:*

$$\begin{array}{ccccc}
\underline{\mathrm{Born}}^{\mathrm{op}} \times \underline{\ell^\infty} & \xrightarrow{\ \eta \times 1\ } & (\underline{\ell^\infty})^{\mathrm{op}} \times \underline{\ell^\infty} & \xrightarrow{\ \ell^\infty\ } & \underline{\ell^\infty} \\
\Big\downarrow {\scriptstyle 1 \times \iota} & & & & \Big\downarrow {\scriptstyle \iota} \\
\underline{\mathrm{Born}}^{\mathrm{op}} \times \underline{\mathrm{Born}} & & \xrightarrow{\ \mathrm{Born}\ } & & \underline{\mathrm{Born}}
\end{array}$$

Proof. Since this result is not used in the following, we only indicate that one uses the adjunction between η and ι (1.2.4), the cartesian closedness of $\underline{\mathrm{Born}}$ and $\underline{\ell^\infty}$ and the fact that η commutes with finite products. □

Remark. The essential consequence of this proposition is the fact that for a given ℓ^∞-space Y the bornology of $\mathrm{Born}(X, \iota Y)$ only depends on ηX (or equivalently: only on the bornological real-valued functions on X).

1.3 Difference quotients

Difference quotients provide simple characterizations of differentiability properties of functions of one or several variables (cf. (1.3.22), (1.3.28)). They are also used in numerical analysis because they yield direct algebraic approximations of (partial) derivatives of higher order (cf. the remark after (1.3.15)). We first give the fundamental definitions, proceed with the results for functions of one variable and terminate with the corresponding results for functions of several variables.

1.3.1 Definition. By $\mathbb{N}:=\{1,2,\ldots\}$ we denote the set of natural numbers. And we will use the sets $\mathbb{N}_0:=\mathbb{N}\cup\{0\}$, $\mathbb{N}_\infty:=\mathbb{N}\cup\{\infty\}$ and $\mathbb{N}_{0,\infty}:=\mathbb{N}\cup\{0,\infty\}$.

Let $D\subseteq\mathbb{R}$ be an arbitrary subset and $f\colon D\to E$ a function with values in a vector space E, $k\in\mathbb{N}_0$.

The natural domain of definition of the difference quotient of order k of f is $D^{\langle k\rangle}:=\{(t_0,\ldots,t_k)\in D^{k+1};\ t_i\neq t_j \text{ for } i\neq j\}$. On $D^{\langle k\rangle}$ we always consider the bornology induced by the compact bornology of D^{k+1}, cf. (1.2.7).

And the *difference quotient* $\delta^k f\colon D^{\langle k\rangle}\to E$ of order k of f is recursively defined by:

$$\delta^0 f:=f$$

$$\delta^k f(t_0,\ldots,t_k):=\frac{k}{t_0-t_k}\cdot(\delta^{k-1}f(t_0,\ldots,t_{k-1})-\delta^{k-1}f(t_1,\ldots,t_k)).$$

The solution of this recursive definition can be expressed explicitly:

1.3.2 Proposition. *$\delta^k f(t_0,\ldots,t_k)=k!\ \sum_{i=0}^k\beta_i\cdot f(t_i)$ where the coefficients $\beta_i=\beta_i(t_0,\ldots,t_k):=\prod_{0\le r\le k,r\neq i}(t_i-t_r)^{-1}$ are independent of f.*

Proof. By induction on k. □

1.3.3 Corollary. *$\delta^k f$ is symmetric in its $k+1$ arguments.*

Next we give the corresponding definitions for functions of several variables.

1.3.4 Definition. Let $D\subseteq\mathbb{R}^m$ be an arbitrary subset and $f\colon D\to E$ a function with values in a vector space E. With $\kappa:=(k_1,\ldots,k_m)\in(\mathbb{N}_0)^m$ we denote a *multi-index of degree* $|\kappa|:=k_1+\cdots+k_m$.

The natural domain of definition of the difference quotient of order κ of f is $D^{\langle\kappa\rangle}:=D^\kappa\cap(\mathbb{R}^{\langle k_1\rangle}\times\ldots\times\mathbb{R}^{\langle k_m\rangle})\subseteq\mathbb{R}^{k_1+1}\times\ldots\times\mathbb{R}^{k_m+1}$, where $D^\kappa:=\{(x^1,\ldots,x^m)\in\mathbb{R}^{k_1+1}\times\ldots\times\mathbb{R}^{k_m+1};\ (x^1_{i_1},\ldots,x^m_{i_m})\in D \text{ for } i_j=0\ldots k_j\}$. On $D^{\langle\kappa\rangle}$ we will always consider the bornology induced by the compact bornology of D^κ, cf. (1.2.7).

In order to define the *difference quotient* $\delta^\kappa f\colon D^{\langle\kappa\rangle}\to E$ of order κ of f we use instead of a recursive definition the analogon to the explicit description given in (1.3.2):

$$\delta^\kappa f(x^1,\ldots,x^m):=\sum_{i_1=0}^{k_1}\cdots\sum_{i_m=0}^{k_m}\beta_{i_1}(x^1)\ldots\beta_{i_m}(x^m)\cdot f(x^1_{i_1},\ldots,x^m_{i_m})$$

where $\beta_i(x):=k!\prod_{0\le j\le k,\, j\ne i}(x_i-x_j)^{-1}$ for $x=(x_0,\ldots,x_k)\in\mathbb{R}^{\langle k\rangle}$.

In the special case where κ has all components equal to 0 except the ith one being equal to k, we define the *ith partial difference quotient $\delta_i^k f$ of order k of f* to be:

$\delta_i^k f(\ldots,x^i,\ldots):=\delta^\kappa f(\ldots,x^i,\ldots)=\delta^k(f(\ldots,x^{i-1},\_,x^{i+1},\ldots))(x^i)$, with $x^i\in\mathbb{R}^{\langle k\rangle}$

and $x^j\in\mathbb{R}$ for $j\ne i$.

1.3.5 Remark. (i) In case $m=1$, i.e. $\kappa=(k)$ with $k\in\mathbb{N}_0$, the general definitions reduce to the special ones in the following way:

$$D^{\langle\kappa\rangle}=D^{\langle k\rangle},\quad D^\kappa=D^{(k)}=D^{k+1},\quad \delta^\kappa f=\delta_1^k f=\delta^k f.$$

(ii) In the case where $D=D_1\times\ldots\times D_m$ one has $D^{\langle\kappa\rangle}=D_1^{\langle k_1\rangle}\times\ldots\times D_m^{\langle k_m\rangle}$.

(iii) The partial difference quotients $\delta_i^k f$ even make sense for $D\subseteq X_1\times\ldots\times X_{i-1}\times\mathbb{R}\times X_{i+1}\times\ldots\times X_m$ for arbitrary sets X_j.

(iv) For $k=1$ we will write $\delta_i^k f=\delta_i^1 f=\delta_i f$ and for $\kappa=(0,\ldots,0)$ we have $D^{\langle\kappa\rangle}=D^\kappa=D$ and $\delta^\kappa f=f$.

(v) We have the following recursion for the difference quotients:

$\delta^\kappa=\delta_{l(1)}^{k_{l(1)}}\circ\cdots\circ\delta_{l(m)}^{k_{l(m)}}$ for any permutation l of $\{1,\ldots,m\}$. To interpret the right side correctly one should remark that $\delta^\kappa f$ is defined on a subset of $\mathbb{R}^{\langle k_1\rangle}\times\ldots\times\mathbb{R}^{\langle k_m\rangle}$ and then use remark (iii) (for the proof it suffices and is easy to show that $\delta^\kappa f=\delta_p^{k_p}(\delta^{\kappa'}f)$ where $\kappa':=(k_1,\ldots,0,\ldots,k_m)$), i.e. κ' is obtained from κ by replacing the pth entry by 0.

We return now to functions of one variable.

1.3.6 Lemma. *For any $f\colon D\to E$ and any $(t_0,\ldots,t_k)\in D^{\langle k\rangle}$ one has*

$$f(t_k)=f(t_0)+\frac{1}{1!}(t_k-t_0)\,\delta^1 f(t_0,t_1)+\cdots$$

$$+\frac{1}{k!}(t_k-t_0)\cdots(t_k-t_{k-1})\,\delta^k f(t_0,\ldots,t_k).$$

Proof. For $k=0$ this is trivial. Suppose the formula holds for $k-1$ instead of k and let $(t_0,\ldots,t_k)\in D^{\langle k\rangle}$.

Applying the induction hypothesis to $(t_0, \ldots, t_{k-2}, t_k) \in D^{\langle k-1 \rangle}$ one has $f(t_k) = f(t_0) + \cdots + \frac{1}{(k-1)!}(t_k - t_0) \ldots (t_k - t_{k-2}) \delta^{k-1} f(t_0, \ldots, t_{k-2}, t_k)$. The claimed formula follows by replacing the last term in the above sum by

$$\frac{1}{(k-1)!}(t_k - t_0) \ldots (t_k - t_{k-2}) \delta^{k-1} f(t_0, \ldots, t_{k-2}, t_{k-1})$$

$$+ \frac{1}{k!}(t_k - t_0) \ldots (t_k - t_{k-1}) \delta^k f(t_0, \ldots, t_k),$$

according to the recursive definition of $\delta^k f$ and its symmetry. □

The interpolation formula of Newton follows from the lemma:

1.3.7 Proposition. *Let $f: D \to E$ and $(t_0, \ldots, t_k) \in D^{\langle k \rangle}$.*

(i) *The unique polynomial function (the interpolation polynomial) $p: \mathbb{R} \to E$ of degree at most k satisfying $p(t_i) = f(t_i)$ for $i = 0, \ldots, k$ is given by*

$$p(t) := f(t_0) + \frac{1}{1!}(t - t_0) \delta^1 f(t_0, t_1) + \cdots$$

$$+ \frac{1}{k!}(t - t_0) \ldots (t - t_{k-1}) \delta^k f(t_0, \ldots, t_k).$$

(ii) *For any $t \in D \backslash \{t_0, \ldots, t_k\}$ the remainder is given by*

$$f(t) - p(t) = \frac{1}{(k+1)!}(t - t_0) \ldots (t - t_k) \delta^{k+1} f(t_0, \ldots, t_k, t).$$

Proof. The uniqueness of such a polynomial function follows from the theory of linear equations; the respective determinant is that of Vandermonde and hence different from zero. The function p as described above is polynomial of degree at most k; and $p(t_i) = f(t_i)$ follows from the lemma (1.3.6) (with i instead of k). The formula for $f - p$ is a direct application of (1.3.6) with $k+1$ instead of k and $t_{k+1} = t$. □

1.3.8 Lemma. *Let $f: D \to E$; $(t_0, \ldots, t_k) \in D^{\langle k \rangle}$; and $i_1, i_2, i_3 \in \{0, \ldots, k\}$ three different integers. Then*

$$0 = \sum_{\sigma} (t_{\sigma(i_1)} - t_{\sigma(i_2)}) \cdot \delta^{k-1} f(t_0, \ldots, \overset{\sqcap}{t}_{\sigma(i_3)}, \ldots, t_k),$$

where the sum goes over the three cyclic permutations σ of $\{i_1, i_2, i_3\}$ and where the symbol $\sqcap$ above a term indicates that it has to be omitted.

Proof. By the symmetry and the recursive definition the first of the three terms is equal to

$$(k-1)(\delta^{k-2}f(\ldots,\widehat{t_{i_1}},\ldots,\widehat{t_{i_3}},\ldots)-\delta^{k-2}f(\ldots,\widehat{t_{i_2}},\ldots,\widehat{t_{i_3}},\ldots)).$$

So one gets six terms which obviously cancel two by two. □

1.3.9 Corollary. *Let* $(t_0,\ldots,t_k)\in D^{\langle k\rangle}$, $0<i<k$ *and* $t_0<t_i<t_k$. *Then there exist positive reals* α *and* β *with* $\alpha+\beta=1$ *and* $\delta^{k-1}f(t_0,\ldots,\widehat{t_i},\ldots,t_k)=\alpha\delta^{k-1}f(t_0,\ldots,t_{k-1})+\beta\delta^{k-1}f(t_1,\ldots,t_k)$ *for any* $f\colon D\to E$.

Proof. According to (1.3.8) $\alpha:=(t_i-t_0)/(t_k-t_0)$ and $\beta:=(t_k-t_i)/(t_k-t_0)$ suffice. □

1.3.10 Lemma. *Let* $t_0<t_1<\ldots<t_n$ *and* $0=i_0<i_1<\ldots<i_k=n$. *There exist positive reals* β_r *for* $r=0,\ldots,n-k$ *with* $\beta_0+\cdots+\beta_{n-k}=1$ *such that for any function* $f\colon D\to E$ *with* $t_i\in D$ *for all* i *one has*

$$\delta^k f(t_{i_0},\ldots,t_{i_k})=\sum_{r=0}^{n-k}\beta_r\delta^k f(t_r,t_{r+1},\ldots,t_{r+k}).$$

Proof. One successively replaces a term $\delta^k f(t_{j_0},\ldots,t_{j_k})$ for which $j_k-j_0>k$ by a linear combination with positive coefficients α and β $(\alpha+\beta=1)$ of two terms $\delta^k f(t_{l_0},\ldots,t_{l_k})$ with $l_k-l_0=j_k-j_0-1$ using (1.3.9). □

This lemma is responsible for the fact that one can restrict in many situations to equidistant difference quotients which are defined as follows:

1.3.11 Definition. For $s\neq 0$ one puts $\delta^k_{\text{eq}}f(t;s):=\delta^k f(t,t+s,\ldots,t+ks)$.

1.3.12 Proposition. *Let* $f\colon I\to E$ *where* $I\subseteq\mathbb{R}$ *is an interval; suppose that* $A\subseteq E$ *is convex and closed for a topology on* E *for which* f *is continuous. If for each* t *and* s *with* $s\neq 0$ *and* $t,t+ks\in I$ *one has* $\delta^k_{\text{eq}}f(t,s)\in A$, *then* $\delta^k f(t_0,\ldots,t_k)\in A$ *for all* $(t_0,\ldots,t_k)\in I^{\langle k\rangle}$.

Proof. Since $\delta^k f$ is symmetric, we can suppose that $t_0<\ldots<t_k$.

(i) Special case: $(t_i-t_0)/(t_k-t_0)\in\mathbb{Q}$ for $i=0\ldots k$.

Let m be the smallest common denominator of these positive rational numbers and put $n_i/m:=(t_i-t_0)/(t_k-t_0)$. Then we write $t_i=t_0+n_is$, i.e. the points $t_0,\ldots,t_k$ are among the points $t_0,t_0+s,\ldots,t_0+n_ks$. Therefore, by (1.3.10), $\delta^k f(t_0,\ldots,t_k)$ is in the convex hull of the values $\delta^k f(t_0+rs,t_0+(r+1)s,\ldots,t_0+(r+k)s)=\delta^k_{\text{eq}}f(t_0+rs,s)$ which by assumption lie in A. So $\delta^k f(t_0,\ldots,t_k)\in A$ since A is assumed to be convex.

(ii) General case. Because f is continuous, $\delta^k f(t_0,\ldots,t_k)$ can be obtained as limit of values $\delta^k f(s_0,\ldots,s_k)$ where the $s_0,\ldots,s_k$ are as in case (i). So $\delta^k f(s_0,\ldots,s_k)\in A$, which implies $\delta^k f(t_0,\ldots,t_k)\in A$ since A is assumed to be closed. □

Remark. Without the continuity assumption on f, (1.3.12) fails as shown by the following example (using a basis for $\mathbb{R}$ as vector space over $\mathbb{Q}$). Let $f\colon\mathbb{R}\to\mathbb{R}$ be a function which is additive (i.e. $f(t+s)=f(t)+f(s)$ for all t,s) but not continuous.

Then $\delta^2_{\text{eq}} f=0$ but $\delta^2 f$ is not locally bounded since otherwise f would be even differentiable, cf. (1.3.17).

1.3.13 Lemma. *Let $T \neq S$ be two subsets of D with $k+1$ elements each. There exist enumerations $T=\{t_0, \ldots, t_k\}$ and $S=\{s_0, \ldots, s_k)$ such that $t_i \neq s_j$ for $i \leq j$; and then*

$$\delta^k f(t_0, \ldots, t_k) - \delta^k f(s_0, \ldots, s_k) = \frac{1}{k+1} \sum_{i=0}^{k} (t_i - s_i)\, \delta^{k+1} f(t_0, \ldots, t_i, s_i, \ldots, s_k)$$

for any $f: D \to E$.

Proof. For the enumeration we put the elements of $T \cap S$ at the end in T and at the beginning in S. Using the recursive definition of $\delta^{k+1} f$, the sum on the right can be written as a telescoping sum. □

1.3.14 Corollary. *Let $I \subseteq \mathbb{R}$ be an interval of finite length L and $(s_0, \ldots, s_k) \in I^{\langle k \rangle}$. Then one has*

$$\sup |\delta^k f(I^{\langle k \rangle})| \leq L \cdot \sup |\delta^{k+1} f(I^{\langle k+1 \rangle})| + |\delta^k f(s_0, \ldots, s_k)|$$

for any function $f: I \to \mathbb{R}$.

In particular, $\delta^{k+1} f$ bounded implies $\delta^k f$ bounded.

Remark. The last statement fails if I is unbounded: If $f(t) := t^{k+1}$, then $\delta^{k+1} f$ is bounded on $\mathbb{R}^{\langle k+1 \rangle}$ since it has the constant value $(k+1)!$, but $\delta^k f(t_0, \ldots, t_k) = k!(t_0 + \cdots + t_k)$ is not bounded on $\mathbb{R}^{\langle k \rangle}$.

1.3.15 Proposition. (*1st Mean Value Theorem.*) *Let $I \subseteq \mathbb{R}$ be an open interval, $0 \leq j \leq k$, $x \in I^{\langle k \rangle}$, and $f: I \to \mathbb{R}$ j-times differentiable. Then there exists a $\xi \in I^{\langle k-j \rangle}$* such that $\delta^k f(x) = \delta^{k-j} f^{(j)}(\xi)$.

Proof. It is obviously enough to prove this for $j=1$. Let $x=(t_0, \ldots, t_k)$ and without loss of generality $t_0 < \ldots < t_k$. Set $r := f - p$ where p is the interpolation polynomial of (1.3.7) for f and the points $t_0, \ldots, t_k$. Hence $r(t_i)=0$ for $i=0, \ldots, k$ and by Rolle's theorem there exists a ξ_i, s.t. $t_i < \xi_i < t_{i+1}$ and $r'(\xi_i)=0$, i.e. $f'(\xi_i) = p'(\xi_i)$. Thus p' is the interpolation polynomial for f' and the points $\xi_0, \ldots, \xi_{k-1}$. Comparing the highest term of p and p' according to (1.3.7) yields the desired result. □

Remark. This lemma has the well-known consequence, that for a function $f: I \to \mathbb{R}$ of class C^k, $f^{(k)}(t)$ is the limit of $\delta^k f(t_0, \ldots, t_k)$ where all the t_i tend towards t. In particular one then has $f^{(k)}(t) = \lim_{s \to 0} \delta^k_{\text{eq}} f(t; s)$.

1.3.16 Proposition. (*2nd Mean Value Theorem.*) *Let $I \subseteq \mathbb{R}$ be an open interval, $0 \leq k$, $x=(t_0, \ldots, t_k) \in I^{\langle k \rangle}$, and $f: I \to \mathbb{R}$ continuous on I and differentiable at all t_i. Then*

(i) $\delta^{k+1}f(x,t_i):=\lim_{t\to 0}\delta^{k+1}f(x,t_i+t)$ *exists;*

(ii) $\delta^k f^{\cdot}(x)=\dfrac{1}{k+1}\sum_{i=0}^{k}\delta^{k+1}f(x,t_i)$;

(iii) *there exists a* $\xi\in I$ *with* $\delta^k f^{\cdot}(x)=\delta^{k+1}f(x,\xi)$.

Proof. (i) Follows from the symmetry of $\delta^{k+1}f$ and (1.3.6).

(ii) Is proved by induction. One uses that with the definition (i) above the recursion formula (1.3.1) remains valid if two points coincide, provided that f is differentiable at the respective point.

(iii) Consider the continuous map $g:I\to\mathbb{R}$ defined by $g(t):=\delta^{k+1}f(t_0,\ldots,t_k,t)$. It takes on I the value $\delta^k f^{\cdot}(t_0,\ldots,t_k)$ since by (ii) this is the mean value of $g(t_0),\ldots,g(t_k)$. □

1.3.17 Lemma. *Let* $I\subseteq\mathbb{R}$ *be an open interval;* $f:I\to\mathbb{R}$*; and suppose* $\delta^2 f$ *is bounded on* $I^{\langle 2\rangle}$*. Then* f *is differentiable.*

Proof. Since

$$\frac{f(t+s)-f(t)}{s}-\frac{f(t+s')-f(t)}{s'}=\frac{1}{2}(s-s')\,\delta^2 f(t,t+s,t+s')$$

the Cauchy condition for the existence of $f^{\cdot}(t)$ is satisfied. □

1.3.18 Lemma. *Let* $I\subseteq\mathbb{R}$ *be an open interval,* $f:I\to\mathbb{R}$*, and* $M\in\mathbb{R}$*. Suppose* $|\delta^{k+1}f|$ *is bounded on* $I^{\langle k+1\rangle}$ *by* M*; then* f *is* k*-times differentiable and* $|\delta^1 f^{(k)}|$ *is bounded on* $I^{\langle 1\rangle}$ *by* M.

Proof. One uses induction. For $k=0$ one has nothing to prove. So let us assume it holds for $k-1$ and let $|\delta^{k+1}f|$ be bounded by M. Using (1.3.14) we obtain that $|\delta^2 f|$ is bounded on $J^{\langle 2\rangle}$ for every bounded subinterval J of I. Thus by (1.3.17) $g:=f^{\cdot}:I\to\mathbb{R}$ exists and by (ii) in (1.3.16) $\delta^k g$ is bounded by M. The induction hypothesis implies that g is $(k-1)$-times differentiable and $\delta^1 g^{(k-1)}$ is bounded by M, as to be shown. □

1.3.19 Definition. Let E,F be normed spaces; $D\subseteq E$; and $f:D\to F$.

(i) f is called *Lipschitzian* (on D) if there exists an $M\in\mathbb{R}$ such that $\|f(x)-f(y)\|\le M\|x-y\|$ for all $x,y\in D$. The function f is called *locally Lipschitzian* if every point of D has a neighborhood U in D such that $f|_U:U\to F$ is Lipschitzian.

(ii) Let $D\subseteq\mathbb{R}^m$ be open and $f:D\to\mathbb{R}$. Then f is called *k-times Lipschitz differentiable* if f is k-times differentiable and has a locally Lipschitzian derivative $f^{(k)}$ of order k.

1.3.20 Proposition. *Let* $U\subseteq\mathbb{R}^m$ *be open;* F *a normed space; and* f: $U\to F$*. Then the following statements are equivalent:*

(1) f *is locally Lipschitzian;*

(2) *f is Lipschitzian on every compact subset of U;*
(3) *the partial difference quotients $\delta_i f$ $(i=1 \ldots m)$ of f are bornological maps;* cf. (1.3.4) *and* (iv) *of* (1.3.5).

Proof. (1⇒2) Suppose f is not Lipschitzian on the compact set K. Then there exist for every $n\in\mathbb{N}$ points $x_n\neq y_n$ in K with $\|f(x_n)-f(y_n)\|>n\|x_n-y_n\|$, and we may assume that $\lim_{n\to\infty}x_n=x$ and $\lim_{n\to\infty}y_n=y$. If $x\neq y$ we get a contradiction with the continuity of f at x and y, the continuity being implied by the hypothesis. And if $x=y$ a contradiction with the fact that f is Lipschitzian in a neighborhood of x results.

(2⇒3) By (1.2.7) it is enough to show that $\delta_i f$ is bounded on sequences $n\mapsto(t_1^n;\ldots;t_i^n,t_i'^n;\ldots;t_m^n)$ of $U^{\langle 0,\ldots,1,\ldots,0\rangle}$ that converge in $U^{(0,\ldots,1,\ldots,0)}$. For such a sequence the points $(t_1^n,\ldots,t_i^n,\ldots,t_m^n)$ and $(t_1^n,\ldots,t_i'^n,\ldots,t_m^n)$ all stay in a compact subset of U, hence the assertion follows.

(3⇒1) One chooses for any given point in U a box-shaped compact neighborhood V. For $x\neq y\in V$ one writes

$$f(x)-f(y)=\sum_{j=1}^{m}(f(\ldots,x_j,y_{j+1},\ldots)-f(\ldots,x_{j-1},y_j,\ldots))$$

In the sum we drop all terms which are zero, divide by $\|x-y\|$ and use $\|x-y\|\geq|x_i-y_i|$. Thus we obtain

$$\frac{\|f(x)-f(y)\|}{\|x-y\|}\leq\sum_j\|\delta_j f(x_1;\ldots;x_j,y_j;\ldots;y_m)\|.$$ □

1.3.21 Lemma. *Let $U\subseteq\mathbb{R}$ be open, $k\geq 1$, $g\colon U\times U^m\to\mathbb{R}$ be a k-times Lipschitz differentiable map. Then $\delta_1 g\colon U^{\langle 1\rangle}\times U^m\to\mathbb{R}$ extends to a $(k-1)$-times Lipschitz differentiable map on $U^2\times U^m=U^{m+2}$.*

Proof. Since such an extension has to be locally unique, it is enough to show its existence locally. So let $(a,a;b)\in(U^2\setminus U^{\langle 1\rangle})\times U^m$. Choose an open interval I with $a\in I\subseteq U$. Since $g|_{I\times U^m}$ is k-times Lipschitz differentiable, $\bar{\delta}_1 g$, defined by $\bar{\delta}_1 g(t,t';x):=\int_0^1\partial_1 g(t+s(t'-t),x)\,ds$, yields a $(k-1)$-times Lipschitz differentiable map which extends $\delta_1 g$, as integration by means of the substitution $s':=t+s(t'-t)$ shows. □

1.3.22 Theorem. *Let $U\subseteq\mathbb{R}$ be open, $f\colon U\to\mathbb{R}$, and $0\leq j\leq k$. Then the following statements are equivalent:*

(1) *f is k-times Lipschitz differentiable,* cf. (1.3.19);
(2) *Every point of U has a neighborhood $V\subseteq U$ such that $\delta^{k+1}f$ is bounded on $V^{\langle k+1\rangle}$;*
(3) *$\delta^{k+1}f\colon U^{\langle k+1\rangle}\to\mathbb{R}$ is a bornological function,* cf. (1.3.1) *for the bornology of $U^{\langle k+1\rangle}$;*
(4_j) *$\delta^j f\colon U^{\langle j\rangle}\to\mathbb{R}$ has a $(k-j)$-times Lipschitz differentiable extension $\bar{\delta}^j f\colon U^{j+1}\to\mathbb{R}$.*

Proof. Obviously (4_0) equals (1). We show first that $(4_j \Rightarrow 4_{j+1})$ for $0 \le j < k$. Thus we are given a $(k-j)$-times Lipschitz differentiable extension $\bar{\delta}^j f$ of $\delta^j f: U^{\langle j \rangle} \to \mathbb{R}$ to U^{j+1}.

We have to find a $(k-j-1)$-times Lipschitz differentiable map $\bar{\delta}^{j+1} f: U^{j+2} \to \mathbb{R}$ extending $\delta^{j+1} f: U^{\langle j+1 \rangle} \to \mathbb{R}$. For this it is enough to extend $\delta_1(\bar{\delta}^j f): U^{\langle 1 \rangle} \times U^j \to \mathbb{R}$ to U^{j+2}. This is guaranteed by (1.3.21).

Hence $(1) = (4_0) \Rightarrow (4_j) \Rightarrow (4_k)$.

$(4_k \Rightarrow 3)$ Since $\bar{\delta}^k f$ is locally Lipschitzian we obtain that its partial difference quotients are bornological, i.e. $\delta^{k+1} f = \delta_1(\bar{\delta}^k f)|_{U^{\langle k+1 \rangle}}$ is bornological.

$(3 \Rightarrow 2)$ One chooses a compact neighborhood $V \subseteq U$.

$(2 \Rightarrow 1)$ Since (1) is a local property this is just (1.3.18). □

1.3.23 Remark. (i) From the remark after (1.3.15) it follows that under the equivalent conditions of the theorem one has: $f^{(k)}(t) = \bar{\delta}^k f(t, \ldots, t)$.

(ii) Conversely one can prove the following formula for the extension:

$$\bar{\delta}^k f(t_0, \ldots, t_k) = \int\limits_{\Delta_0} \cdots \int\limits_{\Delta_p} \partial_0^{k_0} \cdots \partial_p^{k_p} \delta^{p+1} f(x^0, \ldots, x^p) \, \mathrm{d}x^p \ldots \mathrm{d}x^0.$$

Here $(\Delta_0, \ldots, \Delta_p)$ denotes a partition of $(t_0, \ldots, t_k)$, each Δ_i consisting of $k_i + 1$ elements and such that all t_j contained in Δ_i are in the same connected component of U. The integrals on the right side are defined by

$$\int\limits_{\Delta} f(x) \, \mathrm{d}x := \int\limits_0^{r_{k-1}} \cdots \int\limits_0^{r_0} \int\limits_0^1 f(r_0 + s_0(r_1 - r_0) + \cdots + s_k(r_k - r_{k-1})) \, \mathrm{d}s^k \ldots \mathrm{d}s^0$$

$$\text{for } \Delta = (r_0, \ldots, r_k).$$

(iii) Another interpretation of $\bar{\delta}^k f(t_0, \ldots, t_k)$, where the t_i should be equal in groups Δ_j, is as highest coefficient p_k of the polynomial $p(t) := p_0 + (t - t_0)\, p_1 + \ldots + (t - t_0) \ldots (t - t_{k-1})\, p_k$ that agrees with f at the points $t_i \in \Delta_j$ up to order card (Δ_j); e.g. $\delta^2 f(a, a, b)$ is the coefficient of t^3 in the polynomial p of degree at most 3 satisfying $p(a) = f(a)$, $p^{\cdot}(a) = f^{\cdot}(a)$, $p(b) = f(b)$.

1.3.24 Theorem. *Let $U \subseteq \mathbb{R}$ be open and $f: U \to \mathbb{R}$. Then the following statements are equivalent:*

(1) *f is smooth, i.e. has derivatives of all orders;*
(2) *Every point in U has a neighborhood $V \subseteq U$ such that $\delta^k f$ is bounded on $V^{\langle k \rangle}$ for all k (or infinitely many k);*
(3) *For all k (or infinitely many k), $\delta^k f: U^{\langle k \rangle} \to \mathbb{R}$ is a bornological function, see* (1.3.1) *for the bornology on $U^{\langle k \rangle}$;*
(4) *For all k (or infinitely many k) $\delta^k f$ admits a smooth (a locally Lipschitzian) extension to U^{k+1}.*

Proof. This follows immediately from (1.3.22). □

We turn now towards functions of several variables. First we rewrite the recursive definition of δ^k:

1.3.25 Lemma. *Let $D \subseteq \mathbb{R}, f: D \to E$ be a map with values in a vector space E and let $j, k \geq 0$. Then $\delta^{j+k} f = \binom{j+k}{j} \delta_1^j(\delta^k f)$, where $\delta^k f$ has to be considered as map defined on the subset $D^{\langle k \rangle}$ of $D \times D^k$.*

Proof. This follows by induction, using that by the recursion formula in (1.3.1) one has $\delta^{k+1} f = (k+1)\, \delta_1(\delta^k f)$. □

1.3.26 Lemma. *Let $I_1 \times \ldots \times I_m$ be an open box in $\mathbb{R}^m$, $\kappa = (k_1, \ldots, k_m)$, $\kappa' = (k_1 + 1, k_2, \ldots, k_m)$, $f: I_1 \times \ldots \times I_m \to \mathbb{R}$ a map for which the first partial derivative $\partial_1 f$ exists and $\delta^{\kappa'} f$ is bounded. Then $\delta^\kappa \partial_1 f$ is bounded.*

Proof. By the second mean value theorem (1.3.16) we have

$$\delta^\kappa \partial_1 f(x^1, \ldots, x^m) = \delta^{k_1}(\partial_1 \delta^{(k_2, \ldots, k_m)} f(\_, x^2, \ldots, x^m))(x^1) =$$
$$\delta^{k_1+1}(\delta^{(k_2, \ldots, k_m)} f(\_, x^2, \ldots, x^m))(\xi, x^1) = \delta^{\kappa'} f((\xi, x^1), x^2, \ldots, x^m)$$

for some $\xi \in I_1$. In the case where ξ equals some coordinate x_j^1 of x^1, the 'difference quotients' have to be interpreted in the sense of (i) in (1.3.16). □

1.3.27 Corollary. *Let $I_1 \times \ldots \times I_m$ be an open box in $\mathbb{R}^m$, $k \geq 0$, $f: I_1 \times \ldots \times I_m \to \mathbb{R}$ a map for which $\delta^\kappa f$ is bounded for all κ with $|\kappa| = k+1$. Then the partial derivatives $\partial^\kappa f$ of order κ with $|\kappa| \leq k$ exist and all $\delta_i \partial^\kappa f$ are bounded for $i \leq m$.*

Proof. By the recursion formula (v) of (1.3.5) for $\delta^\kappa f$ and (1.3.14) we obtain that $\delta^\kappa f$ is bounded for all $|\kappa| \leq k+1$. Thus by (1.3.17) the partial derivatives $\partial_i f$ exist and by (1.3.26) $\delta^\kappa \partial_i f$ is bounded for all i and all $|\kappa| \leq k$. Applying this argument inductively we obtain the existence of all partial derivatives $\partial^\kappa f$ for $|\kappa| \leq k$ and the boundedness of their difference quotients $\delta_i \partial^\kappa f$. □

1.3.28 Theorem. *Let $U \subseteq \mathbb{R}^m$ be open, $f: U \to \mathbb{R}$ a map, $k \in \mathbb{N}$ and $0 \leq j \leq k$. Then the following statements are equivalent:*

(1) *f is k-times Lipschitz differentiable, cf.* (1.3.19);
(2) *Every point in U has a neighborhood $V \subseteq U$, such that $\delta^\kappa f(V^{\langle \kappa \rangle})$ is bounded for each multi-index κ with $|\kappa| = k+1$;*
(3) *$\delta^\kappa f: U^{\langle \kappa \rangle} \to \mathbb{R}$ is bornological for all κ with $|\kappa| = k+1$, cf.* (1.3.4) *for the bornology of $U^{\langle \kappa \rangle}$;*
(4_j) *$\delta^\kappa f: U^{\langle \kappa \rangle} \to \mathbb{R}$ has a $(k-j)$-times Lipschitz differentiable extension to U^κ for all κ with $|\kappa| = j$.*

Proof. (Compare with the analogous theorem for one variable (1.3.22).)

$(4_j \Rightarrow 4_{j+1})$ for $0 \leq j < k$. We have to extend $\delta^\kappa f$ for all κ with $|\kappa| = j+1$. At least one k_i has to be greater than 0, without loss of generality $k_1 > 0$. Let $\kappa' := (k_2, \ldots, k_m)$. Since the extension problem is a local question, we may assume that $U = I \times W$ with $I \subseteq \mathbb{R}$, $W \subseteq \mathbb{R}^{m-1}$ open. By (4_j) we have a $(k-j)$-times Lipschitz differentiable extension $\overline{\delta}^{(k_1-1,\kappa')} f$ of $\delta^{(k_1-1,\kappa')} f$ to $U^{\langle k_1-1,\kappa'\rangle} = I^{k_1} \times W^{\kappa'} = I \times I^{k_1-1} \times W^{\kappa'}$. By (1.3.21) we obtain a $(k-j-1)$-times Lipschitz differentiable extension of $\delta_1(\overline{\delta}^{(k_1-1,\kappa')} f)$ to $I^2 \times I^{k_1-1} \times W^{\kappa'} = I^{k_1+1} \times W^{\kappa'} = U^{(k_1,\kappa')} = U^\kappa$, which obviously extends at the same time $\delta^\kappa f$.

Hence $(1) = (4_0) \Rightarrow (4_j) \Rightarrow (4_k)$.

$(4_k \Rightarrow 3)$ $\delta^{(k_1,\kappa)} f = \delta_1(\delta^{(k_1-1,\kappa)} f)$ is bornological since $\delta^{(k_1-1,\kappa)} f$ has a locally Lipschitzian extension for $k_1 - 1 + |\kappa| = k$.

$(3 \Rightarrow 2)$ One chooses any compact neighborhood.

$(2 \Rightarrow 1)$ Since (1) is a local property this is obtained by using (1.3.27) and (1.3.20). □

1.3.29 Theorem. *Let $U \subseteq \mathbb{R}^m$ be open, $f: U \to \mathbb{R}$ a map. Then the following statements are equivalent:*

(1) *f is smooth, i.e. has (partial) derivatives of all orders;*
(2) *For every point in U and for every multi-index κ there is a neighborhood $V \subseteq U$, such that $\delta^\kappa f$ is bounded on $V^{\langle\kappa\rangle}$;*
(3) *For all κ the map $\delta^\kappa f: U^{\langle\kappa\rangle} \to \mathbb{R}$ is bornological, cf.* (1.3.4);
(4) *For all κ the difference quotient $\delta^\kappa f: U^{\langle\kappa\rangle} \to \mathbb{R}$ has a locally Lipschitzian (a smooth) extension to U^κ.*

Proof. This follows immediately from (1.3.28). □

The following proposition will be used in section 3.7 in order to determine the differentiable multilinear maps.

1.3.30 Proposition. *Let m: $E_1 \times \ldots \times E_m \to F$ be a multilinear map between vector spaces and c_i: $\mathbb{R} \to E_i$ curves. Then*

$$\delta^k(m \circ (c_1, \ldots, c_m))(t_0, \ldots, t_k)$$
$$= \sum \frac{k!}{k_1! \ldots k_m!} m(\delta^{k_1} c_1(t_0, \ldots, t_{k_1}), \delta^{k_2} c_2(t_{k_1}, \ldots, t_{k_1+k_2}),$$
$$\ldots, \delta^{k_m} c_m(t_{k_1+\cdots+k_{m-1}}, \ldots, t_k)),$$

where the sum is to be taken over all $k_i \geq 0$ with $k_1 + \cdots + k_m = k$.

Proof. One first proves the case $m = 2$. In this case the formula reduces to

$$\delta^k(m \circ (c_1, c_2))(t_0, \ldots, t_k) = \sum_{i=0}^{k} \binom{k}{i} m(\delta^{k-i} c_1(t_0, \ldots, t_{k-i}), \delta^i c_2(t_{k-i}, \ldots, t_k))$$

and is proved by induction on k.

The general formula is then derived by induction on m. Here instead of the multilinear map $\mathscr{m}: E_1 \times \ldots \times E_m \to F$ one considers the associated map $\mathscr{m}^\vee: E_1 \times \ldots \times E_{m-1} \to \mathrm{Lin}(E_m, F)$, where $\mathrm{Lin}(E, F)$ denotes the space of linear maps from E to F. Define $c(t) := \mathscr{m}^\vee(c_1(t), \ldots, c_{m-1}(t))$ and let $\mathrm{ev}: \mathrm{Lin}(E_m, F) \times E_m \to F$ be the bilinear evaluation map. Then $\mathscr{m} \circ (c_1, \ldots, c_m) = \mathscr{m}^\vee(c_1(t), \ldots, c_{m-1}(t))(c_m(t)) = \mathrm{ev}(c(t), c_m(t))$. Now apply step 1 of the proof.

□

1.4 Lipschitz structures and smooth structures

The notion of an $\mathcal{M}$-map introduced in (1.1.1) is particularly useful in order to generalize classical notions of Banach space calculus such as 'locally Lipschitzian', 'k-times Lipschitz differentiable' and 'smooth'. For this we shall use as $\mathcal{M}$ one of the sets $\mathscr{Lip}$, $\mathscr{Lip}^k$ or C^∞ defined as follows:

1.4.1 Definition. (i) $\mathscr{Lip}$ denotes the set of locally Lipschitzian function $\mathbb{R} \to \mathbb{R}$; cf. (1.3.19).

(ii) $\mathscr{Lip}^k$ denotes, for $k \in \mathbb{N}_0$, the set of k-times differentiable functions $\mathbb{R} \to \mathbb{R}$ for which the derivative of order k belongs to $\mathscr{Lip}$. In particular $\mathscr{Lip}^0$ is the same as $\mathscr{Lip}$.

(iii) C^∞ denotes the set of smooth functions $\mathbb{R} \to \mathbb{R}$. This set C^∞ will also be denoted by $\mathscr{Lip}^\infty$.

(iv) With $\underline{\mathscr{Lip}}$, $\underline{\mathscr{Lip}}^k$ and $\underline{C^\infty}$ we denote the respective category obtained according to (1.1.1). Hence notions like $\mathscr{Lip}$-structure, $\mathscr{Lip}^k$-space and $\mathscr{Lip}^k$-map make sense. $\mathscr{Lip}$-structures are also called Lipschitz structures, C^∞-structures also smooth structures.

Any Banach space E, and more generally any vector space E with a given linear subspace E' of its algebraic dual has a natural $\mathscr{Lip}^k$-structure for any $k \in \mathbb{N}_{0,\infty}$, namely the one generated (in the sense of (1.1.3)) by E'. We first show that a map between normed spaces is a $\mathscr{Lip}$-map exactly if it is locally Lipschitzian in the classical sense, cf. (1.3.19). This will be fundamental for proving in Chapter 4 that a map between Banach spaces is a $\mathscr{Lip}^k$-map (respectively a C^∞-map) exactly if it is k-times differentiable with locally Lipschitzian kth derivative (respectively smooth) in the classical sense.

1.4.2 Theorem. *Let E, F be normed spaces, $U \subseteq E$ open and $g: U \to F$ be a map. Then the following statements are equivalent:*

(1) *g is a $\mathscr{Lip}$-map (with respect to the natural Lipschitz structures generated by the continuous linear functionals);*
(2) *g is locally Lipschitzian;*
(3) *g is Lipschitzian on every compact subset of U.*

Proof. (Compare with the similar statement for finite-dimensional E, (1.3.20).)

(1⇒2) Assume that at a point $p \in U$ property (2) fails. Then there exists for each $n \in \mathbb{N}$ points x_n, y_n with $\|x_n - p\| < 1/n^2$, $\|y_n - p\| < 1/n^2$ and $\|g(x_n) -$

$g(y_n)\| > n \cdot \|x_n - y_n\|$. We consider the following curve $c: \mathbb{R} \to E$: $c(t) := x_1$ for $t \le 0$; c runs with constant speed of norm 1 from x_1 to y_1 for $t_1 := 0 \le t \le \|x_1 - y_1\| =: s_1$; similarly it runs from y_1 to x_2 for $s_1 \le t \le s_1 + \|y_1 - x_2\| =: t_2$; and so on. Since $t_\infty := \sum_{n=1}^{\infty} \|x_n - y_n\| + \sum_{n=1}^{\infty} \|x_{n+1} - x_n\|$ is finite, $c(t)$ tends to $p = \lim x_n = \lim y_n$ as t increases towards t_∞; so we define $c(t) := p$ for $t \ge t_\infty$. By construction one has for any $t, s \in \mathbb{R}$: $\|c(t) - c(s)\| \le |t - s|$ and therefore for all $\ell \in E': |(\ell \circ c)(t) - (\ell \circ c)(s)| \le \|\ell\| \cdot |t - s|$, where $\|\ell\| := \sup\{|\ell(x)|; \|x\| \le 1\}$ is the operator norm of ℓ. This shows that c is a $\mathscr{Lip}$-curve in E, and thus at least locally around t_∞ stays in U.

On the other hand we have $\|c(t_n) - c(s_n)\| = \|x_n - y_n\| = |t_n - s_n|$, hence $\|(g \circ c)(t_n) - (g \circ c)(s_n)\| = \|g(x_n) - g(y_n)\| > n \cdot |t_n - s_n|$. This shows that the set

$$\left\{\frac{(g \circ c)(t_n) - (g \circ c)(s_n)}{t_n - s_n}; n \in \mathbb{N}\right\}$$

is unbounded in the norm. Hence it is unbounded under some element $\ell \in F'$ (cf. (i) in (2.1.21)), i.e.

$$\left\{\frac{(\ell \circ g \circ c)(t_n) - (\ell \circ g \circ c)(s_n)}{t_n - s_n}; n \in \mathbb{N}\right\}$$

is unbounded. Thus $\ell \circ g \circ c \notin \mathscr{Lip}$, and this is in contradiction to property (1).

(2⇒3) This is proved in the same way as (1⇒2) in (1.3.20).

(3⇒1) Let $c: \mathbb{R} \to V$ be a $\mathscr{Lip}$-curve; $I \subseteq \mathbb{R}$ a bounded closed interval. Since for any $\ell \in E'$ one has $\ell \circ c \in \mathscr{Lip}$, the set

$$\left\{\ell\left(\frac{c(t) - c(s)}{t - s}\right); t \ne s,\ t, s \in I\right\}$$

is bounded. By the uniform boundedness principle (cf. [Jarchow, 1981, p. 220]) there exists a constant N_1 such that $\|c(t) - c(s)\| / |t - s| \le N_1$ for $t \ne s$; $t, s \in I$. Since c is continuous, $c(I)$ is compact and so using (3) we find a constant N_2 with $\|g(x) - g(y)\| \le N_2 \cdot \|x - y\|$ for $x, y \in c(I)$. Putting this together we get $\|(g \circ c)(t) - (g \circ c)(s)\| \le N_1 N_2 |t - s|$ and hence for any $\ell \in F': |(\ell \circ g \circ c)(t) - (\ell \circ g \circ c)(s)| \le N_1 N_2 \|\ell\| \cdot |t - s|$.

This shows that $\ell \circ g \circ c \in \mathscr{Lip}$ and hence g is a $\mathscr{Lip}$-map. □

The main result of this section, which states that the category $\underline{C^\infty}$ of smooth spaces is cartesian closed, was first proved in [Lawvere, Schanuel, Zame, 1981]. In (4.4.44) we will show that $\underline{C^\infty}$ contains all classical and even many infinite-dimensional smooth manifolds.

1.4.3 Theorem. *The category $\underline{C^\infty}$ of smooth spaces is cartesian closed. The smooth structure of $C^\infty(\mathbb{R}, \mathbb{R})$ is generated by all functions of the form $\ell \circ \delta^k: C^\infty(\mathbb{R}, \mathbb{R}) \to \mathbb{R}$ where $\delta^k: C^\infty(\mathbb{R}, \mathbb{R}) \to \ell^\infty(\mathbb{R}^{\langle k \rangle}, \mathbb{R})$ consists in taking the difference quotient of order k and ℓ runs through the linear ℓ^∞-maps $\ell^\infty(\mathbb{R}^{\langle k \rangle}, \mathbb{R}) \to \mathbb{R}$. In particular the smooth structure of $C^\infty(\mathbb{R}, \mathbb{R})$ is linearly generated; cf.* (1.1.3).

Proof. In order to apply theorem (1.1.7) we have to study the set $\mathscr{C}_{C^\infty}:=\{c:\mathbb{R}\to C^\infty;\hat{c}\in\underline{C^\infty}(\mathbb{R}\sqcap\mathbb{R},\mathbb{R})\}$. Since $\mathbb{R}$ has to be considered with its natural C^∞-structure (C^∞, C^∞) one gets:

$$\mathscr{C}_{C^\infty}=\{c:\mathbb{R}\to C^\infty;\quad \hat{c}\circ(\sigma,\tau)\in C^\infty \text{ for all } \sigma,\tau\in C^\infty\}.$$

Since according to a theorem of [Boman, 1967] (of which a new proof is given in (4.3.30)) a function $\mathbb{R}^2\to\mathbb{R}$ is smooth if and only if it is smooth along smooth curves, this gives:

$$\mathscr{C}_{C^\infty}=\{c:\mathbb{R}\to C^\infty;\quad \hat{c}:\mathbb{R}^2\to\mathbb{R} \text{ is smooth}\}.$$

Again one considers the associated function-set:

$$\Phi\mathscr{C}_{C^\infty}=\{f:C^\infty\to\mathbb{R};\quad f\circ c\in C^\infty \text{ for all } c\in\mathscr{C}_{C^\infty}\}.$$

We first show that it contains at least the following functions:

For any $k\in\mathbb{N}_0$ and any linear $\underline{\ell^\infty}$-morphism $\ell:\ell^\infty(\mathbb{R}^{\langle k\rangle},\mathbb{R})\to\mathbb{R}$, the function $\ell\circ\delta^k:C^\infty\to\mathbb{R}$ belongs to $\Phi\mathscr{C}_{C^\infty}$.

According to (1.3.24) this can be proved by showing that for any $c\in\mathscr{C}_{C^\infty}$ one has: $\delta^j(\ell\circ\delta^k\circ c)$ is an $\underline{\ell^\infty}$-morphism for all $j\geq 0$. For $c\in\mathscr{C}_{C^\infty}$ the function $\hat{c}:\mathbb{R}^2\to\mathbb{R}$ is smooth and hence by (1.3.29): $\delta_1^j\,\delta_2^k\,\hat{c}:\mathbb{R}^{\langle j\rangle}\times\mathbb{R}^{\langle k\rangle}\to\mathbb{R}$ is an ℓ^∞-map. According to the cartesian closedness of $\underline{\ell^\infty}$ (1.2.8) we conclude: $(\delta_1^j\,\delta_2^k\,\hat{c})^\vee:\mathbb{R}^{\langle j\rangle}\to\ell^\infty(\mathbb{R}^{\langle k\rangle},\mathbb{R})$ is an $\underline{\ell^\infty}$-morphism, where $(\,)^\vee$ is defined by $h^\vee(x)(y):=h(x,y)$. Since one has the identity $\delta^j(\delta^k\circ c)=(\delta_1^j\,\delta_2^k\hat{c})^\vee$ we obtain, using the linearity of ℓ: $\delta^j(\ell\circ\delta^k\circ c)=\ell\circ\delta^j(\delta^k\circ c)=\ell\circ(\delta_1^j\,\delta_2^k\,\hat{c})^\vee$.

As composite of two $\underline{\ell^\infty}$-morphisms this is an $\underline{\ell^\infty}$-morphism, and this implies that $\ell\circ\delta^k$ belongs to $\Phi\mathscr{C}_{C^\infty}$.

Now let c: $\mathbb{R}\to C^\infty$ be such that $f\circ c\in C^\infty$ for all $f\in\Phi\,\mathscr{C}_{C^\infty}$ which are obtained according to the lemma, i.e. for $f=\ell\circ\delta^k$ with ℓ as above. We have to show that $\hat{c}$: $\mathbb{R}^2\to\mathbb{R}$ is smooth. By (1.3.29) this is equivalent with

$$\delta_1^j\,\delta_2^k\,\hat{c}:\mathbb{R}^{\langle j\rangle}\times\mathbb{R}^{\langle k\rangle}\to\mathbb{R} \text{ is an } \ell^\infty\text{-map for all } j,k\geq 0.$$

Which by (1.2.8) is again equivalent with

$$(\delta_1^j\,\delta_2^k\,\hat{c})^\vee:\mathbb{R}^{\langle j\rangle}\to\ell^\infty(\mathbb{R}^{\langle k\rangle},\mathbb{R}) \text{ is an } \underline{\ell^\infty}\text{-morphism for all } j,k\geq 0.$$

Since by (1.2.10) the ℓ^∞-structure of $\ell^\infty(\mathbb{R}^{\langle k\rangle},\mathbb{R})$ is linearly generated, we can test this last map by composing with functions ℓ as in the lemma. We use again the identity $\ell\circ(\delta_1^j\,\delta_2^k\,\hat{c})^\vee=\delta^j(\ell\circ\delta^k\circ c)$. By assumption, $\ell\circ\delta^k\circ c$ is smooth, so $\delta^j(\ell\circ\delta^k\circ c)$ is an $\underline{\ell^\infty}$-morphism by (1.3.24) and the theorem is proved. □

1.4.4 Proposition. *Let X be a smooth space and E any vector space with a smooth structure that is generated by a set $\mathscr{S}$ of linear functions. Then the following families of linear morphisms are initial:*

(*i*) $c^*:C^\infty(X,E)\to C^\infty(\mathbb{R},E)\quad(c\in C^\infty(\mathbb{R},X))$;
(*ii*) $\ell_*:C^\infty(X,E)\to C^\infty(X,\mathbb{R})\quad(\ell\in\mathscr{S})$;
(*iii*) $C^\infty(c,\ell):C^\infty(X,E)\to C^\infty(\mathbb{R},\mathbb{R})\quad(c\in C^\infty(\mathbb{R},X),\ \ell\in\mathscr{S})$.

Proof. This result is analogous to (1.2.9) and is an immediate consequence of (1.1.8).

1.4.5 Corollary. *Let X be any smooth space and E any vector space with a linearly generated smooth structure. Then the smooth structure of $C^\infty(X, E)$ is also linearly generated.*

We give now some standard consequences of cartesian closedness.

1.4.6 Proposition. *For any smooth spaces X, Y, Z the evaluation* ev: $C^\infty(X, Y)\sqcap X \to Y$ *and the composition* comp: $C^\infty(Y, Z)\sqcap C^\infty(X, Y) \to C^\infty(X, Z)$ *are C^∞-maps.*

Proof. Since $\mathrm{ev} = (\mathrm{id}_{C^\infty(X,\,Y)})^\wedge$, the first part is obvious. So is the second part since $\mathrm{comp}^\wedge(g, f, x) = \mathrm{ev}(g, \mathrm{ev}(f, x))$ for appropriately chosen evaluation maps. □

1.4.7 Definition. A *smooth group* G is a smooth space for which the underlying set has a given group structure such that the group multiplication $m: G\sqcap G \to G$ and the inversion $\nu: G \to G$ are C^∞-maps.

1.4.8 Proposition. *Let X be any smooth space. If one puts on the group* Diff(X) *of all C^∞-diffeomorphisms of X the initial smooth structure induced by the two maps i, j:* Diff$(X) \to C^\infty(X, X)$, *where $i(f) := f$ and $j(f) := f^{-1}$, then* Diff(X) *is a smooth group.*

Proof. One has the following identity for the group multiplication: $i \circ m = \mathrm{comp} \circ (i \sqcap i)$. Hence, using (1.4.6), $i \circ m$ is a C^∞-map. Similarly $j \circ m = \mathrm{comp} \circ (j \sqcap j) \circ \varphi$ where $\varphi(f, g) := (g, f)$ shows that $j \circ m$ is a C^∞-map. Together this shows that m is a C^∞-map. For the inversion map ν it is even simpler, since $i \circ \nu = j$ and $j \circ \nu = i$. □

1.4.9 Definition. A smooth action of a smooth group G on a smooth space X is a C^∞-map $f: G\sqcap X \to X$ such that

(i) $f(g_1 g_2, x) = f(g_1, f(g_2, x))$ for $g_1, g_2 \in G$, $x \in X$.
(ii) $f(e, x) = x$ for $x \in X$, e the neutral element in G.

Act(G, X) shall denote the smooth space formed by the C^∞-actions of G on X with the smooth structure induced by its inclusion in $C^\infty(G\sqcap X, X)$.

1.4.10 Proposition. *There is a natural isomorphism between the space* Act(G, X) *of smooth actions of G on X and the space* Hom$(G$, Diff$(X))$ *formed by the C^∞-homomorphisms $G \to$* Diff(X), *where* Hom$(G$, Diff$(X))$ *has the smooth structure induced by its inclusion in $C^\infty(G$,* Diff$(X))$ *and* Diff(X) *the one used in* (1.4.8).

Proof. To $f \in \mathrm{Act}(G, X)$ one associates $f^\vee : G \to C^\infty(X, X)$ and verifies that for any $g \in G$ one has $f^\vee(g) \in \mathrm{Diff}(X)$ and that $f^\vee : G \to \mathrm{Diff}(X)$ is a C^∞-homomorphism. To $h \in \mathrm{Hom}(G, \mathrm{Diff}(X))$ one associates $\hat{h} : G \sqcap X \to X$ and verifies that $\hat{h} \in \mathrm{Act}(G, X)$. Obviously these maps define a bijection between $\mathrm{Act}(G, X)$ and $\mathrm{Hom}(G, \mathrm{Diff}(X))$. That the bijection is a $\underline{C}^\infty$-isomorphism also follows from cartesian closedness, i.e. the universal property of the function space structures one works with. □

2 CONVENIENT VECTOR SPACES

In some sense convenient vector spaces are the most general linear spaces for which a differentiation theory upholding the basic classical properties is possible. For calculus one obviously needs limits and therefore certain separation and completeness conditions. These conditions will be specified in sections 2.5 and 2.6, but are not imposed for the preliminary considerations which concern a bigger class of spaces called preconvenient vector spaces. Various types of structures can be used to describe preconvenient vector spaces. We shall discuss these structures carefully in sections 2.1–2.3 and show in section 2.4 how for a preconvenient vector space they determine each other.

In section 2.1 we summarize classical material on locally convex and convex bornological spaces. In section 2.2 convenient vector spaces are identified, by means of Mackey convergence, with certain convergence vector spaces. The Mackey convergence of filters and nets is discussed and it is shown that for most results one can stick to Mackey convergence of ordinary sequences. In the same section also the topology associated to the Mackey convergence structure, called the Mackey closure topology, is introduced. Its main interest is due to the fact that it is, as shown in section 2.3, the final topology induced by various families of curves, e.g. the smooth ones. It is not compatible with the vector addition in the classical sense but it is compatible if considered as arc-generated topology. For calculus on vector spaces the structures as considered in Chapter 1, i.e. ℓ^∞-, $\mathcal{L}ip^k$- or smooth structures are important. Since all the structures mentioned so far determine a dual we also introduce vector spaces structured by specifying a subspace of the algebraic dual. The relations between the various structures are investigated. An appropriate way to do this is in terms of (adjoint) functors. In many cases a non-categorical reformulation is added. In particular (2.4.4) is such a reformulation of (2.4.3) summarizing the various characterizations of preconvenient vector spaces. If the reader is willing to accept this result he can start reading this chapter there.

In sections 2.5 and 2.6 the separation and completeness conditions are described in terms of the various structures, and it is shown that they are not only sufficient but also necessary for the uniqueness and existence of the desired limits. To every non-separated preconvenient vector space one associates a separated one having the usual universal property. Similarly to every non-complete preconvenient vector space an associated complete one will be constructed. The separation and completion functors so obtained are fundamental in order to show in Chapter 3 categorical properties for the convenient vector spaces.

2.1 Locally convex and convex bornological spaces

For many classes of linear spaces, in particular those mentioned in the title, one has natural dual spaces. Since we do not suppose a separation condition it is useful to work with the following

2.1.1 Definition. (i) By a *dualized vector space E* we shall understand a (real) vector space (denoted also by E) together with a given subspace E' of the algebraic dual of E.

(ii) The category $\underline{DVS}$ has as objects the dualized vector spaces; the morphisms from E_1 to E_2 are those linear maps $m: E_1 \to E_2$ which satisfy $m^*(E_2') \subseteq E_1'$.

Remark. In the special case where the functions in E' separate points of E, (E, E') is a dual pair in the usual sense.

We next recall the definition of (convex) bornological vector spaces; cf. [Hogbe-Nlend, 1977, p. 19].

2.1.2 Definition. (i) A bornology $\mathscr{B}$ (cf. (1.2.1)) on a vector space E is called *vector bornology* if the addition $E \times E \to E$ and the scalar multiplication $\mathbb{R} \times E \to E$ are bornological maps with respect to the product bornologies (cf. (1.2.2)), $\mathbb{R}$ being taken with its standard bornology. A *bornological vector space* is a vector space together with a vector bornology.

(ii) A *convex bornological space* is a bornological vector space for which the convex hull of each bounded set is also bounded.

(iii) $\underline{Born\,VS}$ denotes the category of bornological vector spaces with the linear bornological maps as morphisms; $\underline{CBS}$ denotes the full subcategory formed by the convex bornological spaces.

(iv) A bornological vector space E is called *separated* iff $\{0\}$ is the only bounded subspace, or equivalently if for $0 \neq x \in E$ the subspace $\mathbb{R} \cdot x$ is unbounded.

2.1.3 Lemma. *A bornology $\mathscr{B}$ on a vector space E is a vector bornology if and only if it has the following two properties:*

(*i*) $B \in \mathscr{B} \Rightarrow B + B \in \mathscr{B}$;
(*ii*) $B \in \mathscr{B} \Rightarrow \bigcup_{|t| \leq 1} t \cdot B \in \mathscr{B}$.

Proof. Trivial verification. □

2.1.4 Lemma. *A bornology $\mathscr{B}$ on a vector space is a convex vector bornology if and only if it has the following three properties:*

(*i*) $B \in \mathscr{B} \Rightarrow -B \in \mathscr{B}$;
(*ii*) $B \in \mathscr{B} \Rightarrow 2 \cdot B \in \mathscr{B}$;
(*iii*) $B \in \mathscr{B} \Rightarrow \langle B \rangle \in \mathscr{B}$, *where* $\langle \_ \rangle$ *denotes the convex hull.*

Proof. Trivial verification. □

2.1.5 Proposition. *CBS is a reflective and coreflective subcategory of Born VS.*

Proof. Let $\mathscr{B}$ be a vector bornology on a vector space E. Using the lemma above one easily verifies that the following are convex vector bornologies on E:

$$\mathscr{B}_1 := \{B_1 \subseteq E;\ B_1 \subseteq \langle B \rangle \text{ for some } B \in \mathscr{B}\};$$

$$\mathscr{B}_2 := \{B_2 \subseteq E;\ \langle B_2 \rangle \in \mathscr{B}\}.$$

Replacing $\mathscr{B}$ by $\mathscr{B}_1$ resp. $\mathscr{B}_2$ gives two functors *Born VS*$\to$*CBS*; the first is left, the the second right adjoint to the inclusion functor. □

2.1.6 Remark. The categories $\mathscr{X}$ of linear spaces which we shall consider will always contain $\mathbb{R}$ as an object and for any $\mathscr{X}$-object X, the set $\mathscr{X}(X, \mathbb{R})$ will be a linear subspace of the algebraic dual of X. One therefore obtains a dualized vector space δX having the same underlying vector space and $(\delta X)' := \mathscr{X}(X, \mathbb{R})$. This extends to a functor $\delta: \mathscr{X} \to$ *DVS* preserving the underlying vector spaces and the underlying maps. It will be called the duality functor for $\mathscr{X}$ and will be denoted by the same δ for various choices of $\mathscr{X}$.

In most cases $\mathscr{X}$ has initial structures with respect to the forgetful functor $\mathscr{X} \to$ *VS*. We can then associate to any dualized vector space E an $\mathscr{X}$-object σE as follows: it has the same underlying vector space and the initial structure induced by the family $\ell: E \to \mathbb{R}$ ($\ell \in E'$). This extends to a functor $\sigma:$ *DVS*$\to \mathscr{X}$ preserving the underlying vector spaces and the underlying maps. σ will be used with an index indicating the respective choice of $\mathscr{X}$.

2.1.7 Proposition. *The duality functor δ: CBS$\to$DVS for the category of convex bornological spaces has a right adjoint σ_b: DVS$\to$CBS (given by scalar boundedness) and both preserve the underlying vector spaces and the underlying maps.*

Proof. Since the verifications to be made are trivial, we only describe explicitly the structure of $\sigma_b E$ according to the general definition given in (2.1.6): $B \subseteq \sigma_b E$

is bounded iff B is scalarly bounded, i.e. iff $\ell(B) \subseteq \mathbb{R}$ is bounded for all $\ell \in E'$. This is the initial bornology induced by all $\ell \in E'$, where $\mathbb{R}$ is considered with the usual bornology. □

2.1.8 Definition. $\underline{LCS}$ denotes the category of (not necessarily separated) locally convex spaces, cf. [Jarchow, 1981, p. 108], with the linear continuous maps as morphisms.

2.1.9 Proposition. *For the category of locally convex spaces the duality functor $\delta: \underline{LCS} \to \underline{DVS}$ has a left adjoint $\mu: \underline{DVS} \to \underline{LCS}$ (given by the Mackey topology) and a right adjoint $\sigma_t: \underline{DVS} \to \underline{LCS}$ (given by the weak topology); both preserve the underlying vector spaces and the underlying maps and they satisfy $\delta \circ \mu = \delta \circ \sigma_t = Id$.*

Proof. Let us first, for a dualized vector space E, describe μE; it is obtained by supplying E with the finest locally convex topology with the property that E' becomes the topological dual. That this topology exists and behaves functorially is well known in the separated case; it is called the Mackey topology (cf. [Jarchow, 1981, p. 58, p. 61]). For arbitrary E one considers the associated separated dualized vector space $E_1 := E/E_0$ where $E_0 := \{x \in E; \ell(x) = 0$ for all $\ell \in E'\}$, with $(E_1)' := \{\ell_1 : E_1 \to \mathbb{R};\ \ell_1$ linear and $\ell_1 \circ \pi \in E')$, $\pi: E \to E_1$ being the canonical projection. One structures E_1 with the Mackey topology and then E with the initial topology induced by π and calls the result μE, or $\mu(E, E')$. One shows, using Hahn–Banach, that E' becomes the topological dual of μE, and that this gives the finest locally convex topology on E with that property. The adjunction follows since by construction $\delta\mu E = E$ for every dualized vector space; and id: $\mu\delta F \to F$ is continuous for every locally convex space F. We recall that a (separated) locally convex space F is said to be a Mackey space iff $F = \mu\delta F$.

The right adjoint σ_t is obtained according to the general construction of (2.1.6): One puts on E the initial topology, denoted by $\sigma(E, E)$, induced by all $\ell: E \to \mathbb{R}$ of E', called the weak topology, and one verifies that this is the coarsest locally convex topology on E yielding E' as topological dual, cf. [Jarchow, 1981, p. 147]. □

We next recall the classical relation between locally convex and convex bornological spaces, expressed by a pair of adjoint functors $\underline{LCS} \leftrightarrows \underline{CBS}$.

2.1.10 Proposition. (*i*) *For any locally convex space E the subsets B which are absorbed by every 0-neighborhood U (that means there exists a $t \in \mathbb{R}$ with $B \subseteq t \cdot U$) form a convex vector bornology (usually called the von Neumann bornology of E), and thus one obtains a functor $\beta: \underline{LCS} \to \underline{CBS}$.*

(*ii*) *This functor β has a left adjoint $\gamma: \underline{CBS} \to \underline{LCS}$ (one takes as 0-neighborhood basis the bornivorous absolutely convex subsets, i.e. those that absorb bounded* *sets); and both functors preserve the underlying vector spaces and the underlying maps.*

Proof. (i) is trivial.

(ii) On a convex bornological space E there exists a unique locally convex topology having as 0-neighborhood basis the absolutely convex bornivorous subsets of E, cf. [Jarchow, 1981, p. 33]. The so obtained space γE has the finest locally convex topology having as von Neumann bornology the original one, or, in other words, such that $\mathrm{id}: E \to \beta\gamma E$ is bornological. This implies that $\mathrm{id}: \gamma\beta F \to F$ is continuous for any locally convex space F. We thus have described unit and co-unit of the stated adjunction. □

2.1.11 Remark. We shall often meet, as in the above proposition, functors $\mathscr{X} \underset{\psi}{\overset{\varphi}{\rightleftarrows}} \mathscr{Y}$ between concrete categories which preserve the underlying spaces and the underlying maps. It is then a trivial consequence that φ and ψ induce isomorphisms between the full subcategories $\mathscr{X}^* := \{X \in \mathscr{X}; X = \psi\varphi X\}$ and $\mathscr{Y}^* := \{Y \in \mathscr{Y};\ Y = \varphi\psi Y\}$. If in addition ψ is left adjoint to φ and unit and counit of adjunction are formed by the respective identity maps, then $\mathscr{X}^*$ is coreflective in $\mathscr{X}$ (with $\psi\varphi$ as right adjoint to the inclusion); $\mathscr{Y}^*$ is reflective in $\mathscr{Y}$ (with $\varphi\psi$ as left adjoint to the inclusion) and $\varphi\psi\varphi = \varphi$, $\psi\varphi\psi = \psi$. In the case of the functors β and γ of (2.1.10) one finds as $\mathscr{X}^*$ and $\mathscr{Y}^*$ well known subcategories.

We recall now the classical terminology and the consequences of (2.1.10) according to the general remark above.

2.1.12 Definition. (i) $\underline{bLCS}$ denotes the full subcategory of $\underline{\mathrm{LCS}}$ formed by the *bornological locally convex spaces*, i.e., the locally convex spaces F satisfying $F = \gamma\beta F$; this equation means: the bornivorous absolutely convex subsets form a 0-neighborhood basis of the topology. For any locally convex space F, the locally convex space $\gamma\beta F$ is called the *bornologification* of F.

(ii) $\underline{tCBS}$ denotes the full subcategory of $\underline{\mathrm{CBS}}$ formed by the so-called *topological convex bornological spaces*, i.e. the convex bornological spaces E (cf. (2.1.2)) satisfying $E = \beta\gamma E$; this equation means: the subsets that are absorbed by the bornivorous absolutely convex sets are the bounded ones.

2.1.13 Corollary.

(i) $\underline{bLCS} \cong \underline{tCBS}$;
(ii) $\underline{bLCS}$ *is coreflective in* $\underline{LCS}$;
(iii) $\underline{tCBS}$ *is reflective in* $\underline{CBS}$;

Locally convex topologies, convex vector bornologies and the functors β, γ of (2.1.10) can also be described by means of seminorms.

2.1.14 Definition. A *seminorm* on a vector space E is a function $p: E \to \mathbb{R}$ with the properties:

(i) $p(x) \geq 0$ *for all* $x \in E$;
(ii) $p(tx) = |t| p(x)$ for all $x \in E$ and $t \in \mathbb{R}$;
(iii) $p(x+y) \leq p(x) + p(y)$ for all $x, y \in E$.

A non-empty family $\mathscr{P}$ of seminorms on a vector space E determines a locally convex topology on E (one takes the sets $\{x \in E; p(x) < 1/n\}$ for $p \in \mathscr{P}$ and $n \in \mathbb{N}$ as subbasis for the 0-neighborhood filter) and a convex vector bornology (one takes as bounded sets those which are bounded under each $p \in \mathscr{P}$).

2.1.15 Lemma. *Let $B \neq \emptyset$ be an absolutely convex subset of a vector space E. Then $\bigcup_{n \in \mathbb{N}} nB$ is the linear subspace of E generated by B, and one obtains a seminorm on it (called the Minkowsky functional of B) by defining $\|x\|_B := \inf\{s > 0; x \in sB\}$. With E_B one denotes the so obtained seminormed space.*

2.1.16 Remark. Suppose $B_j \neq \emptyset$ is an absolutely convex subset of a vector space E_j and set $E := \prod_{j \in J} E_j$ and $B := \mathsf{X}_{j \in J} B_j \subseteq E$. Then it is easy to show that $E_B \cong \prod_{j \in J} (E_j)_{B_j}$. Here we use the product of normed spaces F_j; it is the vector space of those elements of the cartesian product of the vector spaces F_j for which the norm $x \mapsto \|x\|_\infty := \sup\{\|\mathrm{pr}_j(x)\|; j \in J\}$ is finite, cf. [Hogbe-Nlend, 1977, p. 10]. In the case where J is finite, the underlying vector space of $\prod_{j \in J} F_j$ is the cartesian product of the spaces F_j and the norm $\|\_\|_\infty$ is equivalent to the norm $\|\_\|_p$ for any positive real number p where $(\|x\|_p)^p := \sum_j \|\mathrm{pr}_j(x)\|^p$.

2.1.17 Remark. If U is an absolutely convex 0-neighborhood in a locally convex space E, then $E_U = E$ as vector space and $\|\_\|_U$ is continuous. By taking all U in a 0-neighborhood basis $\mathscr{U}_0$ one obtains a family of seminorms that determines the topology of E. By defining $U_1 \succ U_2$ iff $U_1 \subseteq U_2$ the set $\mathscr{U}_0$ becomes directed (cf. (i) of (2.2.1)), $\mathrm{id}: E_{U_1} \to E_{U_2}$ is continuous for $U_1 \succ U_2$, and E is the projective limit in <u>LCS</u> of the so obtained projective system, cf. (8.3.3). One easily deduces that a locally convex space can always be represented as a projective limit in <u>LCS</u> of seminormed spaces. By replacing the maps $E_{U_1} \to E_{U_2}$ described above by those induced on the separated quotients one deduces that every separated locally convex space can be represented as a projective limit in <u>LCS</u> of normed spaces.

We shall use the analogous results for convex bornological spaces. There the situation is dual to the one above.

2.1.18 Proposition. *Let E be a convex bornological space and let $\mathscr{B}_0$ be a basis of the bornology of E consisting of absolutely convex sets. By defining $B_1 \succ B_2$ iff $B_1 \supseteq B_2$ the set $\mathscr{B}_0$ becomes directed and for $B_1 \succ B_2$ the inclusion $E_{B_2} \subseteq E_{B_1}$ is bornological. E is inductive limit in <u>CBS</u> of the so obtained inductive system of seminormed spaces, cf.* (8.3.4).

Proof. Since $B_1 \succ B_2$ implies $\|x\|_{B_1} \leq \|x\|_{B_2}$ for all $x \in E_{B_2}$ the inclusion $E_{B_2} \to E_{B_1}$ is bornological. One obviously has $E = \bigcup_{B \in \mathscr{B}_0} E_B$, and the inclusions

$E_B \to E$ are bornological, since $A \subseteq E_B$ bounded implies that there exists an n with $\|x\|_B < n$ for all $x \in A$, hence $A \subseteq nB$ is bounded in E. Conversely, if $A \subseteq E$ is bounded then there exists a $B \in \mathscr{B}_0$ with $A \subseteq B$, hence $A \subseteq E_B$ and $\|A\|_B \leq 1$, i.e. A is bounded in E_B. The universal colimit property, cf. (8.3.1), is an immediate consequence. □

2.1.19 Proposition. *Let E be a locally convex space and $\mathscr{B}_0$ a family of absolutely convex subsets forming a basis of the von Neumann bornology. Then the colimit in* <u>LCS</u> *of the inductive system E_B ($B \in \mathscr{B}_0$) is the bornologification of E (i.e. is $\gamma\beta E$).*

Proof. A 0-neighborhood basis for the inductive limit is given by all absolutely convex sets U for which $U \cap E_B$ is a 0-neighborhood in the normed space E_B for all $B \in \mathscr{B}_0$, cf. (3.1.1). Clearly $U \cap E_B$ is a 0-neighborhood in E_B iff U contains εB for some $\varepsilon > 0$, i.e. iff U absorbs B. So the 0-neighborhood basis of the inductive limit is formed by the absolutely convex bornivorous subsets, i.e. is exactly the 0-neighborhood basis of the bornologification of E, cf. (2.1.12). □

2.1.20 Remarks. (i) If the topology of a locally convex space E is determined by a family $\mathscr{P}_0$ of seminorms, then a subset is bounded in βE iff every $p \in \mathscr{P}_0$ is bounded on it. Thus the family $\mathscr{P}_0$ determines the bornology of βE (i.e. the von Neumann bornology of E).

(ii) If E is a convex bornological space and $\mathscr{P}$ the family of all bornological seminorms on E, then the topology determined by $\mathscr{P}$ is that of γE. In fact, for any bornological seminorm p and $\varepsilon > 0$ the set $\{x \in E;\ p(x) < \varepsilon\}$ is absolutely convex and bornivorous, i.e. a 0-neighborhood in γE. Conversely, if U is a 0-neighborhood in γE then its Minkowsky functional is a bornological seminorm and $\{x;\ \|x\|_U < 1\} \subseteq U$.

However, a family $\mathscr{P}_0$ of bornological seminorms on E that determines the bornology does not always determine the topology of γE. For example take a Banach space E; its bornology is determined by the seminorms $x \mapsto |\ell(x)|$ ($\ell \in E'$) and these determine the weak topology of E, but that of γE is the norm-topology.

(iii) As consequence we obtain:

(a) For a locally convex space E the topology of the associated bornological locally convex space $\gamma\beta E$ is determined by all seminorms of E which are bounded on the (von Neumann) bounded sets. In particular, a locally convex space E is bornological (i.e. $E = \gamma\beta E$) iff its continuous seminorms coincide with its bornological seminorms.

(b) For a convex bornological space E the bornology of the associated topological convex bornological space $\beta\gamma E$ is determined by the family of bornological seminorms of E. In particular we have, cf. (2.1.24): the bornology of E is topological iff $B \subseteq E$ is bounded provided $p(B)$ is bounded for every bornological seminorm p. Thus the bornologies determined by families of seminorms on E are exactly the topological convex bornologies.

(iv) Every metrizable locally convex space is bornological, see for example [Horváth, 1966, p. 223] or [Jarchow, 1981, p. 273].

(v) A (separated) locally convex space is bornological iff it can be represented as an inductive limit of seminormable (normable) spaces.

2.1.21 Proposition. *The following triangles commute:*

(i) $$\begin{array}{ccc}\underline{\mathrm{LCS}} & \xrightarrow{\beta} & \underline{\mathrm{CBS}}\\ \delta\big\downarrow & \nearrow{\scriptstyle\sigma_b} & \\ \underline{\mathrm{DVS}} & & \end{array}$$

(ii) $$\begin{array}{ccc}\underline{\mathrm{CBS}} & \xrightarrow{\delta} & \underline{\mathrm{DVS}}\\ \gamma\big\downarrow & \nearrow{\scriptstyle\delta} & \\ \underline{\mathrm{LCS}} & & \end{array}$$

(iii) $$\begin{array}{ccc}\underline{\mathrm{DVS}} & \xrightarrow{\sigma_b} & \underline{\mathrm{CBS}}\\ \mu\big\downarrow & \nearrow{\scriptstyle\beta} & \\ \underline{\mathrm{LCS}} & & \end{array}$$

(iv) $$\begin{array}{ccc}\underline{\mathrm{CBS}} & \xrightarrow{\gamma} & \underline{\mathrm{LCS}}\\ \delta\big\downarrow & \nearrow{\scriptstyle\mu} & \\ \underline{\mathrm{DVS}} & & \end{array}$$

In non-categorical language this means:

(i) For any locally convex space the bounded and the scalarly bounded subsets coincide.

(ii) On any convex bornological space a linear function is bornological iff the inverse image of]-1,1[*is bornivorous.*

(iii) For any dualized vector space the scalarly bounded sets coincide with the sets bounded in the Mackey topology.

(iv) For any convex bornological space an absolutely convex set is bornivorous iff it is a 0-neighbourhood in the Mackey topology determined by the bornological linear functions.

Proof. (i) For separated locally convex spaces, this is a classical theorem, see e.g. [Jarchow, 1981, p. 151]; similarly as in the proof of (2.1.9) the general case can be reduced to the separated one.

(ii) This is a direct consequence of the adjunction stated in (ii) of (2.1.10).

(iii) Using (i) we get $\beta\mu=\sigma_b\delta\mu=\sigma_b$; cf. (2.1.9).

(iv) (a) Using (ii) we have $\mu\delta=\mu\delta\gamma$ and (since $\mu\delta$ refines the topology) id: $\mu\delta E\to\gamma E$ is continuous.

(b) Using the previous results we have $\gamma=\gamma\beta\gamma=\gamma\sigma_b\delta\gamma=\gamma\sigma_b\delta=\gamma\beta\mu\delta$, hence (since $\gamma\beta$ refines the topology) id: $\gamma E\to\mu\delta E$ is continuous. □

2.1.22 Corollary. *A locally convex space E is bornological (i.e. $E=\gamma\beta E$) iff E has the Mackey topology (i.e. $E=\mu\delta E$) and every linear function $\ell: E\to\mathbb{R}$ which is bornological (with respect to the von Neumann bornology) is continuous on E (i.e. $\delta\beta E=\delta E$).*

Proof. ($\Leftarrow$) Using (iv) of (2.1.21) one obtains: $\gamma\beta E=\mu\delta\beta E=\mu\delta E=E$.
($\Rightarrow$) is trivial. □

Using (2.1.21) we also get a simple characterization of the topological convex bornological spaces defined in (ii) of (2.1.12):

2.1.23 Proposition. *A vector space E with a bornology $\mathscr{B}$ is a topological convex bornological space iff a subset $B\subseteq E$ belongs to $\mathscr{B}$ provided that $\ell(B)$ is bounded for all linear bornological $\ell: E\to\mathbb{R}$ (i.e. iff scalarly bounded is equivalent with bounded).*

Proof. Since $\delta\mu=Id$ we obtain, using (i) and (iv) of (2.1.21): $\beta\gamma=\sigma_b\delta\mu\delta=\sigma_b\delta$. 'Topological' for convex bornological spaces is defined as $\beta\gamma$-invariant, while the given condition means $\sigma_b\delta$-invariant. □

2.1.24 Corollary. *For a convex bornological space E the following statements are equivalent:*

(1) *E is a topological convex bornological space;*
(2) *The bornology of E comes from a dualized vector space structure (i.e. $E=\sigma_b F$ for some $F\in|\underline{DVS}|$);*
(3) *The bornology of E comes from a locally convex topology (i.e. $E=\beta F$ for some $F\in|\underline{LCS}|$);*
(4) *A subset is bounded iff all bornological seminorms are bounded on it.*

2.2 The Mackey convergence and the Mackey closure topology

Mackey convergence gives an embedding of the category BornVS of bornological vector spaces into the category LimVS of convergence vector spaces; cf. [Frölicher, Kriegl, 1985]. We start by recalling the basic definitions concerning filters and convergence structures.

2.2.1 Definition. (i) A *directed set* $\mathscr{J}$ is a set J together with a relation $\succ$ which is reflexive, transitive and such that for any $j_1,j_2\in J$ there exists a $j\in J$ with $j\succ j_1$ and $j\succ j_2$. For $j_2\succ j_1$ one usually says: j_2 comes after j_1.

(ii) A *filter basis* on a set X is a non-empty collection of subsets of X which is directed by inclusion (i.e. by the relation $A\succ B$ iff $A\subseteq B$).

(iii) A *filter* on a set X is a filter basis $\mathscr{H}$ with the additional property that $A\in\mathscr{H}$ and $A\subseteq B\subseteq X$ implies $B\in\mathscr{H}$.

(iv) If $\mathscr{G}$, $\mathscr{H}$ are two filter bases on X, $\mathscr{G}$ is called *finer* than $\mathscr{H}$ and we write $\mathscr{G}\leq\mathscr{H}$ iff for every $H\in\mathscr{H}$ there exists a $G\in\mathscr{G}$ with $G\subseteq H$. If $\mathscr{G}$ is a filter then this is obviously equivalent to $\mathscr{H}\subseteq\mathscr{G}$.

2.2.2 Remark. (i) With the relation 'finer' all filters on a set X form a complete lattice. This fails if one restricts the notion of filter (as is often done) by the condition $\emptyset \notin \mathscr{H}$, a condition which eliminates just one filter on X, namely the set of all subsets of X. This filter is also called the zero-filter, since it is the smallest element of the lattice. As usual the lattice operations are denoted by $\wedge$ and $\vee$; i.e. $\mathscr{H} \vee \mathscr{G} = \sup(\mathscr{H}, \mathscr{G})$ and $\mathscr{H} \wedge \mathscr{G} = \inf(\mathscr{H}, \mathscr{G})$.

(ii) If $\mathscr{H}_0$ is a filter basis on X one obtains a filter $\mathscr{H}$ on X by taking all subsets of X containing some set of $\mathscr{H}_0$; $\mathscr{H}$ is called the *filter generated by the filter basis* $\mathscr{H}_0$.

(iii) If $\mathscr{H}$ is a filter on X and $g: X \to Y$ a map then $\{g(H); H \in \mathscr{H}\}$ is a filter basis on Y. The filter generated by it is called the *image of* $\mathscr{H}$ and is shortly denoted by $g(\mathscr{H})$.

(iv) For every $A \subseteq X$ the collection $\{A\}$ consisting of A alone is a filter basis on X. If $A = \{p\}$ for some $p \in X$, then the filter generated by $\{p\}$ is a so-called *ultra-filter*, i.e. only the zero-filter on X is strictly finer.

2.2.3 Definition. (i) A *convergence space* X is a set (also denoted by X) together with a convergence structure on X, i.e. a relation '$\mathscr{H}$ converges to x' between filters $\mathscr{H}$ on X and points x of X, written $\mathscr{H} \xrightarrow[X]{} x$, such that

(a) For all $x \in X$ the (ultra-)filter generated by $\{x\}$ converges to x;

(b) If $\mathscr{H}_1 \xrightarrow[X]{} x$ and $\mathscr{H}_2 \leq \mathscr{H}_1$ then $\mathscr{H}_2 \xrightarrow[X]{} x$;

(c) if $\mathscr{H}_1 \xrightarrow[X]{} x$ and $\mathscr{H}_2 \xrightarrow[X]{} x$ then $\mathscr{H}_1 \vee \mathscr{H}_2 \xrightarrow[X]{} x$.

(ii) Lim denotes the category of convergence spaces, the morphisms being the so-called *continuous* maps $g: X \to Y$, i.e. those for which $\mathscr{H} \xrightarrow[X]{} x$ implies $g(\mathscr{H}) \xrightarrow[Y]{} g(x)$.

(iii) A convergence space is called *separated* or *Hausdorff* if no filter except the zero-filter converges to different points.

Convergence of nets can be described using associated filters. We recall the notion of a net and the relationship to filters.

2.2.4 Lemma. *The final segments $J_j := \{j_1 \in J; j_1 \succ j\}$ of a non-empty directed set $\mathscr{J}$ form a filter basis. The filter generated by it will be called the Fréchet filter of the directed set.*

Proof. Trivial. □

Example. The Fréchet filter of the directed set $(\mathbb{N}, \geq)$ is formed by the subsets of $\mathbb{N}$ which have finite complement.

2.2.5 Definition. A *net* in a set X is a map $x: \mathscr{J} \to X$ from a directed set $\mathscr{J}$ into X. Nets are also called Moore–Smith sequences.

A net $x: \mathscr{J} \to X$ on a convergence space X is said to converge to $x_\infty \in X$ if the associated filter, i.e. the image of the Fréchet filter, converges to x_∞. In the case where x_∞ is uniquely determined by this property we write $x_\infty = \lim_j x(j)$.

On a topological space X one obtains a convergence structure by defining $\mathscr{H} \xrightarrow[X]{} x$ iff $\mathscr{H}$ is finer than the neighborhood filter of x. One thus gets an embedding functor ι: $\underline{\text{Top}} \to \underline{\text{Lim}}$ preserving the underlying sets and the underlying maps.

2.2.6 Proposition. *The embedding functor* ι: $\underline{\text{Top}} \to \underline{\text{Lim}}$ *of the category of topological spaces into that of convergence spaces has a left adjoint* τ, *which preserves the underlying spaces and the underlying maps. For a convergence space* X *the topology of* τX *can be described in the following equivalent ways:*

(i) $O \subseteq \tau X$ *is open iff* $x \in O$ *and* $\mathscr{H} \xrightarrow[X]{} x$ *implies* $O \in \mathscr{H}$;

(ii) $A \subseteq \tau X$ *is closed iff* $\mathscr{H} \xrightarrow[X]{} x$, $A \in \mathscr{H}$, $\emptyset \notin \mathscr{H}$ *implies* $x \in A$.

Proof. One easily verifies that the sets considered in (i) form a topology, that those of (ii) are their complements, that $\tau \circ \iota = \text{Id}$ and finally that id_X: $\iota\tau X \to X$ is continuous for any convergence space X. So the adjunction follows by (8.4.2).

□

2.2.7 Definition. A convergence space is called *first countable* iff for every converging filter there exists a coarser filter with a countable basis converging to the same point.

2.2.8 Proposition. *For a first countable convergence space* X *the following holds:*

(i) *If* $\mathscr{H} \xrightarrow[X]{} p$, $A \in \mathscr{H}$ *and* $\emptyset \notin \mathscr{H}$; *then there exists a sequence of points in* A *converging to* p;

(ii) $A \subseteq \tau X$ *is closed iff* A *is closed under convergent sequences;*

(iii) *The topology of* τX *is the final one induced by the convergent sequences* $\mathbb{N}_\infty \to X$, $\mathbb{N}_\infty := \mathbb{N} \cup \{\infty\}$ *having the usual compact topology.*

Proof. Easy, cf. [Gähler, 1977, p. 254]. □

One can easily show that one has initial and final structures with respect to the forgetful functor $\underline{\text{Lim}} \to \underline{\text{Set}}$ and deduce from this by (8.7.3) that $\underline{\text{Lim}}$ is complete and cocomplete. Moreover $\underline{\text{Lim}}$ is cartesian closed. We shall use only products in $\underline{\text{Lim}}$, so we restrict the consideration to these.

2.2.9 Proposition. *The categorical product of any family* X_j $(j \in J)$ *of convergence spaces exists and can be described as follows: it is the cartesian product of the underlying sets supplied with the convergence structure for which a filter* $\mathscr{H}$ *on the product converges to a point* x *iff* $\text{pr}_j(\mathscr{H})$ *converges to* $\text{pr}_j(x)$ *in* X_j *for all* $j \in J$.

Proof. One easily verifies that this in fact yields a convergence structure having the universal property. □

2.2.10 Definition. (i) A convergence structure on a vector space E is called *compatible* iff the vector space operations $E\sqcap E\to E$ and $\mathbb{R}\sqcap E\to E$ are continuous ($\mathbb{R}$ being considered with the usual convergence).

(ii) A *convergence vector space* is a vector space with a compatible convergence structure; Lim VS denotes the category formed by the convergence vector spaces as objects and the continuous linear maps as morphisms.

(iii) A *Cauchy filter* on a convergence vector space E is a filter $\mathscr{H}$ on E such that $\mathscr{H}-\mathscr{H}\xrightarrow[E]{}0$; $\mathscr{H}-\mathscr{H}$ is the image of the filter $\mathscr{H}\times\mathscr{H}$ on $E\sqcap E$ under the map $(x,y)\to x-y$, hence has a filter basis formed by the sets $H-H:=\{x-y;\, x,y\in H\}$ for $H\in\mathscr{H}$. A *Cauchy net* is a net such that the associated filter is a Cauchy filter.

(iv) A convergence vector space is called *complete* iff every Cauchy filter converges.

(v) A convergence vector space is called *sequentially complete* iff every Cauchy sequence converges.

2.2.11 Remark. The convergence structure of a convergence vector space is determined by the set of filters which converge to zero, since the translations are Lim-morphisms.

2.2.12 Proposition. *Every sequentially complete first countable convergence vector space is complete* [*Gähler*, 1977, *p.* 369].

Proof. Let $\mathscr{G}$ be a Cauchy filter. Then by the countability property there is a filter $\mathscr{H}$ with countable basis $H_1\supseteq H_2\supseteq\ldots$, s.t. $\mathscr{G}-\mathscr{G}\leq\mathscr{H}\to 0$. Choose $G_n\in\mathscr{G}$ and $g_n\in G_n$ with $G_n-G_n\subseteq H_n$, $G_n\supseteq G_{n+1}$. Let $\mathscr{G}_0$ denote the filter associated to the sequence $n\mapsto g_n$. And let $\mathscr{G}_1$ be the filter generated by the G_n. Then $\mathscr{G}_0\leq\mathscr{G}_1$, $\mathscr{G}\leq\mathscr{G}_1$ and $\mathscr{G}_1-\mathscr{G}_1\leq\mathscr{H}\to 0$. Thus $\mathscr{G}_1$ and consequently $\mathscr{G}_0$ are Cauchy filters. And the sequential completeness implies that $\mathscr{G}_0\to x$ for some x. Since $\mathscr{G}_1-x\leq(\mathscr{G}_1-\mathscr{G}_0)+(\mathscr{G}_0-x)\leq(\mathscr{G}_1-\mathscr{G}_1)+(\mathscr{G}_0-x)\to 0$, the filter $\mathscr{G}_1$ and thus also the finer filter $\mathscr{G}$ converges to x. □

We now come to the Mackey convergence structure on a bornological vector space. It will play an important role.

2.2.13 Definition. Let $\mathscr{H}$ be a filter on a bornological vector space E and $x\colon\mathscr{I}\to E$ a net on E.

(i) The filter $\mathscr{H}$ is called *Mackey convergent* (or shortly M-*convergent*) to $p\in E$ and we write $\mathscr{H}\xrightarrow[\xi E]{}p$ or $\mathscr{H}\xrightarrow{M}p$ if there exists a bounded $B\subseteq E$ with $\mathscr{H}-p\leq\mathbb{U}\cdot B$, where $\mathbb{U}$ denotes the filter of 0-neighborhoods on $\mathbb{R}$. The net $x\colon\mathscr{I}\to E$ is called *M-convergent* to p iff the associated filter is M-convergent to p. In the case where p is uniquely determined by this property we will write $p = \text{M-lim}_j x(j)$.

(ii) The filter $\mathscr{H}$ is called *bounded* iff $\mathscr{H}$ contains a bounded set. The net $x: \mathscr{J} \to E$ is called *bounded* iff the associated filter is bounded, i.e. iff there exists an $j_0 \in J$ with $\{x(j); j \succ j_0\} \subseteq E$ bounded.

(iii) The filter $\mathscr{H}$ is called *Mackey–Cauchy* iff it is a Cauchy filter with respect to Mackey convergence. The net $x: \mathscr{J} \to E$ is called *Mackey–Cauchy* iff the associated filter is a Mackey–Cauchy filter.

2.2.14 Remark. An ordinary sequence $x: \mathbb{N} \to E$ (i.e. with $(\mathbb{N}, \geq)$ as directed set) is a bounded sequence iff $x(\mathbb{N})$ is bounded, because the final segments of $\mathbb{N}$ have finite complements and the union of a finite set with a bounded set is bounded.

2.2.15 Proposition. (i) *For any bornological vector space E, Mackey convergence yields a convergence vector space, denoted by ξE, and one thus obtains a functor ξ: $\underline{\text{Born VS}} \to \underline{\text{Lim VS}}$.*

(ii) *This functor has a right adjoint ζ: $\underline{\text{LimVS}} \to \underline{\text{BornVS}}$ satisfying $\zeta \circ \xi = \text{id}$. Both functors preserve the underlying vector spaces and the underlying maps.*

Proof. (i) is trivial. For (ii) let us describe for a convergence vector space G the bornology of ζG: $B \subseteq \zeta G$ bounded iff $\mathbb{U} \cdot B \xrightarrow[G]{} 0$, where as before $\mathbb{U}$ denotes the filter of 0-neighborhoods on $\mathbb{R}$. One verifies functoriality and shows easily that $\zeta\xi = \text{Id}$ and id: $\xi\zeta G \to G$ is continuous. So the adjunction follows by (8.4.2). □

2.2.16 Lemma. *Let $x: \mathscr{J} \to E$ be a net in a separated bornological vector space E. Then x is* M*-convergent to $x_\infty \in E$ iff it can be written in the form $x_j = t_j b_j + x_\infty$ where $t: \mathscr{J} \to \mathbb{R}$ is a zero-converging net and $b: \mathscr{J} \to E$ a bounded net.*

Proof. One trivially reduces the general statement to the case $x_\infty = 0$.

($\Leftarrow$) Choose $j_1 \in J$ such that $B := \{b(j); j \succ j_1\}$ is bounded in E. We claim that the filter $\mathscr{H}$ associated to the net satisfies $\mathscr{H} \leq \mathbb{U}B$. So let $A \in \mathbb{U}B$. Then $A \supseteq [-\delta, \delta]B$ for some $\delta > 0$ and we choose $j_2 \in J$ with $t_j \in [-\delta, \delta]$ for $j \succ j_2$. Using the directedness we obtain a $j_0 \in J$ with $j_0 \succ j_1$ and $j_0 \succ j_2$. Then for $j \succ j_0$ one has $x_j = t_j b_j \in [-\delta, \delta]B \subseteq A$, and this shows that $A \in \mathscr{H}$.

($\Rightarrow$) Suppose $\mathscr{H} \leq \mathbb{U}B$ for some bounded $B \subseteq E$. Since $[-1, 1]B$ is also bounded, we may assume that $B = [-1, 1]B$. In particular $B \in \mathscr{H}$ and thus there exists an $j_1 \in J$ with $x_j \in B$ for all $j \succ j_1$. We define $t: \mathscr{J} \to \mathbb{R}$ as follows:

$$t_j := \begin{cases} \inf\{s > 0;\ x_j \in sB\} & \text{for } j \succ j_1 \\ 1 & \text{otherwise} \end{cases} \qquad \text{(analogous to } \|x_j\|_B \text{ of (2.1.15))}$$

and $b: \mathscr{J} \to E$ by

$$b_j := \begin{cases} t_j^{-1} x_j & \text{for } t_j \neq 0 \\ 0 & \text{for } t_j = 0 \end{cases}$$

Then $x_j = t_j b_j$, since $t_j = 0$ implies that $\mathbb{R}x_j$ is bounded, hence $x_j = 0$ by the separation hypothesis.

We show first that $b: \mathcal{J} \to E$ is bounded by verifying that $\{b_j;\, j \succ j_1\} \subseteq E$ is bounded. For any j we have $x_j \in 2t_j B$, i.e. $b_j \in 2B$ and $2B$ is bounded. We finally show that $t: \mathcal{J} \to \mathbb{R}$ converges to 0. So let $\varepsilon > 0$. Since $[-\varepsilon, \varepsilon]B = \varepsilon B \in \mathcal{H}$ there exists a $j_2 \in J$ with $x_j \in \varepsilon B$ for all $j \succ j_2$. Hence $t_j \leq \varepsilon$ for $j \succ j_0$ where j_0 is chosen such that $j_0 \succ j_1$ and $j_0 \succ j_2$. □

2.2.17 Lemma. *Let $x: \mathcal{J} \to E$ be a net in a separated bornological vector space E. Then x is a Mackey–Cauchy net iff the net $\mathcal{J} \times \mathcal{J} \to E$ defined by $(j_1, j_2) \mapsto x(j_1) - x(j_2)$ can be written in the form $x(j_1) - x(j_2) = t(j_1, j_2) b(j_1, j_2)$ for some zero-converging net $t: \mathcal{J} \times \mathcal{J} \to \mathbb{R}$ and some bounded net $b: \mathcal{J} \times \mathcal{J} \to E$.*

Proof. We first remark that $\mathcal{J} \times \mathcal{J}$ is directed according to $(j_1, j_2) \succ (i_1, i_2)$ iff $j_1 \succ i_1$ and $j_2 \succ i_2$. Let $\mathcal{H}$ be the filter associated to x. Then one verifies easily that the filter associated to the net $(j_1, j_2) \mapsto x(j_1) - x(j_2)$ is exactly $\mathcal{H} - \mathcal{H}$. The result therefore follows from (2.2.16). □

2.2.18 Remark. For an ordinary sequence $x: \mathbb{N} \to E$ the propositions (2.2.16) and (2.2.17) can be formulated equivalently in the following way:

(a) $x_\infty = \text{M-lim}_{n \to \infty} x_n$ iff there exist (positive) reals t_n with $\lim_{n \to \infty} t_n = \infty$ and $\{t_n(x_n - x_\infty);\, n \in \mathbb{N}\} \subseteq E$ bounded.

In case E is a convex bornological space this is also equivalent with the condition that (x_n) converges to x_∞ in the seminormed space E_B for some absolutely convex bounded set $B \subseteq E$.

(b) x is a Mackey–Cauchy sequence iff there exist (positive) reals $t_{n,m}$ with $\lim_{n,m \to \infty} t_{n,m} = \infty$ and $\{t_{n,m}(x_n - x_m);\, n, m \in \mathbb{N}\} \subseteq E$ bounded.

In case E is a convex bornological space this is also equivalent with the condition that (x_n) is a Cauchy sequence in the seminormed space E_B for some absolutely convex bounded set $B \subseteq E$.

For a family of nets $\mathcal{J} \to E$ one can define uniform Mackey convergence. It plays a role for results on commuting double limits. We do not go further into that but we will use uniform M-convergence in the differentiation theory. There it will be shown that very weak differentiability conditions together with a Lipschitz condition on the derivative actually imply very strong differentiability properties which can be expressed in terms of uniform Mackey convergence.

2.2.19 Definition. Let $\mathcal{J}$ be a directed set, X a set, E a separated bornological vector space, $g: \mathcal{J} \times X \to E$ and $f: X \to E$ two maps. One says that $f(x) = \text{M-lim}_j g(j, x)$ uniformly in x iff one can write g in the form $g(j, x) = t_j b(j, x) + f(x)$ with $t: \mathcal{J} \to \mathbb{R}$ converging to 0 and $b: \mathcal{J} \times X \to E$ uniformly bounded, i.e. such that there exists a $j_0 \in \mathcal{J}$ with $\{b(j, x);\, j \succ j_0,\, x \in X\} \subseteq E$ bounded.

We shall now characterize those convergence vector spaces which correspond, by means of Mackey convergence (i.e. by the functor ξ of (2.2.15)) to bornological vector spaces.

2.2.20 Definition. We call a convergence vector space G a *bornological* convergence vector space iff $G = \xi\zeta G$, and we denote the respective full subcategory of Lim VS by bLimVS.

This terminology is justified by the following

2.2.21 Corollary. (*i*) ξ (*i.e. Mackey convergence*) *yields an isomorphism of the category Born VS of bornological vector spaces onto the category bLim VS of bornological convergence vector spaces.*

(*ii*) *bLim VS is a coreflective subcategory of Lim VS.*

We next introduce the so-called Mackey closure topology of a bornological vector space. It should not be confused with the Mackey topology, which was introduced in (2.1.9)!

2.2.22 Definition. The *Mackey closure topology* of a bornological vector space E is the topology of $\tau\xi E$, i.e. the topology associated to the Mackey convergence structure, cf. (2.2.6) and (2.2.8).

If a subset of E is open (closed, dense) with respect to this topology we shall say it is *M-open* (*M-closed*, *M-dense*). In contrast M-convergent or M-continuous will always mean convergent or continuous with respect to the Mackey convergence structure.

2.2.23 Remarks. (i) Since the Mackey convergence structure is obviously first countable, cf. (2.2.7), one can apply (2.2.8) and conclude that 'closed' and 'sequentially closed' is the same for the Mackey closure topology and that the Mackey closure topology is the final one induced by the M-converging sequences. However, the closure of a subset is often strictly larger than the adherence, i.e. the set of limit points of sequences in the subset (or of filters containing the subset); see (6.3.1) for an example.

(ii) Using the remark (2.2.18) one deduces that the Mackey closure topology of a convex bornological space E is the final topology induced by the inclusions of the seminormed spaces E_B for $B \in \mathscr{B}_0$, with $\mathscr{B}_0$ a basis of the bornology of E consisting of absolutely convex sets.

(iii) In general the Mackey closure topology is not a vector space topology since addition is often only partially continuous; see (6.2.8) for various examples.

2.3 ℓ^∞-vector-spaces and $\mathscr{Lip}^k$-vector-spaces

We shall consider vector spaces with an additional $\mathscr{M}$-structure, cf. (1.1.1), for $\mathscr{M} = \ell^\infty$ or $\mathscr{M} = \mathscr{Lip}^k$ with $k \in \mathbb{N}_{0,\infty}$. In these cases the structure functions are real-valued, and $\mathbb{R}$ shall then denote the reals with their natural $\mathcal{M}$-structure, cf. (ii) of (1.1.6).

2.3.1 Definition. An *$\mathcal{M}$-vector-space* E is a vector space together with a compatible $\mathcal{M}$-structure on it, i.e. such that addition $E \sqcap E \to E$ and scalar multiplication $\mathbb{R} \sqcap E \to E$ are $\underline{\mathcal{M}}$-morphisms. By $\underline{\mathcal{M}\text{-}VS}$ we will denote the category of $\mathcal{M}$-vector-spaces, the morphisms being the linear $\underline{\mathcal{M}}$-morphisms.

2.3.2 Proposition. *The duality funtor δ: $\underline{\mathcal{M}\text{-}VS} \to \underline{DVS}$, which associates to an $\mathcal{M}$-vector-space E the dualized vector space δE with $(\delta E)' = \underline{\mathcal{M}\text{-}VS}(E, \mathbb{R})$, has a right adjoint σ: $\underline{DVS} \to \underline{\mathcal{M}\text{-}VS}$. Both functors preserve the underlying vector spaces and the underlying maps, and for a dualized vector space E, the $\mathcal{M}$-structure of σE is the one generated by E'.*

Proof. Functoriality of δ and σ was established in (2.1.6). The definitions imply that id: $\delta\sigma E \to E$ is a $\underline{\text{DVS}}$-morphism. And that id: $E \to \sigma\delta E$ is an $\underline{\mathcal{M}}$-morphism for any $\mathcal{M}$-vector-space E can be easily verified by composing with elements of $(\delta E)'$. The stated adjunction now follows, cf. (8.4.2). □

Let us consider the case $\mathcal{M} = \ell^\infty$ first. In this case we write σ_ℓ instead of σ.

2.3.3 Proposition. *The following diagram commutes:*

$$\begin{array}{ccc} \underline{\text{DVS}} & \xrightarrow{\sigma_b} & \underline{\text{CBS}} \\ \downarrow{\scriptstyle \sigma_\ell} & & \downarrow{\scriptstyle \delta} \\ \underline{\ell^\infty\text{-VS}} & \xrightarrow{\delta} & \underline{\text{DVS}} \end{array}$$

i.e. on any dualized vector space E a linear function is an $\underline{\ell^\infty}$-morphism for the ℓ^∞-structure generated by E' iff it is bounded on the scalarly bounded sets.

Proof. Let E be a dualized vector space. Since the bornology of $\sigma_b E$ coincides with the bornology associated to the ℓ^∞-structure of $\sigma_\ell E$, the assertion follows. □

The analogous result for the case $\mathcal{M} = \mathcal{L}ip^k$, where we shall write σ_k instead of σ, is less obvious and is based on the following lemma, of which a much more general version will be given in (4.2.15).

2.3.4 Proposition. (*Special Curve Lemma.*) *Let a_n be a sequence in a dualized vector space E such that for all $\ell \in E'$ the set $\ell\{n^n a_n;\ n \in \mathbb{N}\}$ is bounded. Then the infinite polygon through the points a_n can be parametrized smoothly: There exists a curve c: $\mathbb{R} \to E$ with the following properties*

(*i*) *$\ell \circ c$ is smooth for all $\ell \in E'$;*
(*ii*) *$c(1/2^n) = a_n$ for all $n \in \mathbb{N}$;*

(*iii*) $c([1/2^{n+1}, 1/2^n])$ *is the segment between* a_{n+1} *and* a_n;
(*iv*) $c(t)=0$ *for* $t\leq 0$.

Proof. Using a fixed smooth function h_0: $\mathbb{R}\to[0,1]$ with $h_0(t)=0$ for $t\leq 0$ and $h_0(t)=1$ for $t\geq 1$ we define a smooth function h: $\mathbb{R}\to[0,1]$ by

$$h(t):=\begin{cases}1-h_0(t) & \text{for} \quad t\geq 0\\ h_0(1+2t) & \text{for} \quad t\leq 0\end{cases}$$

and then the curve c: $\mathbb{R}\to E$ by $c(t):=\sum_{n=1}^{\infty} h(2^n t-1)a_n$.

The sum exists, since for each $t\in\mathbb{R}$ at most two summands are different from 0. Let us verify the stated properties.

(i) For $\ell\in E'$ one has $(\ell\circ c)(t)=\sum f_n(t)$, where $f_n(t):=h(2^n t-1)\cdot\ell(a_n)$. An easy estimation using the boundedness assumption shows that for all k the series $\sum_n f_n^{(k)}(t)$ formed by the kth derivatives converges uniformly with respect to t. Hence $\ell\circ c\in C^\infty$.

(ii) and (iv) are obvious by construction.

(iii) For $2^n t\in[\frac{1}{2},1]$ one has $2^n t-1\leq 0$ and $2\cdot 2^n t-1\geq 0$ and thus $h(2^n t-1)+h(2\cdot 2^n t-1)=1$. □

2.3.5 Proposition. *The following diagram commutes* ($k\in\mathbb{N}_{0,\infty}$):

$$\begin{array}{ccc}\underline{\mathrm{DVS}} & \xrightarrow{\sigma_b} & \underline{\mathrm{CBS}}\\ \downarrow{\scriptstyle\sigma_k} & & \downarrow{\scriptstyle\delta}\\ \underline{\mathscr{Lip}^k\text{-VS}} & \xrightarrow{\delta} & \underline{\mathrm{DVS}}\end{array}$$

i.e. on any dualized vector space E a linear function is a $\mathscr{Lip}^k$*-function for the* $\mathscr{Lip}^k$*-structure generated by E′ iff it is bounded on the scalarly bounded sets.*

Proof. Let E be a dualized vector space. We have to show that for a linear function ℓ_0: $E\to\mathbb{R}$ one has:

$$\ell_0\colon \sigma_b E\to\mathbb{R} \text{ is bornological} \Leftrightarrow \ell_0\colon \sigma_k E\to\mathbb{R} \text{ is } \mathscr{Lip}^k\text{-morphism.}$$

($\Rightarrow$) Let c: $\mathbb{R}\to\sigma_k E$ be a $\mathscr{Lip}^k$-curve and $A\subseteq\mathbb{R}$ be bounded. For $\ell\in E'$ one has $\ell\circ c\in\mathscr{Lip}^k$, hence the difference quotient $\delta^{k+1}(\ell\circ c)(A^{\langle k+1\rangle})$ of order $k+1$ is bounded, cf. (1.3.22). Since δ^{k+1} commutes with ℓ we obtain that $\delta^{k+1}c(A^{\langle k+1\rangle})$ is bounded in $\sigma_b E$. Thus $\ell_0(\delta^{k+1}c(A^{\langle k+1\rangle}))=\delta^{k+1}(\ell_0\circ c)(A^{\langle k+1\rangle})$ is bounded. From (1.3.22) we deduce that $\ell_0\circ c\in\mathscr{Lip}^k$, showing that ℓ_0: $\sigma_k E\to\mathbb{R}$ is a $\mathscr{Lip}^k$-morphism. This proof is for $k<\infty$; the modifications for $k=\infty$ are obvious.
($\Leftarrow$) Assume ℓ_0: $\sigma_b E\to\mathbb{R}$ is not bornological. Then there exists a $B\subseteq\sigma_b E$ bounded with $\ell_0(B)$ unbounded and we can choose $b_n\in B$ with $|\ell_0(b_n)|\geq n^{n+1}$. The sequence a_n defined by $a_n:=n^{-n}b_n$ satisfies the hypothesis of the special curve lemma (2.3.4). Hence there exists a curve c: $\mathbb{R}\to E$ with $c(1/2^n)=a_n$ and

$\ell \circ c \in \mathscr{Lip}^\infty$, so certainly $\ell \circ c \in \mathscr{Lip}^k$ for every $\ell \in E'$. But $|(\ell_0 \circ c)(1/2^n)| \geq n$ shows that $\ell_0 \circ c \notin \mathscr{Lip}^k$ (not even continuous!). So ℓ_0 is not a $\mathscr{Lip}^k$-morphism $\sigma_k E \to \mathbb{R}$. □

2.3.6 Proposition. *For a curve $c: \mathbb{R} \to E$ into a dualized vector space E the following conditions are equivalent ($k \in \mathbb{N}_{0,\infty}$):*

(1) *$c: \mathbb{R} \to \sigma_k E$ is a $\mathscr{Lip}^k$-curve;*
(2) *$\delta^j c: \mathbb{R}^{\langle j \rangle} \to \sigma_b E$ is bornological for $0 \leq j < k+2$.*

Remark. We write $j < k+2$ instead of $j \leq k+1$ in order to include the case $k = \infty$.

Proof. By the definition of the $\mathscr{Lip}^k$-structure of $\sigma_k E$, (1) holds iff $\ell \circ c \in \mathscr{Lip}^k$ for all $\ell \in E'$. For $k = \infty$ this is by (1.3.24) equivalent with $\delta^j(\ell \circ c): \mathbb{R}^{\langle j \rangle} \to \mathbb{R}$ being bornological for all $j < \infty$. For finite k it is by (1.3.22) equivalent with $\delta^{k+1}(\ell \circ c): \mathbb{R}^{\langle k+1 \rangle} \to \mathbb{R}$ being bornological; and by (1.3.14) with $\delta^j(\ell \circ c)$: $\mathbb{R}^{\langle j \rangle} \to \mathbb{R}$ being bornological for all $j \leq k+1$. Using the identity $\delta^j(\ell \circ c) = \ell \circ \delta^j c$ and the fact that a map $g: X \to \sigma_b E$ is bornological iff $\ell \circ g: X \to \mathbb{R}$ is bornological for all $\ell \in E'$ (definition of $\sigma_b E$) one concludes that in both cases (1) is equivalent to (2). □

The special curve lemma (2.3.4) also allows to give still another characterization of the Mackey closure topology of a topological convex bornological space:

2.3.7 Proposition. *The following diagram commutes ($k \in \mathbb{N}_{0,\infty}$):*

$$\begin{array}{ccccc} \underline{\mathrm{DVS}} & \xrightarrow{\sigma_b} & \underline{\mathrm{CBS}} & \xrightarrow{\xi} & \underline{\mathrm{Lim\,VS}} \\ \Big\downarrow{\scriptstyle\sigma_k} & & & & \Big\downarrow{\scriptstyle\tau} \\ \underline{\mathscr{Lip}^k\text{-VS}} & & \xrightarrow{\tau_f} & & \underline{\mathrm{Top}} \end{array}$$

i.e. for any dualized vector space E the Mackey closure topology of $\sigma_b E$ is the final one induced by the $\mathscr{Lip}^k$-curves. The functor τ_f associates to an object the underlying set with the final topology induced by the structure curves and preserves underlying maps. (This topology is called c^∞-topology in [Kriegl, 1982] *and* [Kriegl, 1983].)

Proof. Let $\mathscr{T}_k$ be the final topology induced by the $\mathscr{Lip}^k$-curves; i.e. the topology of $\tau_f \sigma_k E$; $\mathscr{T}_M$ the Mackey closure topology, i.e. the topology of $\tau \xi \sigma_b E$.

($\mathscr{T}_k \leq \mathscr{T}_M$) Let $c: \mathbb{R} \to E$ be a $\mathscr{Lip}^k$-curve and $a \in \mathbb{R}$. Then for any $\ell \in E'$ $\ell \circ c \in \mathscr{Lip}^k \subseteq \mathscr{Lip}$. Hence

$$\left\{ \frac{\ell(c(t)) - \ell(c(a))}{t-a}; 0 < |t-a| \leq 1 \right\}$$

is bounded in $\mathbb{R}$, so $\{(c(t)-c(a))/(t-a); 0<|t-a|\leq 1\}$ is bounded in $\sigma_b E$. This shows that for $t\to 0$ the values $c(t)$ are Mackey convergent to $c(a)$, i.e. that $c: \mathbb{R}\to\xi\sigma_b E$ is continuous. Hence also $c: \mathbb{R}\to\tau\xi\sigma_b E$.

($\mathscr{T}_k \geq \mathscr{T}_M$) Let $A\subseteq E$ be closed for $\mathscr{T}_k$ and suppose (a_n) is a sequence in A which is Mackey convergent to some $a\in E$. We can choose a subsequence a_{n_k} such that $\{k^k(a_{n_k}-a); k\in\mathbb{N}\}$ is bounded in $\sigma_b E$. By the special curve lemma (2.3.4) we obtain a smooth curve c with $c(1/2^k)=a_{n_k}$ and $c(0)=a$. Since c is continuous for $\mathscr{T}_k$ one deduces from $1/2^k\in c^{-1}(A)$, that $0\in c^{-1}(A)$, i.e. $a=c(0)\in A$. According to (ii) of (2.2.8) this shows that A is closed for $\mathscr{T}_M$. □

As remarked in (iii) of (2.2.23) the Mackey closure topology is not a vector space topology in general. In order to discuss continuity of addition it is useful to work in a subcategory of Top. For this purpose the following will suffice:

2.3.8 Definition. By Arc we denote the full subcategory of Top formed by the arc-generated topological spaces, i.e. by those whose topology is the final one induced by their continuous curves.

This subcategory is similar to that formed by the compactly generated spaces; cf. [Engelking, 1968, p. 123]. The inclusion functor has a right adjoint α: Top→Arc; for a topological space X one obtains αX by supplying X with the final topology induced by the continuous curves. According to the adjunction one obtains products in Arc by applying α to the product formed in Top. It is also well known that Arc is cartesian closed, but we will not use this.

2.3.9 Definition. ArcVS denotes the category whose objects are vector spaces with arc-generated topology such that the vector space operations are continuous with respect to the product formed in Arc, and whose morphisms are the linear continuous maps.

Remark. Every arc-generated space is of course compactly generated. Using results of [Kriegl, 1980] it can be shown that every arc-generated vector space is a compactly generated vector space; cf. [Seip, 1979].

2.3.10 Lemma. *Let E be a dualized vector space and $x: \mathbb{N}\to E$ be a sequence. Then the following statements are equivalent:*

(1) *x is convergent to x_∞ with respect to the Mackey closure topology of $\sigma_b E$;*
(2) *every subsequence of x has itself a subsequence which is* M*-convergent to x_∞.*

Proof. (2⇒1) One uses that any M-convergent sequence converges with respect to the M-closure topology of $\sigma_b E$, cf. (iii) of (2.2.8), and that for a topological convergence the Urysohn property holds, i.e. a sequence x converges to x_∞ iff every subsequence of x has itself a subsequence which is convergent to x_∞.

(1⇒2) It is enough to show that x has a subsequence which M-converges to x_∞. We write $\bar{a}$ for the closure in the Mackey closure topology of a singleton $\{a\}$. One easily verifies that $\bar{a}=\{x\in E; \ell(x)=\ell(a)$ for all $\ell\in E'\}$. In the case where $x_n\in\overline{x_\infty}$ for infinitely many n one chooses a subsequence (n_k) with $x_{n_k}\in\overline{x_\infty}$ which is trivially M-convergent to x_∞. Otherwise we can assume that for all $n\in\mathbb{N}$ one has $x_n\notin\overline{x_\infty}$ and we consider $A:=\bigcup_{n\in\mathbb{N}}\overline{x_n}$. Since $x_\infty\notin A$ the set A is not closed in the Mackey closure topology and hence there exists a sequence $y:\mathbb{N}\to A$ which is M-convergent to some $y_\infty\notin A$. Since every finite union $\bigcup_{n\le N}\overline{x_n}$ is closed in the Mackey closure topology and $y_\infty\notin A$, the image $y(\mathbb{N})$ cannot be contained in such a finite union and therefore we can choose strictly increasing sequences (k_i) and (n_i) of natural numbers with $y_{k_i}\in\overline{x_{n_i}}$. Then (x_{n_i}) is convergent to x_∞ and to y_∞ in the Mackey closure topology, which implies $y_\infty\in\overline{x_\infty}$. This subsequence is M-convergent to y_∞ and hence also M-convergent to x_∞. □

Remark. One can prove that there exists a dualized vector space (in fact a convenient vector space) with a sequence that satisfies (2) of the previous lemma but is not M-convergent. This implies that the M-convergence of sequences does not have the Urysohn property and thus is not topological, cf. [Frölicher, Kriegl, 1985].

2.3.11 Corollary. *Let E be a dualized vector space and $c:\mathbb{R}\to E$. Then the following statements are equivalent:*

(1) *c is continuous with respect to the Mackey closure topology of $\sigma_b E$;*
(2) *If $t_n\to t_\infty$ then there exists a subsequence t_{n_k} such that $c(t_\infty)$ is a Mackey limit of $c(t_{n_k})$ for $k\to\infty$.*

2.3.12 Corollary. *For any dualized vector space E one has with respect to the Mackey closure topology of $\sigma_b E$: The sum of converging sequences is converging and the sum of continuous curves is continuous.*

2.3.13 Corollary. *For any dualized vector space E, the Mackey closure topology of $\sigma_b E$ yields an arc-generated vector space.*

Proof. The verification that the addition of E is continuous with respect to the arc-generated product is trivial, since its topology is exactly the Mackey closure topology of the product, as can be seen easily using the lemma above. □

2.3.14 Definition. We denote by τ_M: $\underline{\text{DVS}}\to\underline{\text{ArcVS}}$ the functor which preserves the underlying vector spaces and the underlying maps and for which $\tau_M E$ carries the final topology induced by the smooth curves $\mathbb{R}\to E$, i.e. the Mackey closure topology of $\sigma_b E$, or shortly the topology of $\tau\xi\sigma_b E$.

2.3.15 Proposition. *The following diagram commutes:*

$$\begin{array}{ccc} \underline{\mathrm{DVS}} & \xrightarrow{\sigma_b} & \underline{\mathrm{CBS}} \\ \Big\downarrow{\scriptstyle \tau_M} & & \Big\downarrow{\scriptstyle \delta} \\ \underline{\mathrm{ArcVS}} & \xrightarrow{\delta} & \underline{\mathrm{DVS}} \end{array}$$

i.e. on any dualized vector space E a linear function is bounded on the scalarly bounded subsets iff it is continuous for the Mackey closure topology of $\sigma_b E$.

Proof. Let $\ell : E \to \mathbb{R}$ be a linear function. Then one has:

ℓ is continuous for the Mackey closure topology;
$\Leftrightarrow \ell : \tau\xi\sigma_b E \to \mathbb{R}$ is continuous;
$\Leftrightarrow \ell : \xi\sigma_b E \to \mathbb{R}$ is continuous;
$\Leftrightarrow \ell : \sigma_b E \to \mathbb{R}$ is bornological;
$\Leftrightarrow \ell$ is bounded on the bounded subsets of $\sigma_b E$, i.e. the scalarly bounded subsets of E. The first and the last of the equivalences hold by definition, the others according to the adjunctions (2.2.6) and (2.2.15). □

2.4 Preconvenient vector spaces

For the differentiation theory we shall work with $\mathscr{Lip}^k$-structures on dualized vector spaces, the $\mathscr{Lip}^k$-structures being generated by the duals. We might therefore consider arbitrary dualized vector spaces E. However, since E and $\delta\sigma_b E$ yield the same $\mathscr{Lip}^k$-structure (use $\sigma_k \delta\sigma_k = \sigma_k$ and $\delta\sigma_k = \delta\sigma_b$ according to (2.3.5)) and since the endo-functor $\delta\sigma_b$: $\underline{\mathrm{DVS}} \to \underline{\mathrm{DVS}}$ is idempotent, one can without loss of generality restrict the considerations to $\delta\sigma_b$-invariant dualized vector spaces. This means that among all duals for a given vector space which yield the same $\mathscr{Lip}^k$-structure one chooses the largest one or, equivalently, the one whose set of linear $\mathscr{Lip}^k$-functions is exactly the given dual. The dualized vector spaces so obtained will be called preconvenient. The vector spaces which are convenient for differentiation theory shall be obtained by adding a separation and completeness property.

In the previous sections we described different decompositions of the endofunctor $\delta\sigma_b$. They will be used now in order to show that the category $\underline{\mathrm{Pre}}$ of preconvenient vector spaces embeds in many other categories. This means that preconvenient vector spaces carry in a canonical way many different structures. Since each of these structures determines all the others, any of them could be used in the definition. We choose for the explicit definition the structure as dualized vector space, because it is the simplest one. But others are important too. In particular, as the results of section 1.3 already indicate, the bornological characterization is closely related to differentiation; and the locally convex characterization is important for comparison with classical differentiation theory, e.g. for the case of Fréchet spaces.

2.4.1 Proposition. *The following endo-functors of the category $\underline{DVS}$ of dualized vector spaces are identical:*

(i) $\underline{DVS} \xrightarrow{\sigma_b} \underline{CBS} \xrightarrow{\delta} \underline{DVS}$

(ii) $\underline{DVS} \xrightarrow{\sigma_\ell} \underline{\ell^\infty\text{-}VS} \xrightarrow{\delta} \underline{DVS}$

(iii) $\underline{DVS} \xrightarrow{\sigma_k} \underline{\mathcal{L}ip^k\text{-}VS} \xrightarrow{\delta} \underline{DVS}$

(iv) $\underline{DVS} \xrightarrow{\gamma\beta\mu} \underline{LCS} \xrightarrow{\delta} \underline{DVS}$

(v) $\underline{DVS} \xrightarrow{\xi\sigma_b} \underline{bLim\,VS} \xrightarrow{\delta} \underline{DVS}$

(vi) $\underline{DVS} \xrightarrow{\tau_M} \underline{Arc\,VS} \xrightarrow{\delta} \underline{DVS}$

In (i), (ii), *and* (iii) *the respective duality functor δ is left adjoint to the respective functor σ.*

Proof. The composite functors in (i) and (ii) coincide by (2.3.3); those in (i) and (iii) by (2.3.5); those in (i) and (iv) by (ii) and (iii) in (2.1.21); those in (i) and (vi) by (2.3.15). Finally those in (i) and (v) coincide since by (2.2.15) even the following triangle commutes:

$$\begin{array}{ccc} \underline{\text{Born VS}} & \xrightarrow{\xi} & \underline{\text{Lim VS}} \\ & \searrow^{\delta} & \downarrow{\scriptstyle\delta} \\ & & \underline{\text{DVS}} \end{array}$$

The stated adjunctions were proved in (2.1.7) and (2.3.2). □

2.4.2 Definition. A *preconvenient vector space* is a dualized vector space which is invariant under the endo-functor of $\underline{\text{DVS}}$ described in six equivalent ways in (2.4.1); the respective full subcategory of $\underline{\text{DVS}}$ is denoted by $\underline{Pre}$.

Remarks. (i) The invariance imposed in the above definition means a closure condition for E'. According to the various descriptions of the endo-functor given in (2.4.1) this closure condition can be expressed equivalently in the following ways: a linear function belongs to E' provided it is a morphism with respect to either the bornology or the ℓ^∞-structure or the $\mathcal{L}ip^k$-structure determined by E'; or provided it is continuous with respect to the locally convex topology (i.e. the topology of $\gamma\beta\mu E$) or the Mackey convergence structure or the Mackey closure topology.

(ii) A dualized vector space belongs to Pre iff there exists a (convex) vector bornology on E such that E' becomes the bornological dual; comparison with (2.1.9) shows that in this respect bornologies behave quite different from locally convex topologies.

(iii) For a locally convex space E the following statements are equivalent:

(1) δE, i.e. E structured by its topological dual, is preconvenient;
(2) Every bornological linear function on E is continuous;
(3) The Mackey topology of E coincides with the bornologification of E (i.e. $\mu\delta E = \gamma\beta E$);
(4) The Mackey topology of E is bornological.

Examples of dualized vector spaces that are not preconvenient are obtained by using the classical examples of locally convex spaces not satisfying (2), like the uncountable direct sum of copies of $\mathbb{R}$ with the box topology. Then the bounded sets are the same as for the direct sum topology [Jarchow, 1981, p. 80], thus the function 'sum over all coordinates' is bornological; but it is not continuous with respect to the box topology. Another example is ℓ^∞ with the Mackey topology with respect to ℓ^1 [Jarchow, 1981, p. 223].

2.4.3 Theorem. *The functors with source* DVS *in the list of* (2.4.1) *induce embeddings* (*i.e. full and faithful functors*) *as follows* (*in parentheses we recall the functors*):

(*i*) $\underline{Pre} \xrightarrow{\sigma_b} \underline{CBS}$ (*bornology associated to* E');

(*ii*) $\underline{Pre} \xrightarrow{\sigma_\ell} \underline{\ell^\infty\text{-}VS}$ (ℓ^∞*-structure generated by* E');

(*iii*) $\underline{Pre} \xrightarrow{\sigma_k} \underline{\mathcal{L}ip^k\text{-}VS}$ ($\mathcal{L}ip^k$*-structure generated by* E');

(*iv*) $\underline{Pre} \xrightarrow{\mu} \underline{LCS}$ (*Mackey topology*);

(*v*) $\underline{Pre} \xrightarrow{\xi\sigma_b} \underline{bLim\,VS}$ (*Mackey convergence structure*);

(*vi*) $\underline{Pre} \xrightarrow{\tau_M} \underline{ArcVS}$ (*Mackey closure topology*);

The respective full subcategories which are thereby isomorphic to Pre *are the following*:

(*i*) *The category* tCBS *of topological convex bornological spaces*;
(*ii*) *The category of linearly generated* ℓ^∞*-vector-spaces*;
(*iii*) *The category of linearly generated* $\mathcal{L}ip^k$*-vector-spaces*;
(*iv*) *The category* bLCS *of bornological locally convex spaces*;
(*v*) *The category of* $\xi\sigma_b\delta$*-invariant convergence vector spaces*;
(*vi*) *The category of* $\tau_M\delta$*-invariant arc-determined vector spaces.*

The embeddings (*i*), (*ii*), (*iii*) *and* (*v*) *are reflective*; *the one in* (*iv*) *as well as the inclusion* $\underline{Pre} \to \underline{DVS}$ *are coreflective.*

Proof. The embeddings follow from the functors of (2.4.1) according to the general procedure mentioned in (2.1.11). We only remark that in (iv) we put μ instead of $\gamma\beta\mu$, since applied to a preconvenient vector space E it gives the same: using $E=\delta\sigma_b E$, $\gamma\beta\gamma=\gamma$ and $\gamma=\mu\delta$ (as previously proved) one has $\gamma\beta\mu E=\gamma\beta\mu\delta\sigma_b E=\gamma\beta\gamma\sigma_b E=\gamma\sigma_b E=\mu\delta\sigma_b E=\mu E$. In words this means: the Mackey topology of a preconvenient vector space is always bornological.

The categories isomorphic to $\underline{\text{Pre}}$ are in all cases formed by the objects invariant under the respective endo-functors; they have already been discussed for the cases (i) to (iv).

The embeddings in (i), (ii) and (iii) are reflective as a consequence of the adjunctions stated in (2.4.1); that of (v) because according to (2.2.21) it is up to the isomorphism ξ the same as that in (i). For the coreflexivity in (iv) one remarks that for any locally convex space E one has $\gamma\beta\mu\delta E=\gamma\sigma_b\delta E=\gamma\beta E$, hence id: $\gamma\beta\mu\delta E\to E$ is continuous. Coreflexivity of $\underline{\text{Pre}}$ in $\underline{\text{DVS}}$ is implied by the adjunction of (2.1.7) together with (2.1.11). □

Remarks. (i) We saw that $\underline{\text{Pre}}\to\underline{\text{bLim VS}}$ is reflective, while $\underline{\text{bLim VS}}\to\underline{\text{Lim VS}}$ is coreflective. The composed embedding of $\underline{\text{Pre}}$ in $\underline{\text{Lim VS}}$ has neither a left nor a right adjoint, cf. (7.2.10). The same holds for $\underline{\text{Pre}}\to\underline{\text{Arc VS}}$. The reason is that the Mackey convergence structure and the Mackey closure topology behave in a complicated way with respect to categorical constructions in $\underline{\text{Pre}}$.

(ii) Since, as shown in (2.6.5), the convenient vector spaces form a reflective subcategory of $\underline{\text{Pre}}$, those of the embeddings of $\underline{\text{Pre}}$ which are also reflective yield reflective embeddings of the category of convenient vector spaces.

(iii) Convenient vector spaces were introduced independently according to their characterization as smooth vector spaces (embedding (iii) above for $k=\infty$) in [Frölicher, 1982] and according to their characterization as locally convex spaces (embedding (iv)) in [Kriegl, 1982]; that they can be identified was proved later in [Frölicher, Gisin, Kriegl, 1983].

(iv) $\underline{\text{Pre}}$ has still other embeddings. For example, in any reflective subcategory $\mathscr{X}$ of $\underline{\text{LCS}}$ containing $\mathbb{R}$ and such that the reflector λ: $\underline{\text{LCS}}\to\mathscr{X}$ changes only the topologies but preserves the underlying vector spaces and the underlying maps. One only has to consider the composition $\lambda\circ\gamma\beta\mu$: $\underline{\text{DVS}}\to\underline{\text{LCS}}\to\mathscr{X}$. Composing it with δ: $\mathscr{X}\to\underline{\text{DVS}}$ gives again the same endo-functor as in (2.4.1), since adjointness of λ yields $(\delta\lambda E)'=\mathscr{X}(\lambda E,\mathbb{R})=\underline{\text{LCS}}(E,\mathbb{R})=(\delta E)'$. As examples one can take as $\mathscr{X}$ the category of nuclear locally convex spaces or that of locally convex spaces with weak topology.

We add now a recapitulation of the main results in non-categorical terms. We shall also simplify henceforth the writing by suppressing the various embedding functors.

2.4.4 Summary (non-categorical version of theorem (2.4.3)).

(i) *A preconvenient vector space E has the following structures:*

(0) *a subspace E′ of the algebraic dual, called its dual;*
(1) *a topological convex vector bornology (cf. (ii) in (2.1.12)), called its bornology;*
(2) *a linearly generated (cf. (1.1.3))* ℓ^∞*-structure, called its* ℓ^∞*-structure;*
(3) *a linearly generated* $\mathcal{Lip}^k$*-structure for each* $k \in \mathbb{N}_{0,\infty}$*, called its* $\mathcal{Lip}^k$*-structure;*
(4) *a bornological locally convex topology (cf. (i) in (2.1.12)), called its locally convex topology;*
(5) *a convergence structure (cf. (i) in (2.2.3)), called its Mackey convergence structure;*
(6) *an arc-determined topology (cf. (2.3.8)), called its Mackey closure topology (cf. (2.2.22) and (2.3.7)).*

(ii) *Each of these structures determines all the others. We recall how one gets them from each other:* $B \subseteq E$ *is bounded iff* $\ell(B)$ *is bounded for all* $\ell \in E'$*; the* ℓ^∞*-structure resp. the* $\mathcal{Lip}^k$*-structure is generated by E′; the locally convex topology is the finest one yielding E′ as topological dual; a sequence* (a_n) *is Mackey convergent to 0 if there exist reals* $t_n \to \infty$ *with* $\{t_n a_n; n \in \mathbb{N}\}$ *bounded in E; the Mackey closure topology is the final one induced by the smooth (or* $\mathcal{Lip}^k$*-) curves. The dual E′ is obtained from any of the other structures by forming the set of linear functions respecting the structure in question.*

(iii) *For a linear map* $g: E \to F$ *between preconvenient vector spaces the following statements are equivalent:*

(0) $g^*(F') \subseteq E'$;
(1) *g is bornological;*
(2) *g is an* ℓ^∞*-map;*
(3) *g is a* $\mathcal{Lip}^k$*-map;*
(4) *g is continuous for the locally convex topologies;*
(5) *g is continuous for the Mackey convergence structures;*
(6) *g is continuous for the Mackey closure topologies;*

Remark. For these and further equivalent conditions for linear maps see also [Kriegl, 1982].

On a preconvenient vector space *E* one can consider other locally convex topologies (e.g. the nuclear one or the weak one, as mentioned above) and other bornologies (e.g. that generated by the bornologically compact subsets, which will be introduced later). But if we shortly speak of the locally convex topology or the bornology of *E* we shall always mean those specified above.

2.4.5 Remark. It is, however, important to specify for convergence of filters or even sequences on a preconvenient vector space which structure one considers. One has the following implications for a filter:

(a) convergent for the Mackey convergence structure ⇒
(b) convergent for the Mackey closure topology ⇒

(c) convergent for the locally convex topology ⇒
(d) convergent for the weak topology.

(a) is equivalent to (c) iff the locally convex topology of E is semi-normable.

Every metrizable locally convex space can be considered as preconvenient vector space according to the remark (iv) in (2.1.20). For these spaces (a) and (c) become equivalent for sequences, cf. [Jarchow, 1981, p. 197]. Hence the given locally convex topology of such a space is equal to its Mackey closure topology (i.e. is the final one induced by the $\mathcal{Lip}^k$-curves for any $0 \leq k \leq \infty$, cf. (2.3.7)); in other words: (b) and (c) become equivalent (one uses that both topologies are the final ones induced by their converging sequences, cf. (2.2.23)). For a different proof of the last statement see (i) of (6.1.4).

2.4.6 Definition. An initial $\underline{\text{Pre}}$-morphism is a $\underline{\text{Pre}}$-morphism which is initial with respect to the forgetful functor $\underline{\text{Pre}} \to \underline{\text{VS}}$. An injective initial $\underline{\text{Pre}}$-morphism is called a $\underline{\text{Pre}}$-embedding; cf. (8.8.1).

2.4.7 Remark. For any linear map $m: E \to F$ from a vector space E into a preconvenient vector space F one gets an initial $\underline{\text{Pre}}$-morphism if one supplies E with the initial ℓ^∞-structure, i.e. the ℓ^∞-structure generated by $m^*(\mathcal{S})$ where $\mathcal{S} \subseteq F'$ is some set that generates the ℓ^∞-structure of F, cf. (1.1.4). The bounded sets of E are then those having their image bounded in F. The $\mathcal{Lip}^k$-curves of E are those for which the composite with m is a $\mathcal{Lip}^k$-curve of F.

2.5 Separation

In the following two sections we are going to specify the convenient vector spaces among the preconvenient ones. The reason for restricting the spaces a little more is that we want to do calculus in those spaces. And the most important construction in calculus is that of forming derivatives, which are obtained as limits of certain expressions like difference quotients. For uniqueness of these limits some separation condition is necessary and for existence some completeness property. We will restrict the spaces in a minimal way by imposing conditions which are necessary and sufficient in order that differentiable curves have unique derivatives. In Chapter 4 we will see that these conditions also suffice to get the analogous results for maps in full generality.

2.5.1 Definition. Let $c: \mathbb{R} \to E$ be a curve into a preconvenient vector space. We say that a point $\dot{c}(t) \in E$ is a *weak derivative* of c at t iff the derivative $(\ell \circ c)^{\cdot}(t)$ exists and equals $\ell(\dot{c}(t))$ for all $\ell \in E'$. We say that a point $\int_t^s c \in E$ is a *weak integral* of c iff the integral $\int_t^s (\ell \circ c)(\tau)d\tau$ exists and equals $\ell(\int_t^s c)$ for all $\ell \in E'$.

Note that we already use the usual notation for derivatives and integrals although they are not necessarily unique. But this will be remedied immediately by characterizing the necessary separation condition in different ways:

2.5.2 Theorem. *Let E be a preconvenient vector space, $k \in \mathbb{N}_{0,\infty}$. Then the following statements are equivalent:*

(1) *E' separates points;*
(2) *Every $\mathcal{Lip}^{k+1}$-curve has at most one weak derivative (say at 0);*
(3) *Every $\mathcal{Lip}^{k}$-curve has at most one weak integral (say from 0 to 1);*
(4) *The bornology is separated, i.e. $\{0\}$ is the only bounded linear subspace;*
(5) *The locally convex topology is Hausdorff;*
(6) *The* M*-convergence structure is Hausdorff;*
(7) *The Mackey closure topology is Hausdorff;*
(8) *For every absolutely convex bounded subset B, the seminormed space E_B (cf. (2.1.15)) is a normed space.*

Proof. (1⇒2) and (1⇒3) are trivial.

(2⇒1), (3⇒1) Assume $\ell(a)=0$ for all $\ell \in E'$. Then a and 0 are both weak derivatives at 0 resp. weak integrals from 0 to 1 of the constant smooth curve $c: t \mapsto 0$.

The scheme of proof for the remaining equivalences is:

$$\begin{array}{ccccc} 1 & \Longrightarrow & 5 & \Longrightarrow & 7 \\ \Uparrow & & & & \Downarrow \\ 4 & \Longleftarrow & 8 & \Longleftarrow & 6 \end{array}$$

(1⇒5) is clear, since $\underline{\mathrm{LCS}}(E, \mathbb{R})=E'$ and functionally separated implies separated.

(5⇒7⇒6) By means of the embedding ι: $\underline{\mathrm{Top}} \to \underline{\mathrm{Lim}}$ one has successively finer convergence structures, cf. (2.4.5). Since a topological space X is Hausdorff iff ιX is, the implications follow.

(6⇒8) We have to check only: $\|x\|_B=0 \Rightarrow x=0$. But if $\|x\|_B=0$ then $n \cdot x \in B$ for all $n \in \mathbb{N}$ and thus $x=(1/n) \cdot nx$ is M-convergent to 0 for $n \to \infty$.

(8⇒4) For any bounded subspace F, $\|\_\|_F=0$.

(4⇒1) Suppose $\ell(x)=0$ for all $\ell \in E'$. Then the linear subspace generated by x is bounded. □

Remark. For non-topological convex bornological spaces the separation condition (4) does not imply that the bornological dual separates points; for an example see [Hogbe-Nlend, 1977, p. 128].

2.5.3 Definition. A preconvenient vector space E is called *separated* iff E satisfies one and hence all conditions of the proposition above. With $\underline{sPre}$ we denote the full subcategory of $\underline{\mathrm{Pre}}$ formed by all separated preconvenient vector spaces.

We will discuss general products of preconvenient vector spaces in section 3.3. For the moment we only need products of copies of $\mathbb{R}$.

2.5.4 Lemma. *Let J be a set. Then the vector space $\Pi_J \mathbb{R}$ with the ℓ^∞-structure generated by the projections $\mathrm{pr}_j: \Pi_J \mathbb{R} \to \mathbb{R}$ $(j \in J)$ is a separated preconvenient vector space.*

Proof. Since the projections pr_j are linear, the considered ℓ^∞-structure is linearly generated, and thus the result follows by (2) of (2.4.4). Obviously the projections separate points. □

2.5.5 Proposition. (*Special Embedding Lemma.*) *For any preconvenient vector space E the canonical map $\iota_E: E \to \Pi_{E'} \mathbb{R}$ defined by $\iota_E(x) := (\ell(x))_{\ell \in E'}$ is an initial Pre-morphism.*

Proof. This follows from (2.4.7) by using as $\mathcal{S}$ the family of projections $\{\mathrm{pr}_\ell; \ell \in E'\}$, which generates the structure of the product according to (2.5.4). □

It is possible to associate to any preconvenient vector space in a natural way a separated one.

2.5.6 Proposition. *For any preconvenient vector space E there exists a morphism onto a separated preconvenient vector space ωE, such that any morphism into a separated preconvenient vector space factors uniquely over ωE. By* (8.4.3) *one thus obtains a functor ω: Pre→sPre which is left adjoint to the inclusion ι: sPre→Pre. Explicitly ωE can be constructed as the Pre-subspace $\iota_E(E)$ of $\Pi_{E'} \mathbb{R}$, cf.* (2.5.5). *The kernel of $\iota_E: E \to \omega E$ is exactly the closure of $\{0\}$ with respect to either the Mackey closure topology or the locally convex topology. It is the largest bounded subspace. Furthermore ι_E is a retraction in Pre, i.e. admits a right-inverse in Pre.*

Proof. As Pre-subspace of the separated preconvenient vector space $\Pi_{E'} \mathbb{R}$, cf. (2.5.5), the described space ωE is also separated. Let $g: E \to F$ be a Pre-morphism with F separated. Then $\iota_F: F \to \Pi_{F'} \mathbb{R}$ is an injective and initial linear ℓ^∞-morphism. Obviously $g: E \to F$ extends to a Pre-morphism $\bar{g}: \Pi_{E'} \mathbb{R} \to \Pi_{F'} \mathbb{R}$ characterized by $\mathrm{pr}_\ell \circ \bar{g} = \mathrm{pr}_{\ell \circ g}$ for all $\ell \in F'$. From $\bar{g}(\omega E) = \bar{g}(\iota_E(E)) = \iota_F(g(E)) \subseteq \iota_F(F)$ it follows that $\bar{g}$ restricts to a Pre-morphism $\omega E \to F$. It is unique since $\iota_E: E \to \omega E$ is surjective.

Denote the kernel of $\iota_E: E \to \omega E$ by E_0. Then E_0 is equal to $\bigcap_{\ell \in E'} \ell^{-1}(0)$. Hence E_0 is bounded, and any other bounded subspace has to be included in E_0 since it gets annihilated by all $\ell \in E'$. Being a kernel, E_0 is obviously closed with respect to the locally convex topology, hence also with respect to the Mackey closure topology. On the other hand the constant sequence 0 converges Mackey to any $x \in E_0$. Thus E_0 is the closure of $\{0\}$ in both topologies. Finally E_0 is the largest linear subspace F with trivial dual, since $F' = \{0\}$ implies that F is bounded.

The map $\iota_E: E \to \omega E$ is a retraction, since by the initiality of ι_E every linear right inverse map is a morphism. □

2.6 Completion

Now we turn to the question of existence of certain limits which are needed for calculus. Again we give equivalent characterizations of the respective completeness property, using various of the structures on a separated preconvenient vector space. For the case of Mackey convergence we need a lemma:

2.6.1 Lemma. *Let $\mathcal{G}$ be a Mackey–Cauchy filter on a preconvenient vector space E (with $\mathcal{G}-\mathcal{G}\leq \mathbb{U}B$ for some weakly closed B).* If *$\mathcal{G}$ converges weakly to $x\in E$, then $\mathcal{G}$ is* M-*convergent to x (with $\mathcal{G}-x\leq \mathbb{U}B$).*

Proof. By definition a Mackey–Cauchy filter satisfies $\mathcal{G}-\mathcal{G}\leq \mathbb{U}B$ for some bounded B, which may be assumed to be weakly closed, since the weak closure of a bounded set is bounded. Then there exists for any $\varepsilon>0$ a $G\in\mathcal{G}$ with $G-G\subseteq[-\varepsilon,\varepsilon]B$ and so $[-\varepsilon,\varepsilon]B\supseteq g-G\in g-\mathcal{G}$ for every $g\in G$. Since $[-\varepsilon,\varepsilon]B$ is weakly closed and $g-\mathcal{G}$ converges weakly to $g-x$ one concludes that $g-x\in[-\varepsilon,\varepsilon]B$. Therefore $G-x\subseteq[-\varepsilon,\varepsilon]B$, showing that $\mathcal{G}-x\leq\mathbb{U}B$. □

2.6.2 Theorem. *Let E be a separated preconvenient vector space, $k\in\mathbb{N}_{0,\infty}$. Then the following statements are equivalent:*

(1) *The Mackey convergence structure is complete;*
(2) *Every Mackey–Cauchy sequence converges* (*weakly*);
(3) *The bornology is complete* (*cf.* [*Hogbe-Nlend*, 1977, p. 42]), *i.e. every bounded set is contained in an absolutely convex bounded set B, such that E_B is a Banach space;*
(4) *For every $\mathcal{L}ip^k$-curve c the weak integral $\int_0^1 c$ exists;*
(5) *For every $\mathcal{L}ip^{k+1}$-curve c the weak derivative $\dot{c}(0)$ exists;*
(6) *For every $\mathcal{L}ip^{k+1}$-curve c the difference quotient $\delta^1 c:\mathbb{R}^{\langle 1\rangle}\to E$ has a $\mathcal{L}ip^k$-extension to $\mathbb{R}^2$;*
(7) *If $\{x_n; n\in\mathbb{N}\}\subseteq E$ is bounded and $(t_n)\in\ell^1$ then $\Sigma_n t_n x_n$ converges* (*weakly*);
(8) *For every* <u>Pre</u>*-embedding of E into a preconvenient vector space the image is* M-*closed.*

Proof. We will show the implications, cf. [Kriegl, 1982]:

$$\begin{array}{ccccccccc} & & 1 & \Longleftarrow & 2 & \Longrightarrow & & & 7\\ & & \Downarrow & & \Uparrow & & & & \Downarrow\\ 6 & \Leftrightarrow & 5 & \Longrightarrow & 8 & \Longleftarrow & 4 & \Longleftarrow & 3\end{array}$$

(1⇒5) We first show that

$$\left\{\frac{1}{t-s}\left(\frac{c(t)-c(0)}{t}-\frac{c(s)-c(0)}{s}\right); t,s\in[-1,1]\setminus\{0\}, t\neq s\right\}$$

is bounded. By composing with $\ell \in E'$ one reduces this to the case $E=\mathbb{R}$. In this case $t\mapsto(c(t)-c(0))/t$ has a $\mathscr{Lip}^k$-extension to $\mathbb{R}$ by (1.3.22), hence is locally locally Lipschitzian and the claimed boundedness follows. This boundedness implies the Mackey–Cauchy condition for the net $t\mapsto(c(t)-c(0))/t$ $(t\neq 0)$, where $t_1 \succ t_2$ iff $|t_1|\leq|t_2|$. So by (1) the Mackey limit and hence the weak limit exists.

(5⇒6) One defines the extension $\overline{\delta^1 c}$ of the difference quotient by $\overline{\delta^1 c}(t,t):=\dot c(t)$. Since $\ell\circ\delta^1 c=\delta^1(\ell\circ c)$ has a $\mathscr{Lip}^k$-extension to $\mathbb{R}^2$ by (1.3.22) it has to coincide with $\ell\circ\overline{\delta^1 c}$, and thus $\overline{\delta^1 c}$ is $\mathscr{Lip}^k$.

(6⇒5) This is obvious since $\overline{\delta^1 c}\,(0,0)$ is the weak derivative $\dot c(0)$ of c at 0.

(5⇒8) Let E be contained as Pre-subspace in some F and let $y\in F$ be the M-limit of a sequence in E. Then there exists a subsequence $x_n\in E$ with $\{n^n(x_n-y);\,n\in\mathbb{N}\}$ bounded. The special curve lemma (2.3.4) yields a smooth parametrization $c\colon\mathbb{R}\to F$ of the infinite polygon through the points x_n-y. Using a smooth monotonic function $h\colon\mathbb{R}\to[-1,1]$ with $h(t)=t$ for t in a neighborhood of 0 we define a smooth curve $e\colon\mathbb{R}\to F$ by $e(t):=t\cdot(c(h(t))+y+c(-h(t)))$. Then $e(0)=0$ and for $t\neq 0$ the point $e(t)$ lies on the segment between tx_n and tx_{n+1} for some $n\in\mathbb{N}$. Hence the smooth curve e lies in E. Its weak derivative at 0 is equal to y and lies in E by assumption (5).

(8⇒2) Let (x_n) be a Mackey–Cauchy sequence in E. Then $\ell(x_n)$ is a Cauchy sequence for every $\ell\in E'$ and hence converges to some number t_ℓ. Consider now the embedding $\iota\colon E\to\Pi_{E'}\mathbb{R}$ of (2.5.5). From (2.6.1) it follows that $\iota(x_n)$ is M-convergent to $(t_\ell)_{\ell\in E'}$. Since, by (8), $\iota_E(E)$ is M-closed in $\Pi\mathbb{R}$ this M-limit has to lie in $\iota_E(E)$. Using that ι_E is an initial injective ℓ^∞-morphism we conclude that x_n converges in E.

(2⇒1) is an immediate consequence of (2.2.12).

(2⇒7) Let B be a bounded convex set containing all x_n. Then $s_{nm}\cdot\Sigma_{i=n}^m t_i x_i$ is contained in B, where $s_{nm}:=(\Sigma_{i=n}^m|t_i|)^{-1}\to\infty$. Hence the series is a Mackey–Cauchy sequence.

(7⇒3) For $A\subseteq E$ bounded take as B the weak closure of the absolutely convex hull of A. Use then the fact that a normed space like E_B is complete if every absolutely summable series is convergent. By definition a series Σy_n is absolutely convergent in E_B iff Σt_n converges where t_n is the E_B-norm of y_n. Define $x_n:=(1/t_n)\,y_n$ for $t_n\neq 0$ and $x_n:=0$ for $t_n=0$. Then $\{x_n;\,n\in\mathbb{N}\}$ is bounded and hence $\Sigma t_n x_n=\Sigma y_n$ converges weakly by (7). Using (2.6.1) we conclude that Σy_n converges in E_B.

(3⇒4) Let c be a $\mathscr{Lip}^0$-curve. Then $c|_{[0,1]}$ is continuous into some E_B, which can be assumed to be complete. Hence $\int_0^1 c$ exists in E_B and thus in E. Another method, cf. (4.1.3), to prove this would be to show that the Riemann sums form a Mackey–Cauchy net and then use (3).

(4⇒8) Let F, x_n, y, c and h as in (5⇒8). One defines a smooth curve $e\colon\mathbb{R}\to F$ by $e:=(c\circ h)^{\cdot}$. Then $e(0)=0$ and for $t\neq 0$ the point $e(t)$ lies in E since $(c\circ h)^{\cdot}(t)$ is a multiple of some difference $x_{n+1}-x_n$. Hence e is a smooth curve in E. Its weak integral from 0 to 1 is equal to $(c\circ h)(1)-(c\circ h)(0)=c(h(1))$ and lies in E. Since $c(h(1))$ lies on the segment between x_n-y and $x_{n+1}-y$ for some $n\in\mathbb{N}$, one concludes that y lies in E. □

Remark. Note that we gave no characterization of completeness in terms of the locally convex topology. A locally convex space E is called *Mackey complete* (or *locally complete*, cf. [Jarchow, 1981, p. 196]) iff the associated preconvenient vector space βE (cf. (2.1.10) and (1) of (2.4.4)) is complete. Since every Mackey–Cauchy sequence is also a Cauchy sequence in the locally convex topology, the sequential completeness of this topology is enough to ensure Mackey completeness. Although it is quite likely that the converse is false, we do not know any counter-example, i.e. a bornological (!) locally convex space that is Mackey complete but not sequentially complete. An example showing that Mackey completeness does not imply (quasi-)completeness of the locally convex topology can be found in (7.4.3). For metrizable locally convex spaces E, however, Mackey completeness is equivalent to completeness, cf. [Jarchow, 1981, p. 197]. In fact, for a Mackey complete metrizable locally convex space the completion is metrizable, cf. [Jarchow, 1981, p. 60]; hence one can apply (8) of the previous theorem to conclude that E coincides with this completion.

For non-topological convex bornological spaces Mackey completeness does not imply bornological completeness; cf. (7.4.4).

2.6.3 Definition. Any separated preconvenient vector space that satisfies one and hence all conditions of the proposition above is called *complete* and the complete separated preconvenient vector spaces will shortly be called *convenient vector spaces.* With $\underline{Con}$ we denote the full subcategory of $\underline{\text{Pre}}$ formed by all convenient vector space.

By the previous remark and (iv) of (2.1.20) Fréchet spaces, i.e. complete metrizable locally convex spaces, are convenient.

Let us give some useful information on M-closed subspaces of convenient vector space.

2.6.4 Proposition. (*Closed Embedding Lemma.*) *Let $m: F \to E$ be a $\underline{PRE}$-embedding into a convenient vector space E with M-closed image. Then F is a convenient vector space and the Mackey closure topology of F is the trace topology of the Mackey closure topology of E.*

Proof. In order to show that F is complete, let x_n be a Mackey–Cauchy sequence in F, hence in E. But in E it has to converge Mackey towards an $x \in E$. Since F is assumed to be M-closed we obtain that $x \in F$ and $x_n \xrightarrow{M} x$ in E.

The spaces E and F with their Mackey closure topologies are again denoted $\tau_M E$ and $\tau_M F$. Obviously the injection $\tau_M F \to \tau_M E$ is continuous. Conversely let A be closed in $\tau_M F$ and $a_n \in A$ a sequence converging Mackey to $x \in E$. Since F is M-closed, $a_n \xrightarrow{M} x$ in F. Therefore $x \in A$, showing that A is also closed in $\tau_M E$. □

Remark. For an arbitrary $\underline{\text{Pre}}$-subspace F, the Mackey closure topology of F is not always the trace topology of that of E, cf. (6.3.3).

We now show that for every preconvenient vector space there exists a completion having the usual universal property.

2.6.5 Theorem. *For any preconvenient vector space E there exists a convenient vector space $\tilde{E}$ which is the universal solution for extending bornological linear maps into convenient vector spaces. One thus obtains a functor $\bar{\omega}$: $\underline{\text{Pre}} \to \underline{\text{Con}}$ which is left adjoint to the inclusion $\underline{\text{Con}} \to \underline{\text{Pre}}$. Explicitly $\bar{\omega}E = \tilde{E}$ can be constructed as the M-closure of the image of E under the initial $\underline{\text{Pre}}$-morphism $\iota_E: E \to \Pi_{E'} \mathbb{R}$, cf. (2.5.5). This $\tilde{E}$ will be called the (separated) completion of E and $\bar{\omega}$ the completion functor.*

Proof. We first show that $\Pi_J \mathbb{R}$ as defined in (2.5.4) is convenient. For any Mackey–Cauchy sequence (x^n) the coordinates x_j^n from a Cauchy sequence in $\mathbb{R}$ and thus converge to some x_j^∞. Let $x^\infty \in \Pi_J \mathbb{R}$ be the point with coordinates x_j^∞. By construction x^n converges weakly to x^∞, thus $\Pi_J \mathbb{R}$ is convenient by (2.6.1).

From this and (2.6.4) it follows that the M-closure of $\iota_E(E)$ in $\Pi_J \mathbb{R}$ is convenient.

Let now $g: E \to F$ be a $\underline{\text{Pre}}$-morphism into an arbitrary convenient vector space F. Then F is M-closed in $\Pi_{F'} \mathbb{R}$ by property (8) of (2.6.2) and g extends to a bornological linear map $\bar{g}: \Pi_{E'} \mathbb{R} \to \Pi_{F'} \mathbb{R}$. Hence $\bar{g}$(M-closure of E) $\subseteq$ M-closure of $\bar{g}(E) \subseteq$ M-closure of $F = F$. Furthermore this extension is unique, since $\iota_E(E)$ is M-dense in $\tilde{E}$ as (2.6.4) implies. □

Remark. The separated completion of E can also be obtained in two steps: first one forms the associated separated space ωE and then the convenient vector space $\bar{\omega}(\omega E)$ associated to ωE. More formally: $\bar{\omega} = \bar{\omega} \circ \iota \circ \omega$, where ι: $\underline{\text{sPre}} \to \underline{\text{Pre}}$ denotes the inclusion.

One might believe that the completion of E can be obtained by taking the M-closure of E in any convenient vector space F that contains E as a $\underline{\text{Pre}}$-subspace. However this is not true in general; see (6.3.2) for an example. We will show in (2.6.7) that it holds under an additional assumption. The following lemma will be useful.

2.6.6 Lemma. (*Linear Extension Lemma.*) *Let X be a set, F a preconvenient vector space, G a convenient vector space; $f: X \to F$ and $g: X \to G$ two maps. Suppose that the linear subspace $\langle f(X) \rangle$ generated by $f(X)$ is dense in F with respect to the Mackey closure topology of F. If for every $\ell \in G'$ there exists an $\ell_F \in F'$ with $\ell \circ g = \ell_F \circ f$, then there exists a unique $\underline{\text{Pre}}$-morphism $\tilde{g}: F \to G$ with $g = \tilde{g} \circ f$.*

Proof. Let $\iota_G: G \to \Pi_{G'} \mathbb{R}$ be the $\underline{\text{Pre}}$-embedding of (2.5.5); $\tilde{m}$: $F \to \Pi_{G'} \mathbb{R}$ the $\underline{\text{Pre}}$-morphism with $\text{pr}_\ell \circ \tilde{m} = \ell_F$ for all $\ell \in G'$. Then $\text{pr}_\ell \circ \tilde{m} \circ f = \ell_F \circ f = \ell \circ g = \text{pr}_\ell \circ \iota_G \circ g$ for $\ell \in G'$, hence $\tilde{m} \circ f = \iota_G \circ g$, which implies $\tilde{m}(f(X)) \subseteq \iota_G(G)$. Using that $\tilde{m}$ is linear and continuous with respect to the Mackey closure topologies one obtains: $\tilde{m}(F)=\tilde{m}(\overline{\langle f(X)\rangle})\subseteq\overline{\tilde{m}\langle f(X)\rangle}$ $\subseteq\overline{\langle\tilde{m}(f(X))\rangle}\subseteq\overline{\langle\iota_G(G)\rangle}=\iota_G(G)$. Therefore $\tilde{m}$ factors as $\tilde{m}=\iota_G\circ\tilde{g}$ and by initiality of ι_G one concludes that $\tilde{g}: F\to G$ is a $\underline{\text{Pre}}$-morphism. One has $\ell\circ\tilde{g}\circ f$ $=\mathrm{pr}_\ell\circ\iota_G\circ\tilde{g}\circ f=\mathrm{pr}_\ell\circ\tilde{m}\circ f=\mathrm{pr}_\ell\circ\iota_G\circ g=\ell\circ g$ for all $\ell\in G'$, thus $\tilde{g}\circ f=g$. Uniqueness of $\tilde{g}$ is trivial. □

2.6.7 Proposition. *Let $m: E\to F$ be an injective morphism from a preconvenient vector space E into a convenient vector pace F with M-dense image. Then the following statements are equivalent:*

(1) $m: E\to F$ *is a completion, i.e. every morphism ℓ from E into a convenient vector space G extends uniquely to a morphism $\tilde{\ell}: F\to G$;*
(2) *every element of E' has an extension belonging to F';*
(3) $m: E\to F$ *is an embedding for the locally convex topologies.*

Proof. (1⇒3) We first show that m is a $\underline{\text{Pre}}$-embedding. So let $B\subseteq E$ with $m(B)\subseteq F$ bounded. Then $\ell(B)=\tilde{\ell}(m(B))$ is bounded for $\ell\in E'$, thus B is bounded in E.

Consider on $m(E)$ the trace of the locally convex topology on F. Then the locally convex topology on E is the topology induced by the map m into the bornologification of $m(E)$. So it remains to show that $m(E)$ is bornological. For this it is enough to show that all bornological linear maps into any Banach space are continuous, cf. [Jarchow, 1981, p. 272]. So let g: $m(E)\to G$ be such a map. Then $g\circ m$ is bornological, and so it extends to a morphism $(g\circ m)^\sim$ on F. Thus $g=(g\circ m)^\sim|_{m(E)}$ is continuous with respect to the trace topology.

(3⇒2) This is a direct consequence of the Hahn–Banach theorem.

(2⇒1) This follows from the linear extension lemma (2.6.6). □

2.6.8 Corollary. *Let $m: E\to F$ be an injective $\underline{\text{Pre}}$-morphism into a convenient vector space F. If every $\ell\in E'$ extends to F (i.e. $m^*: F'\to E'$ is surjective) then a completion of E is given by $m: E\to\overline{m(E)}$, where $\overline{m(E)}$ denotes the* M-closure *of the image $m(E)$* in *F.*

Remark. Using this corollary one shows easily that the completion of a separated preconvenient vector space E can be obtained by taking its M-closure in the bidual E'' which will be introduced later.

3 MULTILINEAR MAPS AND CATEGORICAL PROPERTIES

After having demonstrated in the previous chapter that convenient vector spaces have good internal properties which guarantee that derivatives and integrals exist and are unique, we show in this chapter that at the same time they form a category with excellent properties. As for analogous constructions of topological spaces, initial and final structures can be used to show the existence and give explicit descriptions of all categorical limits and colimits. The general procedure is described in section 3.1. The most important special cases, namely subspaces, quotients, products, direct sums and inductive limits, are discussed in sections 3.2 to 3.5, and it is shown that in these cases the general constructions often can be simplified.

Function spaces of linear and multilinear maps are clearly of importance since for a differentiable map the (higher) derivative at a point will be a (multi-)linear map. Section 3.6 starts with the linear case. For convenient vector spaces the linear maps from E to F which satisfy the equivalent conditions of being continuous, bornological or differentiable, i.e. being morphisms, form in a natural way a convenient vector space denoted by $L(E, F)$.

In section 3.7 multilinear morphisms are introduced and characterized in various ways. The function spaces formed by them are discussed and it is shown that they can be identified in the usual way with iterated function spaces of linear morphisms.

In section 3.8 it is proved that for any convenient vector spaces E_1 and E_2 there exists a tensor product, i.e. a convenient vector space $E_1 \tilde{\otimes} E_2$ with the property that the bilinear morphisms from $E_1 \times E_2$ to any third space F are in one-to-one correspondence with the morphisms from $E_1 \tilde{\otimes} E_2$ to F.

Of great importance for many results is the fact that a version of the Banach–Steinhaus theorem, also called the linear uniform boundedness theorem, holds. It states that a subset of the space $L(E, F)$ is bounded on bounded subsets of E provided it is bounded on points of E. It allows to obtain related results for spaces of non-linear functions: a so-called bornological uniform boundedness principle in the same section, a multilinear version in section 3.7 and differentiable ones in Chapter 4.

In the last section the duality functor which associates to a convenient vector space the space $E' = L(E, \mathbb{R})$ is studied. It is proved that any E can be naturally embedded into its bidual E''. Furthermore the duals of products and direct sums are determined.

3.1 Initial and final structures, categorical completeness

Since we not only want to prove the existence but also give explicit descriptions of categorical limits and colimits of convenient vector spaces, we first study some of the ambient categories.

3.1.1 Lemma. *For the following categories initial and final structures exist with respect to the forgetful functor to the category $\underline{VS}$ of vector spaces:*

(i) *the category $\underline{DVS}$ of dualized vector spaces;*
(ii) *the category $\underline{LCS}$ of locally convex spaces;*
(iii) *the category $\underline{CBS}$ of convex bornological spaces.*

They will be called initial and final dualized vector space structures in case (i), initial and final locally convex structures in case (ii) and initial and final convex bornological structures in case (iii) without explicit mention of the forgetful functor to $\underline{VS}$.

Proof. The verifications being trivial, we only describe for the three cases under (a) the initial structure for a given family of linear maps $m_j: E \to E_j$ and under (b) the final structure for a given family of linear maps $m_j: E_j \to E$, where j varies in an arbitrary index set J, $E \in |\underline{VS}|$ and E_j is an object in the corresponding category stated in the lemma.

(i) (a) Take as E' the linear subspace of the algebraic dual of E generated by $\bigcup_{j\in J} m_j^*(E_j')$.

(b) Take $E' := \{\ell : E \to \mathbb{R} \text{ linear}; \ell \circ m_j \in E_j' \text{ for all } j \in J\}$.

(ii) (a) Put on E the initial topology, which has as sub-basis the sets $m_j^{-1}(O_j)$ for $j \in J$ and O_j running through a sub-basis of the topology of E_j.

(b) Put on E the locally convex topology having as basis for the 0-neighborhoods the collection of absolutely convex absorbent subsets U for which $m_j^{-1}(U)$ is a 0-neighborhood in E_j for all $j \in J$; cf. [Jarchow, 1981, p. 108].

(iii) (a) Put on E the initial bornology $\{B \subseteq E; m_j(B) \subseteq E_j \text{ bounded for all } j \in J\}$; cf. [Hogbe-Nlend, 1977, p. 29].

(b) Put on E the convex vector bornology $\mathscr{B}$ generated by the sets $m_j(B_j)$ for $j \in J$ and B_j running through a basis of the bornology of E_j. The bornology $\mathscr{B}$

has as basis the absolutely convex hulls of finite unions of these and the finite sets; cf. [Hogbe-Nlend, 1977, p. 32] and (2.1.4). □

3.1.2 Theorem.

(i) *Initial structures with respect to the forgetful functor* $\underline{Pre} \to \underline{VS}$ *exist. They will be called initial preconvenient structures and have, for given linear maps* $m_j: E \to E_j$, $j \in J$, $E \in |\underline{VS}|$ and $E_j \in |\underline{Pre}|$, *the following descriptions:*

— *the dual of E is obtained by applying the retraction functor* $\delta\sigma$: $\underline{DVS} \to \underline{Pre}$, *cf.* (2.4.1), *to E equipped with the initial dualized vector space structure;*
— *the locally convex topology of E is obtained by bornologifying the initial locally convex structure, i.e. applying the functor* $\gamma\beta: \underline{LCS} \to \underline{bLCS}$;
— *the bornology of E is the initial (convex vector) bornology; cf.* (3.1.1);
— *the* ℓ^∞*-structure of E is the initial* ℓ^∞*-structure;*
— *the* $\mathscr{Lip}^k$*-structure of E is the initial* $\mathscr{Lip}^k$*-structure.*

(ii) *Final structures with respect to the forgetful functor* $\underline{Pre} \to \underline{VS}$ *exist. They will be called final preconvenient structures and have, for given linear maps* $m_j: E_j \to E$, $j \in J$, $E \in |\underline{VS}|$, $E_j \in |\underline{Pre}|$, *the following descriptions:*
— *the dual of E is the final dualized vector space structure;*
— *the locally convex topology of E is the final locally convex topology;*
— *the bornology of E is obtained by topologifying the final convex vector bornology, i.e. applying the functor* $\beta\gamma: \underline{CBS} \to \underline{tCBS}$;
— *the* ℓ^∞*-structure of E is the one generated by those linear functions* $\ell: E \to \mathbb{R}$ *for which* $\ell \circ m_j$ *is an* ℓ^∞*-morphism for all* $j \in J$;
— *the* $\mathscr{Lip}^k$*-structure of E is the one generated by those linear functions* $\ell: E \to \mathbb{R}$ *for which* $\ell \circ m_j$ *is a* $\mathscr{Lip}^k$*-function for all* $j \in J$.

Proof. For the first three structures the result follows from (3.1.1) by using the special case (8.7.5) of the categorical proposition (8.7.4) and (2.4.3). For the ℓ^∞- or $\mathscr{Lip}^k$-structure one verifies directly that the described structures suffice. □

Remark. It will be shown in sections 3.2–3.5 that there are important special cases where certain of the above descriptions become much simpler in the sense that applying a retraction functor like $\delta\sigma$, $\gamma\beta$ or $\beta\gamma$ becomes superfluous.

3.1.3 Corollary. *The category* $\underline{Pre}$ *of preconvenient vector spaces is complete and cocomplete. The forgetful functor to the category* $\underline{VS}$ *of vector spaces has a left and a right adjoint, hence commutes with limits and colimits. Limits (colimits) in* $\underline{Pre}$ *are obtained by forming them in* $\underline{VS}$ *and putting the initial (final) preconvenient structure on them.*

Proof. The categorical proposition (8.7.3) applies directly.

3.1.4 Theorem. *The category Con of convenient vector spaces is complete and cocomplete. Limits are obtained by forming them in Pre. Colimits are obtained by applying the completion functor $\bar{\omega}$ (cf. (2.6.5)) to the colimit formed in Pre.*

Proof. Since the retraction functor $\bar{\omega}$: Pre→Con is left adjoint to the inclusion functor ι: Con→Pre, the categorical proposition (8.5.3) applies. □

Remark. Again we refer to sections 3.2–3.5 for some special cases where colimits taken in Pre are already separated and sometimes even complete, so that the application of $\bar{\omega}$ is superfluous.

3.2 Subspaces and quotients

We saw in (3.1.2) that initial and final Pre-structures do exist; this is not the case for Con-structures, even for families consisting of a single map only. The following proposition (of which only the second part will be used) describes the initial Con-morphisms:

3.2.1 Proposition. *Let $f: E \to F$ be a linear map from a vector space E into a convenient vector space F. Then the initial Con-structure on E induced by f exists (and a basis of its bornology is given by the absolutely convex bounded sets B for which $F_{f(B)}$ is complete) iff f is injective and the ultrabornologification of the trace topology of the locally convex topology of F on the image $f(E)$ is Mackey complete. If in particular the image of f is* M*-closed then this ultrabornologification is Mackey complete and the initial Pre-structure equals the initial Con-structure.*

Proof. We first recall that the ultrabornologification of a locally convex space G has as 0-neighborhood basis those absolutely convex sets that absorb the bounded absolutely convex sets B for which G_B is a Banach space.

($\Leftarrow$) Supply E with the initial topology induced by the locally convex topology of F and take its ultrabornologification. This makes E convenient since ultrabornological implies bornological, separation is trivial and Mackey completeness is assumed. That the mentioned sets form a basis of the bornology of E is immediate by the description of the ultrabornologification.

Let G be a convenient vector space and $g: G \to E$ a linear map for which $f \circ g$ is a morphism. Then $g: G \to E$ is continuous with respect to the locally convex topology of G and the considered initial topology on E. By applying the ultrabornologification functor and using that the locally convex topology of every convenient vector space G is ultrabornological (i.e. invariant under ultrabornologification) one deduces that $g: G \to E$ is a morphism.

($\Rightarrow$) Suppose $f(x) = 0$. Let G be a Banach space and $\ell: G \to \mathbb{R}$ a linear functional that is not continuous. Then $y \mapsto \ell(y) \cdot x$ defines a linear map $g: G \to E$ such that $f \circ g$ is a morphism (since $f \circ g = 0$). By the assumed initiality $g: G \to E$ has to be a morphism. Let $\ell_1 \in E'$ be arbitrary. Then $(\ell_1 \circ g)(y) = \ell(y) \cdot \ell_1(x)$

defines a morphism on G. Thus $\ell_1(x)$ has to be zero, otherwise ℓ would be a morphism on G. Since E is assumed to be separated, x has to be 0. This proves the injectivity of f.

The ultrabornologification of the trace topology of $f(E)$ can be described as colimit in $\underline{LCS}$ of the Banach spaces F_B with bounded $B \subseteq f(E)$. Let g_B be the composite of the inverse of $f: E \to f(E)$ with the inclusion of F_B into $f(E)$. Then g_B is linear and $f \circ g_B$ is the inclusion and thus a morphism. By the initiality assumption g_B has to be a morphism and by the universal property of the colimit the inverse of the restriction of f to the ultrabornologification of $f(E)$ into E has to be a homeomorphism. Thus $f(E)$ is convenient since it is isomorphic as locally convex space to E.

That M-closedness suffices was proved in (2.6.4). □

3.2.2 Remark. Any $\underline{\text{Con}}$-morphism that is an initial $\underline{\text{Pre}}$-morphism is also an initial $\underline{\text{Con}}$-morphism. However an initial $\underline{\text{Con}}$-morphism is in many cases not an initial $\underline{\text{Pre}}$-morphism. For an example see (3.4.3) together with (3.4.4).

For an initial $\underline{\text{Pre}}$-structure (but not in general for an initial $\underline{\text{Con}}$-structure) one has, as shown in (3.1.2), that the bornology is the initial bornology, the ℓ^∞-structure is the initial ℓ^∞-structure, and the $\mathcal{Lip}^k$-structure is the initial $\mathcal{Lip}^k$-structure.

But the locally convex topology is only in special cases the initial locally convex structure. A sufficient condition is the following:

> The inclusion of a $\underline{\text{Pre}}$-subspace of F is also initial for the locally convex topologies if it is of finite codimension in F, or of at most countable codimension provided F is convenient.

That subspaces of finite codimension of bornological locally convex spaces are bornological can be found in [Jarchow, 1981, p. 281]. Subspaces of countable codimension of ultrabornological locally convex spaces are bornological by [Valdivia, 1971]. For subspaces of arbitrary codimension this is in general false, cf. (7.4.2). □

Now we turn towards final morphisms.

3.2.3 Lemma. *Let $(m_j): E_j \to E$ be a final family of $\underline{\text{CBS}}$-morphisms. If every E_j is complete and E is separated then E is complete too.*

Proof. Let $B \subseteq E$ be bounded. By (3.1.1) there are finitely many bounded sets $B_1 \subseteq E_{j_1}, \ldots, B_n \subseteq E_{j_n}$ and a finite set $B_0 \subseteq E$ with B being contained in the absolutely convex hull of the union of $m_{j_1}(B_1), \ldots, m_{j_n}(B_n)$; B_0. Since each E_{j_i} is complete we may choose absolutely convex bounded sets $K_i \subseteq E_{j_i}$ containing B_i and such that $(E_{j_i})_{K_i}$ is a Banach space. Since B_0 is finite the space E_{K_0} is finite dimensional and hence a Banach space, where K_0 denotes the absolutely convex hull of B_0. The map $m_{j_1} + \cdots + m_{j_n} + \mathrm{id}$ defined on the product $\Pi_{i=1}^n (E_{j_i})_{K_i} \Pi E_{K_0}$ is a quotient map onto the normed space E_K, where $K := m_{j_1}(K_1) + \cdots + m_{j_n}(K_n) + K_0$, hence E_K is a Banach space which certainly contains B. □

3.2.4 Proposition. *Let $f: E \to F$ be a linear map from a convenient vector space E into a vector space F. Then the final $\underline{\text{Con}}$-structure on F induced by f exists (and equals the final $\underline{\text{Pre}}$-structure) iff the kernel of f is closed in the locally convex topology and the final locally convex topology on F is Mackey complete. If in particular the final convex bornological structure on F is separated and topological then these conditions are satisfied and the bornology of the final $\underline{\text{Con}}$-structure on E is the final convex bornological structure.*

Proof. ($\Rightarrow$) Every final $\underline{\text{Con}}$-morphism has to be a final $\underline{\text{Pre}}$-morphism and hence to be final for the locally convex topologies. But a final topology is separated only if the kernel is closed; cf. [Jarchow, 1981, p. 76].

($\Leftarrow$) The locally convex topology of the final $\underline{\text{Pre}}$-structure is the final locally convex structure, thus the assumptions imply that it is convenient and hence the final $\underline{\text{Con}}$-structure.

In a case where the final convex bornological structure is topological and separated, it defines a separated preconvenient vector space which is complete since final convex bornological structures inherit completeness by the previous lemma. □

3.2.5 Remark. In (3.1.2) we showed that for the final $\underline{\text{Pre}}$-structure the locally convex topology is the final locally convex topology. This is not the case for the bornology, the ℓ^∞-structure and the $\mathcal{Lip}^k$-structure. An example for the bornology and hence ℓ^∞-structure can be found in [Jarchow, 1981, p. 233]. An example for the $\mathcal{Lip}^\infty$-structure will be provided in (7.3.7). In (7.3.3) and (7.3.4) final $\underline{\text{Pre}}$-morphisms will be described explicitly.

3.3 Products

3.3.1 Proposition. *Let E_j for $j \in J$ be preconvenient (resp. convenient) vector spaces. The categorical product $E = \prod_{j\in J} E_j$ in $\underline{\text{Pre}}$ (resp. in $\underline{\text{Con}}$) has as underlying vector space the product of the underlying vector spaces.*

- *The bornology of E is the product bornology, i.e. $B \subseteq E$ is bounded iff $\text{pr}_j(B)$ is bounded in E_j for all $j \in J$.*
- *The dual of E is obtained by applying the retraction functor $\delta\sigma$, cf. (2.4.1), to the product formed in $\underline{DVS}$ (having as dual the direct sum of the duals, cf. (3.1.1)). (In most cases it is not necessary to apply $\delta\sigma$, cf. (3.9.5)).*
- *The locally convex topology is obtained by bornologification (applying $\gamma\beta$) of the product of the locally convex topologies (Again in most cases it is not necessary to apply $\gamma\beta$, cf. (3.3.5)).*
- *The bounded sequences ($\underline{\ell^\infty}$-morphisms) are those $c: \mathbb{N} \to E$ for which the coordinates $\text{pr}_j \circ c: \mathbb{N} \to E_j$ are bounded sequences of E_j for all $j \in J$.*
- *The $\mathcal{Lip}^k$-curves $c: \mathbb{R} \to E$ are those whose coordinates $\text{pr}_j \circ c: \mathbb{R} \to E_j$ are $\mathcal{Lip}^k$-curves of E_j for all $j \in J$.*

Proof. This is an immediate consequence of the general result in (3.1.3) and (3.1.4). □

Remark. If $F_j = F$ for all $j \in J$ we also write F^J for $\Pi_{j \in J} F_j$.

We did not describe the Mackey convergence for general limits. Let us describe it now for finite products.

3.3.2 Proposition. *For a finite product E of (pre-)convenient vector spaces E_i $(i = 1, \ldots, m)$ the Mackey convergence on E is the product convergence structure of those of the factors, i.e. for a filter $\mathscr{H}$ on $E = \Pi_{i=1\ldots m} E_i$ one has $\mathscr{H} \xrightarrow[E]{} p$ iff $\mathrm{pr}_j(\mathscr{H}) \xrightarrow[E_j]{} \mathrm{pr}_j(p)$ for $j = 1 \ldots m$.*

Proof. Using the translation invariance one reduces the proof to the case $p = 0$. Suppose first that $\mathrm{pr}_j(\mathscr{H}) \xrightarrow[E_j]{} 0$ for $j = 1 \ldots m$. Then there exist bounded sets $B_j \subseteq E_j$ such that $\mathrm{pr}_j(\mathscr{H}) \leq \mathbb{U} \cdot B_j$ and we thus obtain $\mathscr{H} \leq \mathrm{pr}_1(\mathscr{H}) \times \ldots \times \mathrm{pr}_m(\mathscr{H}) \leq \mathbb{U} B_1 \times \ldots \times \mathbb{U} B_m \leq \mathbb{U}(B_1 \times \ldots \times B_m)$, and since $B_1 \times \ldots \times B_m$ is bounded in E the assertion $\mathscr{H} \xrightarrow[E]{} 0$ follows. We remark that the first inequality fails for infinite products. Conversely, $\mathscr{H} \xrightarrow[E]{} 0$ implies $\mathrm{pr}_j(\mathscr{H}) \xrightarrow[E_j]{} 0$ because pr_j is continuous with respect to the Mackey convergence structures. □

3.3.3 Remark. The following example shows that (3.3.2) in fact fails already for infinite denumerable products. Choose $E = \mathbb{R}^{\mathbb{N}}$, i.e. the product of countably many factors $\mathbb{R}$. Take as directed set $J := \{k + 1/m;\ k, m \in \mathbb{N}\}$ with its natural order inherited from $\mathbb{R}$ and define a net $x: J \to E$ as follows:

$$\mathrm{pr}_n(x(t)) := \begin{cases} 0 & \text{for} \quad t \geq n \\ m & \quad\quad t = k + m^{-1} < n \end{cases}$$

For every $n \in \mathbb{N}$ the net $\mathrm{pr}_n \circ x: J \to \mathbb{R}$ is trivially M-convergent to zero, but x is not M-converging to 0 in $\mathbb{R}^{\mathbb{N}}$. In fact, otherwise there would exist reals s_t with $s_t \to \infty$ for $t \to \infty$ and such that $\{s_t x(t); t \in J\}$ would be bounded. This is obviously impossible, because for every $t_0 \in J$ even $\{x(t); t \in J, t > t_0\}$ is unbounded.

3.3.4 Remark. Even for a product of two convenient vector spaces the Mackey closure topology is often strictly finer than the product of the Mackey closure topologies of the factors. According to the characterization of multilinear morphisms in (3.7.1) it is enough to find two bornological locally convex spaces E and F that are complete and a bilinear function $E \times F \to \mathbb{R}$ which is bornological but not continuous. Take as F any non-normable Fréchet space (as e.g. $\mathbb{R}^{\mathbb{N}}$) and as E its dual with the strong topology; cf. [Jarchow, 1981, p. 154]. Then the evaluation map $E \times F \to \mathbb{R}$ is bilinear and bornological. But for a non-normable locally convex space F there does not even exist a topology with point absorbing 0-neighborhoods on F' making the evaluation map continuous. To see this, suppose there exists such a topology. Then take a 0-neighborhood U in F' and a 0-neighborhood V in F with $\mathrm{ev}(U \times V) \subseteq [-1, 1]$. The 0-neighborhood V is bounded, since (cf. (i) in (2.1.21)) for every $\ell \in F'$ one has $\ell \in K \cdot U$ for some $K > 0$ and hence $\ell(V) \subseteq K \cdot U(V) \subseteq K[-1, 1] = [-K, K]$; this is a contradiction to the non-normability of F, cf. [Jarchow, 1981, p. 116].

An important special case where the product of the M-closure topologies is the M-closure topology of the product is the following: E and F convenient with E finite dimensional.

It is enough to show this for $E = \mathbb{R}$ since any finite-dimensional E is isomorphic to some $\mathbb{R}^m$. So let $(t, x) \in U$ with U M-open in $\mathbb{R} \times F$. Take a compact interval I containing t such that $I \times \{x\} \subseteq U$. Let $V := \{x' \in F; (t', x') \in U \text{ for all } t' \in I\}$. It remains to show that V is M-open in F. Otherwise there is a sequence $x_n \notin V$ in F with $x_n \xrightarrow{M} x_\infty \in V$. Thus there are $t_n \in I$ with $(t_n, x_n) \notin U$. Since I is compact we may assume that $t_n \to t_\infty \in I$, and hence $(t_\infty, x_\infty) \in U$. This is a contradiction to the assumption that U is M-open.

3.3.5 Theorem. *If the cardinality of the index set J is non-measurable* (*i.e. J does not admit a non-trivial additive $\{0, 1\}$-measure defined on all subsets of J; cf.* [*Jarchow*, 1981, *p.* 282]) *then the locally convex topology of a product $\Pi_{j \in J} E_j$ of (pre-)convenient vector spaces is the product topology of the locally convex topologies of the factors.*

Proof. The topologies of the factors are bornological. Hence by the Mackey–Ulam theorem [Jarchow, 1981, p. 282] so is the product topology. Thus the bornologification according to the general description in (3.3.1) is not necessary. □

Remark. If in some model of set theory a measurable cardinal exists, then the smallest such cardinal has to be strongly inaccessible [Jarchow, 1981, p. 282]. Hence one can restrict set theory in such a way that all cardinals are non-measurable.

3.3.6 Corollary. *For a product $\Pi_{j \in J} E_j$ of (pre-)convenient vector spaces with an index set of non-measurable cardinality the dual $(\Pi_{j \in J} E_j)'$ is the direct sum of the duals.*

Proof. One only has to combine the preceding result with the fact that the dual formed by the continuous linear functionals on any product of locally convex spaces is the direct sum of the duals of the factors; cf. [Jarchow, 1981, p. 165] for the separated case.

3.4 Coproducts or direct sums

3.4.1 Proposition. *Let E_j for $j \in J$ be convenient (preconvenient) vector spaces. The categorical coproduct $E = \coprod_{j\in J} E_j$ in Con (in Pre) has as underlying vector space the direct sum (i.e. the coproduct in VS) of the underlying vector spaces [Jarchow, 1981, p. 17]. The dual of E is the product of the duals E'_j. The locally convex topology of E is the final one induced by the canonical injections $\mathrm{in}_j\colon E_j \to E$ (also called the locally convex sum topology; see [Jarchow, 1981, p. 111]). The bounded sets of E are those sets which are contained in sums of finitely many bounded sets $B_j \subseteq E_j$. The ℓ^∞-structure of E has as structure curves the sequences $c\colon \mathbb{N} \to E$ that are sums of a finite number of bounded sequences $c_j\colon \mathbb{N} \to E_j$. The $\mathcal{Lip}^k$-structure of E has as structure curves those maps $c\colon \mathbb{R} \to E$ which are locally representable as a finite sum of $\mathcal{Lip}^k$-curves $c_j\colon \mathbb{R} \to E_j$.*

Proof. We first consider the case of preconvenient spaces E_j and begin with the description of the bornology of E. Using (2.1.4) it is easily verified that the finite sums of sets of the form $\mathrm{in}_j(B_j)$ with $B_j \subseteq E_j$ bounded form a basis of a convex vector bornology (and it is of course the finest one for which all these sets $\mathrm{in}_j(B_j)$ are bounded); i.e. it is the final convex bornological structure induced by the maps $\mathrm{in}_j\colon E_j \to E$. Remains to verify that it is a topological bornology (otherwise we would have to apply the topologification functor $\beta\gamma$ according to (3.1.2)). This is easy; one uses either (2.1.23), or the fact that it is the von Neumann bornology of the locally convex sum topology. The statements concerning the dual of E and the locally convex topology of E follow directly from the general results, cf. (3.1.2) and (3.1.1); those concerning the ℓ^∞-structure and the $\mathcal{Lip}^k$-structure follow from the given description of the bornology, combined with (1.3.22) in the case of $\mathcal{Lip}^k$-curves, cf. (4.1.12).

For the case of convenient vector spaces E_j it remains to show that the described coproduct E in Pre is already separated and complete and hence is the coproduct in Con. Separation is obvious: the only bounded subspace of E is the 0-subspace. Completeness follows from lemma (3.2.3) applied to the family $\mathrm{in}_i\colon E_i \to \coprod_{j\in J} E_j$. □

Remark. If $F_j = F$ for all $j \in J$ then we also write $F^{(J)}$ for $\coprod_{j\in J} F_j$.

3.4.2 Proposition. *Let E_j $(j \in J)$ be preconvenient vector spaces; $\mathcal{S}_j \subset E'_j$ linear subspaces that generate the bornology of E_j (i.e. $B \subseteq E_j$ is bounded iff $\ell(B)$ is bounded for all $\ell \in \mathcal{S}_j$). Then $\prod_{j\in J} \mathcal{S}_j$ generates the bornology of $\coprod_{j\in J} E_j$.*

Proof. Trivially $B \subseteq \coprod E_j$ bounded implies $\ell(B) \subseteq \mathbb{R}$ bounded for all $\ell \in \prod \mathcal{S}_j$. We have to show the converse. So let $B \subseteq \coprod E_j$ be unbounded. Then by (3.4.1) either B meets some E_j in an unbounded set and one trivially obtains an $\ell \in \prod \mathcal{S}_j$ with $\ell(B)$ unbounded; or there exist $j_n \in J$ $(n \in \mathbb{N})$, all different, for which points $b^n \in B$ with j_nth coordinate $b^n_{j_n} \neq 0$ exist. Since every $b \in B$ has only a finite number of non-zero coordinates we can choose the j_n and b^n such that, in addition, $b^m_{j_n} = 0$ for $m < n$. One chooses $\ell_{j_k} \in \mathscr{S}_{j_k}$ inductively such that $|\ell_{j_1}(b^1_{j_1})| = 1$ and $\ell_{j_k}(b^k_{j_k}) = k \cdot \mathrm{sign}\ (\Sigma^{k-1}_{m=1} \ell_{j_m}(b^k_{j_m}))$. In $\Pi \mathscr{S}_j$ we consider the element ℓ having as j_nth coordinate ℓ_{j_n}, the others being 0. For $x \in \amalg E_j$ one has $\ell(x) = \Sigma_k \ell_{j_k}(x_{j_k})$, the sum being actually finite. For $x := b^n$, the terms of this sum with $k > n$ are zero (by the choice of the indices j_n); the nth term has absolute value n; the sum of the preceding terms has the same signature as the nth term. Therefore $|\ell(b^n)| \geq n$ proving that $\ell(B)$ is unbounded. □

We continue this section by comparing products and coproducts. The inclusion of a coproduct into the corresponding product is of course always a Pre-morphism. But even for convenient vector spaces it forms only in very special cases a Pre-subspace. However, the coproduct of convenient vector spaces is always a Con-subspace of the corresponding product.

3.4.3 Proposition. *Let E_j $(j \in J)$ be preconvenient vector spaces. If only finitely many factors E_j are non-zero then the inclusion $\iota: \amalg_{j \in J} E_j \to \Pi_{j \in J} E_j$ is a Pre-isomorphism and in this case we will also write $\bigoplus_{j \in J} E_j$ for the product and the coproduct. Otherwise the Pre-morphism ι is neither initial nor surjective.*

Proof. Using the given descriptions of the bornologies on products and coproducts one easily verifies all the statements. □

3.4.4 Proposition. *Let E_j $(j \in J)$ be convenient vector spaces. Then the inclusion $\iota: \amalg_{j \in J} E_j \to \Pi_{j \in J} E_j$ is an initial Con-morphism; i.e. $\amalg E_j$ has the coarsest convenient vector space structure for which the projections $\mathrm{pr}_i: \amalg E_j \to E_i$ are morphisms.*

Proof. For this we have to show that for an absolutely convex set $B \subseteq \amalg E_j$, that generates a Banach space and is bounded in ΠE_j, only finitely many projections $\mathrm{pr}_j(B)$ are unequal to $\{0\}$; i.e. B is bounded in $\amalg E_j$. Suppose indirectly that $\mathrm{pr}_j(B) \neq \{0\}$ for infinitely many j, i.e. there exist $b^j \in B$ with $b^j_j \neq 0$. Using that only finitely many coordinates of b^j are non-zero we can choose a sequence $b^n \in B$ and distinct $j_n \in J$ with $b^n_{j_n} \neq 0$ and $b^n_{j_k} = 0$ for $k > n$. Next we want to modify this sequence to obtain a sequence $x^n \in B$ and $\ell_n \in (E_{j_n})'$ with $\ell_k(\mathrm{pr}_{j_k}(x^n)) = \delta^n_k$ (i.e. 1 for $k = n$ and 0 for $k \neq n$). We prove the existence of x^n and ℓ_n by induction. For $n = 1$ let $x^1 := b^1$ and ℓ_1 be chosen such that $\ell_1(b^1_{j_1}) = 1$. Suppose we already have constructed $x^1, \ldots, x^{n-1}$ and $\ell_1, \ldots, \ell_{n-1}$ satisfying the above equations. Define $x^n := t_1 \cdot x^1 + \cdots + t_{n-1} \cdot x^{n-1} + t \cdot b^n$ with appropriately chosen $t_i \in \mathbb{R}$. To ensure that $x^n \in B$ we impose the condition $|t_1| + \cdots + |t_{n-1}| + |t| = 1$. As additional equations we have to satisfy $0 = \ell_k(x^n_{j_k}) = t_1 \cdot \ell_k(x^1_{j_k}) + \cdots + t_{n-1} \cdot \ell_k\ (x^{n-1}_{j_k}) + t \cdot \ell_k(b^n_{j_k})$ for $k = 1 \ldots n-1$. Since by induction hypothesis $\ell_k\ (x^m_{j_k}) = \delta^m_k$ for $k, m \leq n$, the kth equation reduces to $t_k = -t \cdot \ell_k(b^n_{j_k})$. One uses this equation to define t_k, and then has only to choose t according to $|t_1| + \cdots + |t_{n-1}| + |t| = 1$; i.e. $t^{-1} := 1 + \sum_{k<n} |\ell_k(b^n_{j_k})|$. Finally as ℓ_n one can use any $\ell_n \in (E_{j_n})'$ with $\ell_n(x^n_{j_n}) = 1$. Now consider the series $\Sigma 2^{-k} x^k$. It is a Cauchy sequence in the Banach space

generated by B, hence converges to some $x \in \coprod E_j$. On the other hand for the j_kth coordinate of x one obtains $\ell_k(x_{j_k}) = \ell_k(\mathrm{pr}_{j_k}(x)) = \Sigma 2^{-k} \ell_k(\mathrm{pr}_{j_k}(x^k)) = 2^{-k} \neq 0$. This is a contradiction to $x \in \coprod E_j$. □

3.4.5 Proposition. *On any vector space E there exists a finest, convenient vector space structure (called the discrete $\underline{\mathrm{Con}}$-structure) and it is such that the dual is the algebraic dual E^* of E. One thus gets a functor $\psi: \underline{VS} \to \underline{\mathrm{Con}}$ which is left adjoint to the forgetful functor. It preserves the underlying vector spaces and the underlying maps.*

Proof. If J is a basis of E, then according to (3.4.1) the underlying vector space of $\mathbb{R}^{(J)}$ gets identified with E and its dual with E^*. This means that ψE is a dualized vector space isomorphic to $\mathbb{R}^{(J)}$ and hence a convenient vector space. The remaining statements follow from (8.4.3). □

3.4.6 Corollary

(i) *The forgetful functor $\underline{\mathrm{Con}} \to \underline{VS}$ commutes with limits.*
(ii) *The left adjoint functor ψ identifies $\underline{VS}$ with a full coreflective subcategory of $\underline{\mathrm{Con}}$.*
(iii) *A convenient vector space belongs to this subcategory iff it is isomorphic to a coproduct of the form $\mathbb{R}^{(J)}$ for some set J.*

Proof. (i) and (ii) are immediate, cf. (8.5.1) and (8.4.4). For (iii) one uses that every vector space has a basis, or equivalently, is a coproduct in $\underline{\mathrm{VS}}$ of summands equal to $\mathbb{R}$, and that ψ preserves colimits according to (8.5.1). □

3.5 Inductive limits

In section 3.1 we proved that an inductive limit in $\underline{\mathrm{Con}}$ is obtained by forming the limit in $\underline{\mathrm{Pre}}$ and then applying the completion functor $\bar{\omega}$ to it. We come now to the question in which situations the last step can be omitted.

3.5.1 Proposition. *Let E_j ($j \in \mathbb{N}$) be a sequence of convenient vector spaces such that E_j is a $\underline{\mathrm{Pre}}$-subspace of E_{j+1} and closed for the locally convex topology. Then the inductive limit $E = \bigcup_j E_j$ formed in $\underline{\mathrm{Pre}}$ is convenient and hence the inductive limit in $\underline{\mathrm{Con}}$. The bounded sets in the inductive limit are those that are contained and bounded in some E_j. And $\mathcal{Lip}^k$-curves into the limit are locally $\mathcal{Lip}^k$-curves into some E_j.*

Proof. In [Jarchow, 1981, pp. 83] it is shown that the inductive limit formed in $\underline{\mathrm{LCS}}$ is separated and that it is regular, i.e. every bounded set is contained and bounded in some E_j or, equivalently, that the bornology is the final bornology. Thus completeness follows from (3.2.3). □

Remark. Important cases of this situation are countable direct sums $\coprod_{n\in\mathbb{N}} E_n$. In this situation the finite subsums define such a sequence of increasing subspaces and the corresponding inductive limit is the countable coproduct.

Another application is given by the spaces of test functions used to define distributions. Here the space of smooth functions with compact support on a finite-dimensional separable smooth manifold X is the inductive limit of the Fréchet spaces of smooth functions with support contained in some K_n, where the K_n's are assumed to form an increasing sequence of compact sets covering X.

If the connecting maps in an inductive sequence of convenient vector spaces E_n are no longer closed embeddings, then the inductive limit in $\underline{\text{Pre}}$ need not be separated or M-complete. A special situation when this is nevertheless the case is described in the following:

3.5.2 Proposition. *Let $E_j \to E_{j+1}$ ($j \in \mathbb{N}$) be morphisms between Fréchet spaces. Then the inductive limit formed in $\underline{\text{Pre}}$ is convenient (and hence equals the inductive limit formed in $\underline{\text{Con}}$) iff the limit is separated and its von Neumann bornology is the final bornology.*

Proof. ($\Leftarrow$) Completeness follows from (3.2.3).

($\Rightarrow$) Obviously it has to be separated and M-complete. A result of [Floret, 1973] shows that completeness is equivalent to the assumption on the von Neumann bornology. □

3.5.3 Remark. An easy (categorical) consideration shows that for every inductive limit $\text{in}_j: E_j \to E$ ($j \in J$) in $\underline{\text{Pre}}$ the map $\coprod_{j\in J} E_j \to E$, $(x_j) \mapsto \sum_j \text{in}_j(x_j)$ is a quotient map in $\underline{\text{Pre}}$.

3.6 Function spaces of linear maps

We shall show that for any convenient vector spaces E and F the function space $L(E, F)$ of linear morphisms $E \to F$ has a natural convenient vector space structure and that the uniform boundedness principle holds for the corresponding bornology of $L(E, F)$.

3.6.1 Proposition

(i) *One has a functor $\ell^\infty: (\underline{\ell^\infty})^{\text{op}} \times \underline{\text{Pre}} \to \underline{\text{Pre}}$, lifting the functor $\ell^\infty: (\underline{\ell^\infty})^{\text{op}} \times \underline{\ell^\infty} \to \underline{\ell^\infty}$, as follows: the vector space operations on the ℓ^∞-space $\ell^\infty(X, E)$ are defined pointwise, cf.* (1.2.8), *and on morphisms one has $\ell^\infty(f, g): h \mapsto g \circ h \circ f$.*

(ii) *The functor $\ell^\infty: (\underline{\ell^\infty})^{\text{op}} \times \underline{\text{Pre}} \to \underline{\text{Pre}}$ restricts to a functor $\ell^\infty: (\underline{\ell^\infty})^{\text{op}} \times \underline{\text{Con}} \to \underline{\text{Con}}$; i.e. $\ell^\infty(X, E)$ is convenient for convenient E.*

Proof. (i) Since the preconvenient vector spaces can be identified with the linearly generated ℓ^∞-vector spaces (cf (ii) in (2.4.4)), we can reformulate proposition (1.2.10) as follows: for any ℓ^∞-space X and any preconvenient vector space E the function space $\ell^\infty(X,E)$ is also a preconvenient vector space. Furthermore by (1.2.9) its ℓ^∞-structure and therefore its preconvenient vector space structure is the initial one induced by the maps $\ell^\infty(c,\ell)=\ell_*\circ c^*$: $\ell^\infty(X,E)\to\ell^\infty$ for $\ell\in E'$ and $c\in\ell^\infty(\mathbb{N},X)$. Functoriality follows trivially from cartesian closedness of $\underline{\ell^\infty}$, cf. (1.2.8), and from the fact that $\ell^\infty(f,g)$ is linear provided g is linear.

(ii) $\ell^\infty(X,E)$ is separated, since the linear functionals $\ell\circ \mathrm{ev}_x$ are obviously point separating morphisms on $\ell^\infty(X,E)$.

Completeness can be proved along the following pattern: take a Mackey–Cauchy sequence (g_n) in $\ell^\infty(X,E)$. Then the values $g_n(x)=\mathrm{ev}_x(g_n)$ form a Mackey–Cauchy sequence in E for any $x\in X$. Thus the functions g_n converge pointwise (Mackey) to some function g. And one then shows that g is an $\underline{\ell^\infty}$-morphism and is the Mackey limit of the given sequence.

A more efficient proof is based on the classical result that ℓ^∞ with its usual norm is a Banach space. Thus the preconvenient vector space $\ell^\infty=\ell^\infty(\mathbb{N},\mathbb{R})$ is convenient since its bornology is the von Neumann bornology of the Banach space topology; cf. (1.2.12). As mentioned above $\ell^\infty(X,E)$ embeds by means of the maps $\ell_*\circ c^*$ into the product of factors ℓ^∞ taken over all $(\ell,c)\in E'\times\ell^\infty(\mathbb{N},X)$. Since the product of convenient vector spaces is again convenient it is by the closed embedding lemma (2.6.4) enough to show that the image is M-closed in the product. This is trivial since it is formed by the solutions of the equations $x_n^{(\ell,c)}=x_m^{(\ell,e)}$ for all $\ell\in E'$, $c,e\in\ell^\infty(\mathbb{N},X)$ and $n,m\in\mathbb{N}$ with $c(n)=e(m)$, and hence is the intersection of the kernels of the morphisms $\mathrm{ev}_n\circ\mathrm{pr}_{(\ell,c)}-\mathrm{ev}_m\circ\mathrm{pr}_{(\ell,e)}$. □

3.6.2 Proposition. *One has a functor $L:\underline{Pre}^{\mathrm{op}}\times\underline{Pre}\to\underline{Pre}$, lifting the hom-functor, as follows: $L(E,F)$ is the $\underline{Pre}$-subspace of $\ell^\infty(E,F)$ formed by the linear morphisms, and $L(f,g): h\mapsto g\circ h\circ f$.*

Proof. One only has to show that for two linear morphisms $f: E_2\to E_1$ and $g: F_1\to F_2$ one has a linear morphism $L(f,g): L(E_1,F_1)\to L(E_2,F_2)$ defined by $h\mapsto g\circ h\circ f$. Clearly $L(f,g)$ has values in $L(E_2,F_2)$ and is linear. It is a morphism since it is the restriction of the morphism $\ell^\infty(f,g)$ and since the inclusion of $L(E_2,F_2)$ into $\ell^\infty(E_2,F_2)$ is an initial ℓ^∞-morphism. □

3.6.3 Proposition. *The functor $L:\underline{Pre}^{\mathrm{op}}\times\underline{Pre}\to\underline{Pre}$ restricts to a functor $L:\underline{Pre}^{\mathrm{op}}\times\underline{Con}\to\underline{Con}$ and hence to a functor $L:\underline{Con}^{\mathrm{op}}\times\underline{Con}\to\underline{Con}$.*

Proof. We show that for a convenient vector space F also $L(E,F)$ is convenient. Since $L(E,F)$ is defined as $\underline{Pre}$-subspace of the convenient vector space $\ell^\infty(E,F)$ it is by the closed embedding lemma (2.6.4) enough to show that it is M-closed. Again, this is trivial to verify using that this subspace is formed by the solutions of the equations $f(x+t\cdot y)=f(x)+t\cdot f(y)$ with $x,y\in E$ and $t\in\mathbb{R}$ (in fact the equations with $t = 1$ suffice), hence is the intersection of the kernels of the morphisms $\mathrm{ev}_x + t\cdot\mathrm{ev}_y - \mathrm{ev}_{x+ty}$. □

We state now a generalization of the Banach–Steinhaus theorem which will play an important role also for the differentiation theory.

3.6.4 Theorem. (*Linear Uniform Boundedness Principle.*) *Let E be a convenient and F a preconvenient vector space. Then for a subset $B \subseteq L(E, F)$ the following statements are equivalent:*

(1) *B is bounded, i.e. $B(A)$ is bounded for $A \subseteq E$ bounded;*
(2) *B is equicontinuous with respect to the locally convex topologies, i.e. for every 0-neighborhood V in F there is a 0-neighborhood U in E such that $B(U) \subseteq V$;*
(3) *B is pointwise bounded, i.e. $\mathrm{ev}_x(B)$ is bounded for all $x \in E$.*

Proof. ($1 \Rightarrow 2$) Let V be an absolutely convex 0-neighborhood in F. Then $U := \bigcap_{g\in B} g^{-1}(V)$ is absolutely convex in E. We show that U is bornivorous and hence the 0-neighborhood we search for. Let A be bounded in E; then $B(A)$ is bounded in F by (1.2.13) and hence gets absorbed by V, i.e. $B(A) \subseteq K\cdot V$ for some $K > 0$. Thus $K\cdot U = \bigcap g^{-1}(K\cdot V) \supseteq \bigcap g^{-1}(g(A)) \supseteq A$, i.e. U absorbs A.

($2 \Rightarrow 3$) Suppose B is unbounded at some point $x \in E$. Then some 0-neighborhood V in F does not absorb $B(x)$ and hence we can choose points $g_n \in B$ with $g_n(x) \notin n\cdot V$. By equicontinuity there exists a 0-neighborhood U in E with $B(U) \subseteq V$. It follows that $(1/n)x \notin U$ in contradiction with the fact that U is point absorbing.

($3 \Rightarrow 1$) By composing with elements of F' one immediately reduces the general case to the case $F = \mathbb{R}$. It is enough to prove that $B(A)$ is bounded for A belonging to a basis of the bornology of E and thus we may assume that A is absolutely convex and E_A a Banach space (cf. (3) in (2.6.2)). For $B(A)$ only the restrictions $g|_{E_A}$ of the elements $g \in B$ play a role and these form a family of continuous linear operators on the Banach space E_A. By the classical Banach–Steinhaus theorem [Jarchow, 1981, p. 220] we conclude that they are uniformly bounded on the unit ball of E_A which contains A; hence $B(A)$ is bounded. □

3.6.5 Theorem. *Let E be a convenient and F a preconvenient vector space. Then the (pre)convenient structure of $L(E, F)$ is the initial one induced by the evaluation maps $\mathrm{ev}_x\colon L(E, F) \to F$ for $x \in E$. In particular this implies that a curve $g\colon \mathbb{R} \to L(E, F)$ or more generally a map $g\colon X \to L(E, F)$ for any $\mathcal{Lip}^k$-space X is a $\mathcal{Lip}^k$-map iff all its composites $\mathrm{ev}_x \circ g\colon X \to F$ with $x \in E$ are $\mathcal{Lip}^k$-maps.*

Proof. This is a direct consequence of (3.6.4) using the description of initial Pre-structures in (3.1.2). □

3.6.6 Proposition. (*Bornological Uniform Boundedness Principle.*) *Let X be an ℓ^∞-space and E a convenient vector space. The structure of $\ell^\infty(X, E)$ introduced in*

(3.6.1) *is the coarsest convenient vector space structure making all evaluations* ev_x $(x \in X)$ *morphisms. In categorical language this means that* $\mathrm{ev}_x : \ell^\infty(X, E) \to E$ $(x \in X)$ *is an initial source with respect to the forgetful functor* $\underline{Con} \to \underline{VS}$.

Proof. Let F be a convenient vector space and $m: F \to \ell^\infty(X, E)$ a linear map such that $\mathrm{ev}_x \circ m$ is a morphism (i.e. $\mathrm{ev}_x \circ m \in L(F, E)$) for all $x \in X$. We have to show that m is a morphism, i.e. that m is bornological. Let $A \subseteq X$ be bounded. Put, for $x \in X$, $\tilde{m}(x) := \mathrm{ev}_x \circ m \in L(F, E)$ (i.e. $\tilde{m}(x)(y) = m(y)(x)$). So we have a map $\tilde{m}: X \to L(F, E)$. For $y \in F$ one has $\mathrm{ev}_y(\tilde{m}(A)) = \tilde{m}(A)(y) = m(y)(A)$ and since $m(y) \in \ell^\infty(X, E)$ we deduce that $\mathrm{ev}_y(\tilde{m}(A)) \subseteq F$ is bounded. By the linear uniform boundedness principle (3.6.4) this implies that $m(B)(A) = \tilde{m}(A)(B) \subseteq E$ is bounded for any bounded $B \subseteq F$. Since A was an arbitrary bounded subset of X the image $m(B) \subseteq \ell^\infty(X, E)$ is bounded for all bounded B. □

Remark. The name of the theorem is justified, since it shows that for an absolutely convex subset $B \subseteq \ell^\infty(X, E)$ one has: B pointwise bounded and the normed space $\ell^\infty(X, E)_B$ complete implies that B is uniformly bounded on bounded subsets of X, i.e. $B(A)$ is bounded for every bounded $A \subseteq X$.

We conclude this section with another description of the structure of $\ell^\infty(X, E)$ and thereby of $L(E, F)$.

3.6.7 Lemma. *Let X be any bornological space, and βE the bornological space associated to a locally convex space E (cf. (i) in (2.1.10)). Then the bornology of the vector space* $\mathrm{Born}(X, \beta E)$ *is the von Neumann bornology of the topology of uniform convergence in E on bounded subsets of X.*

Proof. Recall that a subset B of $\mathrm{Born}(X, \beta E)$ is bounded iff $B(A)$ is bounded in βE for all bounded $A \subseteq X$. By definition of the bornology of βE this is the case iff $B(A)$ gets absorbed by every (absolutely convex) 0-neighborhood U of E; i.e. B gets absorbed by the sets $\{g; g(A) \subseteq U\}$ with A bounded in X and U an absolutely convex 0-neighborhood in E. But these subsets define a basis of the topology of uniform convergence, cf. [Jarchow, 1981, p. 44]. □

3.6.8 Corollary. *Let X be an ℓ^∞-space and E a (pre)convenient vector space. Then the bornology of $\ell^\infty(X, E)$ is the von Neumann bornology of the topology of uniform convergence on bounded sets.*

It is an immediate consequence of (3.6.8) that the considered topology of uniform convergence is the locally convex topology of the preconvenient vector space $\ell^\infty(X, E)$ provided it is bornological. In general this is not the case. The following proposition gives conditions on X and E under which this holds.

3.6.9 Proposition. *If an ℓ^∞-space X has a countable basis for its bornology and the preconvenient vector space E has a countable basis for the 0-neighborhoods of* *its locally convex topology, then the topology of uniform convergence on the bounded sets is semimetrizable and is the locally convex topology of* $\ell^\infty(X,E)$.

Proof. Let $\mathscr{B}_0$ denote a countable basis of the bornology of X and $\mathscr{U}_0$ a countable basis of 0-neighborhoods of E. Then the sets $\{g\in\ell^\infty(X,E);\ g(B)\subseteq U\}$ (for $B\in\mathscr{B}_0$ and $U\in\mathscr{U}_0$) form a countable basis of 0-neighborhoods of the topology of uniform convergence, which therefore is semi-metrizable [Jarchow, 1981, p. 40] and thus bornological [Jarchow, 1981, p. 273]. □

Remark. A locally convex space E has a countable basis of 0-neighborhoods iff there exists a countable family $\mathscr{P}_0$ of seminorms generating the topology. The locally convex topology of $\ell^\infty(X,E)$ is then generated by the countable family of seminorms $q_{p,B}(g):=\sup\{p(g(x));\ x\in B\}$, where $B\in\mathscr{B}_0$ and $p\in\mathscr{P}_0$.

3.7 Function spaces of multilinear maps

We first characterize those multilinear maps which generalize linear morphisms between preconvenient vector spaces, and which will be called multilinear morphisms, cf. also (3.7.5).

3.7.1 Theorem. (*Characterizations of Multilinear Morphisms.*)
Let $m: E_1 \sqcap \dots \sqcap E_m \to F$ *be an m-linear map between preconvenient vector spaces and* $k\in\mathbb{N}_{0,\infty}$. *Then the following statements are equivalent:*

(1) m *is bornological;*
(2) m *is an* $\underline{\ell^\infty}$*-morphism;*
(3) m *is bounded on sequences* M*-converging to* 0;
(4) m *is continuous with respect to* M*-convergence;*
(5) m *is continuous with respect to the* M*-closure topologies;*
(6) m *is a* $\underline{\mathscr{Lip}^k}$*-morphism.*

Proof. Trivial are $(1\Rightarrow 2\Rightarrow 3)$ and $(4\Rightarrow 5\Rightarrow 3)$.

$(3\Rightarrow 1)$ Suppose m is unbounded on some bounded set B. Then there exists a 0-neighborhood U and $b_k\in B$ with $(1/k)\,m((1/k)\cdot b_k)=(1/k)^{m+1}\,m(b_k)\notin U$. Since $(1/k)b_k$ is M-converging to 0 this is a contradiction to (3).

$(1\Rightarrow 4)$ Let us show continuity at 0 first: a 0-converging filter $\mathscr{H}$ on $E_1\sqcap\dots\sqcap E_m$ is finer than some filter $\mathbb{U}B_1\times\dots\times\mathbb{U}B_m$ with B_j bounded in E_j. Since $m(\mathbb{U}B_1\times\dots\times\mathbb{U}B_m)=\mathbb{U}m(B_1\times\dots\times B_m)$ and since $m(B_1\times\dots\times B_m)$ is by assumption bounded, $m(\mathscr{H})$ is 0-converging.

Now the continuity at an arbitrary point $(a_1,\dots,a_n)$. Using multilinearity one develops $m((\mathbb{U}B_1+a_1)\times\dots\times(\mathbb{U}B_n+a_m))$. One of the resulting terms is $m(a_1,\dots,a_m)$. The others contain at least one $\mathbb{U}B_j$; that they are M-convergent to 0 can be seen by applying the above argument concerning continuity at 0 to the multilinear map obtained by keeping in m those variables

a_i fixed that appear in the given term. Thus $m((\bigcup B_1 + a_1) \times \ldots \times (\bigcup B_n + a_m))$ is M-convergent to $m(a_1, \ldots, a_m)$.

$(1 \Rightarrow 6)$. Let $c_j: \mathbb{R} \to E_j$ be $\mathcal{Lip}^k$-curves and $\ell \in F'$. One has to show that $h := \ell \circ m \circ (c_1, \ldots, c_m): \mathbb{R} \to \mathbb{R}$ is $\mathcal{Lip}^k$. One has $\delta^j h = \ell \circ \delta^j (m \circ (c_1, \ldots, c_m))$. Using (1.3.30) (i.e. $\delta^j (m \circ (c_1, \ldots, c_m))$ is a linear combination of $m \circ (\delta^{j_1} c_1, \ldots, \delta^{j_n} c_n)$ where $\Sigma j_i = j$) and (1.3.22) one finds: $\delta^j h: \mathbb{R}^{\langle j \rangle} \to \mathbb{R}$ is bornological for $j < k+2$. The assertion now follows from (1.3.22) and (1.3.24).

$(6 \Rightarrow 3)$ Suppose m is unbounded on a sequence x that is M-convergent to 0. Passing to a subsequence we may assume, using the special curve lemma (2.3.4), that there exists a smooth curve c such that $c(1/2^n) = x_n$. Then $m \circ c$ is $\mathcal{Lip}^k$ and hence according to (1.3.14) bounded on $[0,1]$ and one reaches a contradiction. □

3.7.2 Definition. Let E_i and F be preconvenient vector spaces. With $L(E_1, \ldots, E_m; F)$ we denote the Pre-subspace of $\ell^\infty(E_1 \sqcap \ldots \sqcap E_m, F)$ formed by the multilinear morphisms $E_1 \sqcap \ldots \sqcap E_m \to F$ described in (3.7.1).

3.7.3 Proposition. *For preconvenient vector spaces E_i and F one has natural isomorphisms: $L(E_1, \ldots, E_m; F) \cong L(E_1, \ldots, E_k; L(E_{k+1}, \ldots, E_m; F))$. If F is assumed to be convenient then $L(E_1, \ldots, E_m; F)$ is convenient as well.*

Proof. Since the category of ℓ^∞-spaces is cartesian closed one has $\ell^\infty(E_1 \sqcap \ldots \sqcap E_m, F) \cong \ell^\infty(E_1 \sqcap \ldots \sqcap E_k, \ell^\infty(E_{k+1} \sqcap \ldots \sqcap E_m, F))$. This isomorphism restricts, of course, to the Pre-subspaces formed by the multilinear maps.

The second statement is obtained inductively by means of the stated isomorphism, using that it holds for $m = 1$ by (3.6.3). □

3.7.4 Theorem. (*Multilinear Uniform Boundedness Principle.*) *Let E_i be convenient vector spaces and F a preconvenient vector space. Then for a subset $B \subseteq L(E_1, \ldots, E_m; F)$ the following statements are equivalent:*

(1) *B is bounded, i.e. $B(A)$ is bounded for all bounded $A \subseteq E_1 \sqcap \ldots \sqcap E_m$;*
(2) *$B(A_1 \times \ldots \times A_m)$ is bounded for all bounded $A_i \subseteq E_i$;*
(3) *$B(x_1, \ldots, x_m)$ is bounded for all $x_i \in E_i$.*

Proof. $(1 \Leftrightarrow 2)$ is deduced easily, using that the products $A_1 \times \ldots \times A_m$ of bounded sets A_i form a basis for the bornology of $E_1 \sqcap \ldots \sqcap E_m$.

$(2 \Rightarrow 3)$ is trivial.

$(3 \Rightarrow 2)$ We proceed by induction. For $m = 1$ this is (3.6.4). Suppose now $B \subseteq L(E_1, \ldots, E_{m+1}; F)$ is pointwise bounded. Set $g^\vee(x_1, \ldots, x_m)(x_{m+1}) := g(x_1, \ldots, x_{m+1})$. Then $B^\vee(x_1, \ldots, x_m) \subseteq L(E_{m+1}, F)$ is pointwise bounded. Thus by (3.6.4) $B^\vee(x_1, \ldots, x_m)$ is bounded in $L(E_{m+1}, F)$. By induction hypothesis this implies that $B^\vee(A_1 \times \ldots \times A_m) \subseteq L(E_{m+1}, F)$ is bounded for $A_j \subseteq E_j$ bounded, i.e. $B(A_1 \times \ldots \times A_{m+1}) = B^\vee(A_1 \times \ldots \times A_m)(A_{m+1})$ is bounded in F for all bounded sets A_{m+1} of E_{m+1}. □

We can now, for the case of convenient vector spaces, add to the characterizations of multilinear morphisms given in (3.7.1) a further one:

3.7.5 Corollary. *Let $m: E_1 \sqcap \ldots \sqcap E_m \to F$ be a multilinear map, $E_1, \ldots, E_m$ being convenient vector spaces. If m is partially bornological (i.e. partially a $\underline{\text{Pre}}$-morphism), then m is bornological.*

Proof. One uses induction on m. For $m=1$ nothing has to be proved. For $m>1$ the assertion is by (3.7.3) equivalent with $m^\vee: E_1 \to L(E_2, \ldots, E_m; F)$ being well defined and bornological. Since m is partially bornological in the last $m-1$ variables, $m^\vee$ has by induction hypothesis values in $L(E_2, \ldots, E_m; F)$. In order to show that $m^\vee(A)$ is bounded for any bounded $A \subseteq E_1$ one applies the previous theorem with $B = m^\vee(A)$. □

3.8 The tensor product

3.8.1 Proposition. *The category $\underline{\text{Pre}}$ of preconvenient vector spaces is symmetric monoidal closed; i.e. there exist functors $L: \underline{\text{Pre}}^{\text{op}} \times \underline{\text{Pre}} \to \underline{\text{Pre}}$ and $\otimes: \underline{\text{Pre}} \times \underline{\text{Pre}} \to \underline{\text{Pre}}$ with natural isomorphisms $L(E_1; L(E_2; E_3)) \cong L(E_1 \otimes E_2; E_3)$; $E_1 \otimes E_2 \cong E_2 \otimes E_1$; $E_1 \otimes (E_2 \otimes E_3) \cong (E_1 \otimes E_2) \otimes E_3$; $E \otimes \mathbb{R} \cong E$.*

Proof. The functor L was already defined in (3.6.2). The tensor product $E_1 \otimes E_2$ has as underlying vector space the algebraic tensor product of the underlying vector spaces and as ℓ^∞-structure the one generated by the following set of functions: $\mathcal{F}_0 := \{h: E_1 \otimes E_2 \to \mathbb{R};\ h$ is linear and $h \circ b: E_1 \sqcap E_2 \to \mathbb{R}$ is an $\underline{\ell^\infty}$-morphism$\}$, where $b: E_1 \sqcap E_2 \to E_1 \otimes E_2$ denotes the canonical bilinear map. With this linearly generated ℓ^∞-structure, $E_1 \otimes E_2$ becomes an object of $\underline{\text{Pre}}$, cf. (ii) in (2.4.4.), b an $\underline{\ell^\infty}$-morphism, and by construction we have $\mathcal{F}_0 \subseteq (E_1 \otimes E_2)'$. In order to show that in fact $\mathcal{F}_0 = (E_1 \otimes E_2)'$ we first remark that for arbitrary bounded sequences $s_1: \mathbb{N} \to E_1$ and $s_2: \mathbb{N} \to E_2$ the sequence $s: \mathbb{N} \to E_1 \otimes E_2$ defined by $s(n) := b(s_1(n), s_2(n))$ is an $\underline{\ell^\infty}$-morphism. This is verified by composing with every $h \in \mathcal{F}_0$. If $m: E_1 \otimes E_2 \to \mathbb{R}$ is linear and $m \circ s \in \ell^\infty$ for all $\underline{\ell^\infty}$-morphisms $s: \mathbb{N} \to E_1 \otimes E_2$, then in particular $m \circ b \circ (s_1, s_2) \in \ell^\infty$ for any s_1, s_2 as before. This implies that $m \circ b$ is an $\underline{\ell^\infty}$-morphism and hence $m \in \mathcal{F}_0$.

Let now $g: E_1 \sqcap E_2 \to E_3$ be any element of $L(E_1, E_2; E_3)$. By the universal property of the algebraic tensor product there exists a unique linear map $\bar{g}: E_1 \otimes E_2 \to E_3$ such that $g = \bar{g} \circ b$. For the proof that $\bar{g}$ is an $\underline{\ell^\infty}$-morphism we have to show that $\ell \circ \bar{g} \in (E_1 \otimes E_2)' = \mathcal{F}_0$ for all $\ell \in E_3'$. This holds since $\ell \circ \bar{g} \circ b = \ell \circ g$ is an $\underline{\ell^\infty}$-morphism. So we get a natural bijection $L(E_1 \otimes E_2; E_3) \cong L(E_1, E_2; E_3)$. That this is an isomorphism follows using cartesian closedness of $\underline{\ell^\infty}$. The first of the claimed isomorphisms now follows since $L(E_1, E_2; E_3) \cong L(E_1; L(E_2; E_3))$ by (3.7.3). The others are trivial. Functoriality of $\otimes$ is a consequence of the universal property; cf. (8.4.3). □

3.8.2 Lemma. *Let $a_1, \ldots, a_n$ be linearly independent points of a separated preconvenient vector space E. Then there exists an $\ell \in E'$ with $\ell(a_1)=1$ and $\ell(a_2)= \cdots =\ell(a_n)=0$.*

Proof. Since E is assumed to be separated the lemma certainly holds for $n=1$. Assume now it is valid for $n-1$, and let $a_1, \ldots, a_n$ be linearly independent. By the induction hypothesis there exists an $\ell_1 \in E'$ with $\ell_1(a_1)=1$ and $\ell_1(a_2)= \ldots =\ell_1(a_{n-1})=0$. Since $a_1, \ldots, a_{n-1}, a_n-\ell_1(a_n)\cdot a_1$ are linearly independent, the induction hypothesis also implies the existence of an $\ell_2 \in E'$ with $\ell_2(a_2)= \ldots =\ell_2(a_{n-1})=0$ and $\ell_2(a_n-\ell_1(a_n)\cdot a_1)=1$. We form $\ell:=s\cdot\ell_1+t\cdot\ell_2$, with $s,t\in\mathbb{R}$ chosen in such a way that $\ell(a_1):=s+t\cdot\ell_2(a_1)=1$ and $\ell(a_n)=s\cdot\ell_1(a_n)+t\cdot\ell_2(a_2)=0$. This is possible since the determinant of this linear system equals 1 by the choice of ℓ_2. □

3.8.3 Proposition. *For separated preconvenient vector spaces E and F also $E \otimes F$ is separated.*

Proof. Let $0\neq z\in E\otimes F$. One can write z in the form of $z=b(x_1,y_1)+ \ldots +b(x_n,y_n)$ with $x_1, \ldots, x_n\in E$ linearly independent and $y_1, \ldots, y_n\in F$; as before $b: E\sqcap F\to E\otimes F$ denotes the canonical bilinear map. Since $z\neq 0$ we may assume $x_1\neq 0$ and $y_1\neq 0$. We choose $h\in F'$ with $h(y_1)=1$, and according to the lemma above, $\ell\in E'$ with $\ell(x_1)=1$ and $\ell(x_2)= \ldots =\ell(x_n)=0$. Then the map $m: E\sqcap F\to\mathbb{R}$ defined by $m(x,y):=\ell(x)\cdot h(y)$ is a bilinear $\underline{\ell^\infty}$-morphism and thus factors as $m=\bar{m}\circ b$, where $\bar{m}: E\otimes F\to\mathbb{R}$ is a linear $\underline{\ell^\infty}$-morphism, i.e. $\bar{m}\in(E\otimes F)'$. This gives $\bar{m}(z)=m(x_1,y_1)+ \ldots +m(x_n,y_n)=\ell(x_1)\cdot h(y_1)=1$. □

Since for separated E and F both $L(E,F)$ and $E\otimes F$ are separated one concludes that by restricting the functors L and $\otimes$ of proposition (3.8.1) the category $\underline{\text{sPre}}$ of separated preconvenient vector spaces is symmetric monoidal closed as well.

3.8.4 Theorem. *The category $\underline{Con}$ of convenient vector spaces is symmetric monoidal closed; i.e. there exist functors $L: \underline{Con}^{op}\times\underline{Con}\to\underline{Con}$ and $\tilde{\otimes}: \underline{Con}\times\underline{Con}\to\underline{Con}$ with natural isomorphisms $L(E_1;L(E_2;E_3))\cong L(E_1\tilde{\otimes}E_2;E_3)$; $E_1\tilde{\otimes}E_2\cong E_2\tilde{\otimes}E_1$; $E_1\tilde{\otimes}(E_2\tilde{\otimes}E_3)\cong(E_1\tilde{\otimes}E_2)\tilde{\otimes}E_3$; $E\tilde{\otimes}\mathbb{R}\cong E$.*

Proof. As shown in (3.6.3) the functor L used in (3.8.1) restricts to convenient vector spaces. However, for convenient vector spaces E and F the space $E\otimes F$ introduced in (3.8.1) is in general not complete. So we define by means of the completion functor $\bar{\omega}$ (2.6.5) the functor $\tilde{\otimes}:=\bar{\omega}\circ\otimes$; i.e. $E\tilde{\otimes}F$ is the completion of the separated preconvenient vector space $E\otimes F$. We further define $\tilde{b}: E\sqcap F\to E\tilde{\otimes}F$ as the composite of $b: E\sqcap F\to E\otimes F$ and the canonical embedding $E\otimes F\to E\tilde{\otimes}F$. Thus $\tilde{b}$ is bilinear and bornological, and one easily verifies that it has the desired universal property: for any bilinear bornological map $m: E\sqcap F\to G$ into a convenient vector space G there exists a unique linear morphism $\tilde{m}: E\tilde{\otimes}F \to G$ with $m = \tilde{m} \circ \bar{b}$. The rest follows from (3.8.1) and the universal property (2.6.5) of the completion functor $\bar{\omega}$. □

Let us now give a consequence of the categorical situation established in this section:

3.8.5 Proposition. *For preconvenient vector spaces one has the following $\underline{\text{Pre}}$-isomorphisms:*

(i) $L(E;\Pi_{j\in J}F_j) \cong \Pi_{j\in J}L(E;F_j)$;
(ii) $L(\amalg_{j\in J}E_j;F) \cong \Pi_{j\in J}L(E_j;F)$.

Proof. (i) The functor $L(E,\_): \underline{\text{Pre}} \to \underline{\text{Pre}}$ has a left adjoint (namely $(\_)\otimes E$), hence commutes with categorical limits and thus in particular with products; cf. (8.5.1).

(ii) The existence of natural isomorphisms $L(E_1;L(E_2;F)) \cong L(E_2;L(E_1;F))$ can be expressed by saying that the functor $L(\_;F): \underline{\text{Pre}}^{\text{op}} \to \underline{\text{Pre}}$ has a left adjoint (namely $L(\_;F): \underline{\text{Pre}} \to \underline{\text{Pre}}^{\text{op}}$), hence commutes with categorical limits. The limits in $\underline{\text{Pre}}^{\text{op}}$ are the colimits in $\underline{\text{Pre}}$. In particular, $L(\_;F)$ carries coproducts in $\underline{\text{Pre}}$ to products in $\underline{\text{Pre}}$. □

3.9 The duality functor

3.9.1 Definition. The functor $L(\_,\mathbb{R}): \underline{\text{Pre}}^{\text{op}} \to \underline{\text{Pre}}$ is called the *duality functor* of the category of preconvenient vector spaces. Since for any preconvenient vector space E the space $L(E,\mathbb{R})$ is always convenient this functor actually has values in $\underline{\text{Con}}$. From now on E' will denote the convenient vector space $L(E,\mathbb{R})$, i.e. the dual of E together with its natural structure as convenient vector space. By restriction one gets the duality functor $L(\_;\mathbb{R}): \underline{\text{Con}}^{\text{op}} \to \underline{\text{Con}}$.

3.9.2 Proposition. *For the dual E' of a convenient vector space E the linear uniform boundedness principle of* (3.6.4) *holds and the preconvenient structure of E' is the initial one induced by the point evaluations* $\text{ev}_x: E' \to \mathbb{R}$ *for* $x \in E$.

Proof. This is (3.6.4) and (3.6.5) for $F = \mathbb{R}$. □

3.9.3 Proposition. *For any preconvenient vector space E the canonical map $\iota_E: E \to E''$ into its bidual is an initial morphism of $\underline{\text{Pre}}$; it is injective iff E is separated. Moreover ι_E is even an initial morphism of $\underline{\text{LCS}}$.*

Proof. The morphism ι_E is obtained by symmetric monoidal closedness as follows: to the morphism id: $E' \to E'$ corresponds a bilinear morphism $E'\sqcap E \to \mathbb{R}$, which corresponds to a bilinear morphism $E \sqcap E' \to \mathbb{R}$ and this corresponds to the morphism $\iota_E: E \to E''$.

A trivial calculation shows that $\iota_E: E \to E''$ composed with the embedding $E'' \to \Pi_{E'}\mathbb{R}$ given by $\ell \in (E')' \mapsto (\ell(f))_{f \in E'}$ is exactly the initial morphism defined in (2.5.5). Thus $\iota_E: E \to E''$ is initial too by (i) in (8.7.2). It is trivial to verify the statement on injectivity.

Now let us show the initiality for the locally convex topologies. Consider a closed absolutely convex 0-neighborhood U of E. Then $U^0 := \{\ell \in E'; \ell(U) \subseteq [-1, 1]\}$ is equicontinuous and therefore bounded in E'. By the bipolar theorem [Jarchow, 1981, p. 149] U is equal to $\{x \in E; |\ell(x)| \leq 1 \text{ for all } \ell \in U^0\}$ which is the trace of $(U^0)^0$ on E. Since $(U^0)^0$ is the polar of a bounded set it is a 0-neighborhood in $(E')'$. This shows that $\iota_E: E \to E''$ is an embedding for the locally convex topologies. □

3.9.4 Proposition. *For preconvenient vector spaces E_j one has a canonical Con-isomorphism: $\Pi_{j \in J} E'_j \cong (\amalg_{j \in J} E_j)'$.*

Proof. This is (ii) in (3.8.5) for $F := \mathbb{R}$. □

3.9.5 Proposition. *For preconvenient vector spaces E_j one has a canonical injective and initial Pre-morphism (of convenient vector spaces): $\amalg_{j \in J} E'_j \to (\Pi_{j \in J} E_j)'$.*

For index sets J of non-measurable cardinality it is an isomorphism.

Proof. The morphism is obtained by the universal property of the coproduct $\amalg_{j \in J} E'_j$ using the morphism $(\mathrm{pr}_j)^*: E'_j \to (\Pi_{j \in J} E_j)'$. Injectivity is trivial. Initiality of this morphism, denoted by m, can be seen as follows: let $B \subseteq \amalg E'_j$ be unbounded. Using (3.4.2) with $\mathcal{S}_j := E_j \subseteq (E'_j)'$ and (3.9.2) one deduces that there exists an $x \in \Pi E_j$ such that $x(B)$ is unbounded. Since $x(B) = m(B)(x) = \mathrm{ev}_x(m(B))$ this proves that $m(B) \subseteq (\Pi E_j)'$ is unbounded.

If J has non-measurable cardinality then the embedding is a surjection by (3.3.6) and thus it is an isomorphism. □

4 CALCULUS IN CONVENIENT VECTOR SPACES

Besides calculus function spaces are the major theme of this chapter. The differentiability classes which are considered yield function spaces which are convenient vector spaces provided the range itself is a convenient vector space or more generally a vector bundle.

In classical approaches k-fold differentiability is defined recursively. Our definition is, however, non-recursive in so far as it is based on the behavior of the map along curves. Thus it is natural to start with curves in section 4.1. We define $\mathscr{Lip}^k$-curves as those curves $c: \mathbb{R} \to E$ into a convenient vector space E for which all composites $\ell \circ c: \mathbb{R} \to \mathbb{R}$ with functionals ℓ of the dual E' are k-times differentiable with a locally Lipschitzian derivative of order k. As in the finite-dimensional case they are also characterized by means of their difference quotients. The result that these curves are Mackey–Riemann integrable constitutes a useful tool in differentiation theory. A mean value theorem is established. It estimates the increment of a $\mathscr{Lip}^1$-curve by means of its derivative in terms of a convex set.

The purpose of section 4.2 is to show that the $\mathscr{Lip}^k$-curves of a convenient vector space form again a convenient vector space, and to give various descriptions of its structure. These spaces are used later for the description of general function spaces. The section ends with the so-called general curve lemma which describes how certain sequences of pieces of smooth curves can be joined by a single smooth curve within a finite parameter interval. It is one of the main tools used for the study of differentiable maps.

In section 4.3 we consider $\mathscr{Lip}^k$-maps. These are the maps for which all composites with $\mathscr{Lip}^k$-curves are $\mathscr{Lip}^k$-curves. We show that $\mathscr{Lip}^k$-maps have derivatives up to order k, establish the chain rule and the symmetry of the higher derivatives, and prove that a map is $\mathscr{Lip}^k$ if and only if it is $\mathscr{Lip}^j$ and its derivative of order j is $\mathscr{Lip}^{k-j}$ for some $j \leq k$. In order to determine $\mathscr{Lip}^k$-ness of a map whose derivatives can be guessed by looking at the composites with the elements of a certain point separating subset $\mathscr{S}$ of the dual of the range we introduce the notion of $\mathscr{S}$-differentiability. Its usefulness is due to the fact that one does not have to know all elements of the dual. In particular for function spaces one obtains such sets $\mathscr{S}$ by means of the point evaluations. On the other hand, it is interesting to know that $\mathscr{Lip}^k$-ness implies rather strong differentiability properties. Together with a Lipschitz condition on the derivatives either k-fold $\mathscr{S}$-differentiability as well as k-fold strong differentiability is equivalent with $\mathscr{Lip}^k$-ness.

Section 4.4 shows that for very general domains, namely $\mathscr{Lip}^k$-spaces (in particular arbitrary subsets of convenient vector spaces and arbitrary classical differentiable manifolds) the $\mathscr{Lip}^k$-functions with values in a convenient vector space form a convenient vector space as well. Based on the respective results for curve spaces the function space structure is described and compared with classical ones in case of manifolds modelled on convenient vector spaces. The so-called differentiable uniform boundedness principle gives an intrinsic description of the function space structure as the coarsest convenient vector space structure making the point evaluations morphisms. Several natural maps between function spaces such as evaluation and composition are proved to be $\mathscr{Lip}^k$. Some considerations on polynomial maps between convenient vector spaces open the way for the study of Taylor polynomials. It is shown that $\mathscr{Lip}^k$-maps admit Taylor expansions of order j for $j \leq k$ and that these give a direct sum decomposition of the respective function spaces.

In section 4.5 we study the relation between differentiability and partial differentiability of a map on a finite product of convenient vector spaces. It is also proved that for a $\mathscr{Lip}^k$-map $f\colon E \sqcap \mathbb{R} \to F$ the function $g(x) := \int_0^1 f(x,t)\mathrm{d}t$ has as derivative the function $g'(x) = \int_0^1 \partial_1 f(x,t)\mathrm{d}t$.

The vector bundles which are considered in section 4.6 have convenient vector spaces as fibres, a $\mathscr{Lip}^k$-map as projection, and triviality is only assumed along $\mathscr{Lip}^k$-curves of the base space.

Spaces of sections of vector bundles appear naturally if one wants to show, as it is done in section 4.7, that certain function spaces of maps between manifolds are manifolds modelled on convenient vector spaces. For any classical smooth manifolds X and Y, where X is supposed to be compact, the following function spaces are manifolds modelled on convenient vector spaces which are even nuclear Fréchet spaces: The space $\mathrm{Diff}(X)$ of all smooth diffeomorphisms of X, the space $\mathrm{Emb}(X, Y)$ of all smooth embeddings of X into Y, the space $\mathrm{Submf}(X, Y)$ of all submanifolds of Y which are diffeomorphic to X. The main result of section 4.7 states that the canonical map $\mathrm{Emb}(X, Y) \to \mathrm{Submf}(X, Y)$ constitutes a smooth principal fibre bundle of Fréchet manifolds with structure group $\mathrm{Diff}(X)$.

There exist many examples showing that an inverse or implicit function theorem in the classical formulation fails even for smooth maps between nuclear Fréchet spaces. However, the results given in section 4.8 show that the respective theorems do not go wrong with respect to the differentiability class of the inverse or implicit functions one is looking for. One can therefore obtain inverse and implicit function theorems for $\mathcal{Lip}^k$-maps by adding tameness conditions, cf. [Hamilton, 1982].

4.1 Differentiable curves

Any subset U of a convenient vector space E has a natural $\mathcal{Lip}^k$-structure, namely the initial one induced by the inclusion $U \subseteq E$. According to the description of initial $\mathcal{M}$-structures the $\mathcal{Lip}^k$-curves of U are those $\mathcal{Lip}^k$-curves of E having their image in U, or equivalently those curves c into U for which the composites $\ell \circ c$ are k-times Lipschitz differentiable for all $\ell \in E'$. For two subsets $U \subseteq E$ and $W \subseteq F$ of convenient vector spaces a map $f: U \to W$ is called a $\mathcal{Lip}^k$-map iff it is a $\mathcal{Lip}^k$-morphism. This is precisely the case when $\iota \circ f: U \to F$ is a $\mathcal{Lip}^k$-morphism (ι being the inclusion $W \subseteq F$), or equivalently if $\ell \circ f \circ c$ is k-times Lipschitz differentiable for every $\ell \in F'$ and every $\mathcal{Lip}^k$-curve $c: \mathbb{R} \to E$ with $c(\mathbb{R}) \subseteq U$.

Since we want to get derivatives of $\mathcal{Lip}^k$-maps for $k \geq 1$, we will restrict to special subsets U as domains; one might require U to be open for the locally convex topology of E, but the weaker condition of U being open in the Mackey-closure topology turns out to be adequate, and in order to get shorter formulations we adopt:

4.1.1 Convention. By $f: E \supseteq U \to F$ we will always mean that E and F are convenient vector spaces, U is M-open in E and f is a function from U to F.

We first consider the integration of $\mathcal{Lip}$-curves.

4.1.2 Definition. Let $a, b \in \mathbb{R}$ with $a < b$.

A *marked partition* $P = (t_0, \ldots, t_n; \tau_1, \ldots, \tau_n)$ of $[a, b]$ consists of real numbers t_i with $a = t_0 < \ldots < t_{j-1} < t_j < \ldots < t_n = b$ and real numbers $\tau_j \in [t_{j-1}, t_j]$.

The *mesh* $\mu(P)$ of a (marked) partition is defined by $\mu(P) := \max\{t_j - t_{j-1};\ j = 1 \ldots n\}$.

The set $\mathcal{P}$ of marked partitions of $[a, b]$ is directed by $P \succ Q$ iff $\mu(P) \leq \mu(Q)$.

Let c: $[a, b] \to E$ be a curve into a preconvenient vector space.

The *Riemann sum* $R_c(P)$ of c with respect to a marked partition P is defined as $R_c(P) := \sum_{j=1}^n c(\tau_j)(t_j - t_{j-1})$.

If the net $P \mapsto R_c(P)$ converges Mackey, then its limit will be called the *Mackey–Riemann integral* of c and denoted by $\int_a^b c$ or $\int_a^b c(t)\,dt$ and c is called *Mackey–Riemann integrable.*

$\int_a^a c$ is defined to be 0 and if $a > b$ then $\int_a^b c$ is defined as $-\int_b^a c$.

4.1.3 Lemma. *Let $c: \mathbb{R} \to E$ be a $\mathcal{Lip}$-curve of a preconvenient vector space E, $[\mathrm{a}, \mathrm{b}] \subseteq \mathbb{R}$ a bounded interval. Then the net formed by the Riemann sums of $c|_{[a, b]}$ is a Mackey–Cauchy net in E.*

Proof. Since c is a $\mathcal{Lip}$-curve there exists for each $\ell \in E'$ an $M_\ell \in \mathbb{R}$ with $|(\ell \circ c)(s) - (\ell \circ c)(t)| \leq M_\ell |s-t|$ for all $s, t \in [a,b]$. Let $P := (t_0, \ldots, t_n; \tau_1, \ldots, \tau_n)$ and $Q := (s_0, \ldots, s_m; \sigma_1, \ldots, \sigma_m)$ be two marked partitions of $[a,b]$. Denote by $a = r_0 < \ldots < r_h = b$ the ordered points of the set $\{t_0, \ldots, t_n, s_0, \ldots, s_m\}$. Decomposing the terms of the Riemann sums for those intervals which are subdivided we get:

$$R_c(P) = \sum (r_k - r_{k-1})\, c(\tau_{i(k)}) \quad \text{and} \quad R_c(Q) = \sum (r_k - r_{k-1})\, c(\sigma_{j(k)})$$

where $|\tau_{i(k)} - \tau_{j(k)}| \leq \mu(P) + \mu(Q)$. Thus $|(\ell \circ c)(\tau_{i(k)}) - (\ell \circ c)(\sigma_{j(k)})| \leq M_\ell(\mu(P) + \mu(Q))$ and $|\ell(R_c(P)) - \ell(R_c(Q))| \leq (b-a) M_\ell |\mu(P) + \mu(Q)|$. Setting $t_{P,Q} := 1/(\mu(P) + \mu(Q))$ gives: $\{t_{P,Q}(R_c(P) - R_c(Q));\ P, Q \in \mathcal{P}\}$ is bounded in E, and this yields the Mackey–Cauchy condition, cf. (2.2.13). □

4.1.4 Proposition. *Let $c: \mathbb{R} \to E$ be a $\mathcal{Lip}^k$-curve into a convenient vector space with $k \in \mathbb{N}_{0,\infty}$. Then:*

(i) *For all $a, b \in \mathbb{R}$ the Mackey–Riemann integral $\int_a^b c$ exists.*

(ii) *For any linear morphism $m: E \to F$ into a convenient vector space one has $\int_a^b (m \circ c) = m(\int_a^b c)$.*

Proof. (i) follows immediately from (4.1.3).

(ii) One uses that for any marked partition P of $[a,b]$ one trivially has $R_{m \circ c}(P) = m(R_c(P))$, and that m is continuous with respect to the Mackey convergence structure. □

4.1.5 Lemma. *Let $U \subseteq \mathbb{R}^2$ be open, with $\mathbb{R} \times [0,1] \subseteq U$, and $f: U \to F$ be a $\mathcal{Lip}^0$-map into a convenient vector space. Then the map $t \mapsto \int_0^1 f(t,s)\,ds$ is $\mathcal{Lip}^0$ from $\mathbb{R}$ to F.*

Proof. Let $c: \mathbb{R} \to F$ be defined by $c(t) := \int_0^1 f(t,s)\,ds$. We have to show that $\delta^1 c$ is bornological.

For this we first consider the case $F = \mathbb{R}$. Then

$$\delta^1 c(t_1, t_2) = \left(\int_0^1 f(t_1, s)\,ds - \int_0^1 f(t_2, s)\,ds \right) \cdot \frac{1}{t_1 - t_2} = \int_0^1 \frac{f(t_1, s) - f(t_2, s)}{t_1 - t_2}\,ds$$

$$= \int_0^1 \delta_1 f(t_1, t_2; s)\,ds.$$

Since f is $\mathcal{Lip}^0$ it is locally Lipschitz by (1.4.2). Hence $\delta_1 f$ is bornological by (1.3.20), i.e. for every bounded $B \subseteq \mathbb{R}$ the set $\delta_1 f(B^{\langle 1 \rangle} \times [0,1])$ is bounded. Thus $\{\int_0^1 \delta_1 f(t_1, t_2; s)\,ds; (t_1, t_2) \in B^{\langle 1 \rangle}\}$ is bounded, i.e. $\delta^1 c$ is bornological.

Now the general case. The curve c is $\mathcal{Lip}^0$ iff $\ell \circ c = \ell \circ \int_0^1 f(\_, s)\,ds = \int_0^1 (\ell \circ f)(\_, s)\,ds$ (cf. (ii) in (4.1.4)) is $\mathcal{Lip}^0$. This holds by the special case applied to the function $\ell \circ f$. □

4.1.6 Lemma. *Let $U \subseteq E$ be* M-*open. The subspace topology on U of the Mackey-closure topology of E is the final one induced by the $\mathscr{Lip}^k$*-curves into *U.*

Proof. Since the inclusion $\iota\colon U \to E$ is by definition a $\mathscr{Lip}^k$-map it is continuous with respect to the mentioned topologies.

Conversely let $W \subseteq U$ be open in U with respect to the $\mathscr{Lip}^k$-curves in U. It is enough to show that W is open with respect to the $\mathscr{Lip}^k$-curves in E. So let $c\colon \mathbb{R} \to E$ be a $\mathscr{Lip}^k$-curve and $c(t) \in W \subseteq U$ for some $t \in \mathbb{R}$. Since $U \subseteq E$ is M-open, we conclude that an $\varepsilon > 0$ exists with $[t-\varepsilon, t+\varepsilon] \subseteq c^{-1}(U)$. Take a smooth function $h\colon \mathbb{R} \to [t-\varepsilon, t+\varepsilon]$ with $h(s) = t+s$ for $2|s| \leq \dot{\varepsilon}$. Then $c \circ h\colon \mathbb{R} \to E$ is a $\mathscr{Lip}^k$-curve into U with $(c \circ h)(0) \in W$ and hence there exists a $\delta > 0$ with $[-\delta, \delta] \subseteq (c \circ h)^{-1}(W)$. We may assume that $2\delta < \varepsilon$, hence $c(t+s) = (c \circ h)(s) \in W$ for $|s| < \delta$. This shows that $c^{-1}(W)$ is open, i.e. W is open in E with respect to the Mackey-closure topology. □

4.1.7 Corollary. *Let $f\colon E \supseteq U \to F$ be a $\mathscr{Lip}^k$-map. Then f is continuous with respect to the Mackey-closure topologies.*

We next give a lemma on extensions of maps which will be used in order to extend difference quotients of differentiable curves.

4.1.8 Lemma. *($\mathscr{Lip}^k$-Extension Lemma.) Let E and F be convenient vector spaces; $U \subseteq E$ be* M-*open; D a dense subset of U (with respect to the Mackey closure topology); and $f\colon D \to F$ a map such that for each $\ell \in F'$ the function $\ell \circ f\colon D \to \mathbb{R}$ has a $\mathscr{Lip}^k$-extension $f_\ell\colon U \to \mathbb{R}$. Then f has a unique $\mathscr{Lip}^k$-extension $\tilde{f}\colon U \to F$.*

Proof. Let $\iota\colon F \to \Pi_{F'}\mathbb{R}$ be the canonical map, which is an embedding by the special embedding lemma (2.5.5). The map $g\colon U \to \Pi_{F'}\mathbb{R}$ characterized by $\mathrm{pr}_\ell \circ g = f_\ell$ for all $\ell \in F'$ is a $\mathscr{Lip}^k$-map and satisfies $g|_D = \iota \circ f$. Therefore $g(D) \subseteq \iota(F)$, and since $\mathscr{Lip}^k$-maps are continuous for the Mackey closure topologies one obtains $g(U) \subseteq g(\bar{D}) \subseteq \overline{g(D)} \subseteq \overline{\iota(F)}$. But $\overline{\iota(F)} = \iota(F)$ since F is complete, cf. (8) of (2.6.2). Hence g factors over ι; let us put $g = \iota \circ \tilde{f}$ for some $\tilde{f}\colon U \to F$. Since ι is initial $\tilde{f}$ is a $\mathscr{Lip}^k$-map, and it satisfies $\tilde{f}|_D = f$ by construction. This proves the existence. Uniqueness is trivial since $\mathscr{Lip}^k$-maps are continuous, cf. (4.1.7), and D is dense in U with respect to the Mackey closure topology. □

4.1.9 Definition. Let $U \subseteq \mathbb{R}$ be open, $c\colon U \to E$ be a curve into a preconvenient vector space, and $\mathscr{S} \subseteq E'$ point separating.

(i) c is called *$\mathscr{S}$-differentiable at t* iff $\lim_{s \to 0} (c(t+s) - c(t))/s$ exists with respect to the weak topology $\sigma(E, \mathscr{S})$, i.e. the initial topology on E induced by the family $\mathscr{S}$. Since the value of this limit (not the existence) is independent of the family $\mathscr{S}$ it is simply denoted by $c^{\cdot}(t)$ and called the *derivative of c at t*.

If c is $\mathscr{S}$-differentiable at every point of U, then c is called *$\mathscr{S}$-differentiable* and $c^{\cdot}$ is called the *derivative of c* (E'-differentiability coincides with weak differentiability as defined in (2.5.1)).

(ii) c is called *strongly differentiable* iff $\text{M-lim}_{r,s\to t,\, r\neq s}(c(r)-c(s))/(r-s)$ exists uniformly for t in compact subsets of U. This implies that c is $\mathscr{S}$-differentiable for any $\mathscr{S}$ and that this M-limit is equal to $c^{\cdot}(t)$.

(iii) c is called *$(k+1)$-times $\mathscr{S}$-differentiable* (*$(k+1)$-times strongly differentiable*) iff it is $\mathscr{S}$-differentiable (strongly differentiable) and $c^{\cdot}$ is k-times $\mathscr{S}$-differentiable (k-times strongly differentiable). The *derivative* $c^{(k+1)}$ of order $k+1$ is defined as derivative of order k of $c^{\cdot}$. Instead of k-times E'-differentiable we also say *k-times weakly differentiable*.

Remark. Since the weak topology $\sigma(E,\mathscr{S})$ depends only on the linear subspace of E' generated by $\mathscr{S}$, the same holds for $\mathscr{S}$-differentiability.

4.1.10 Lemma. *Let $c\colon \mathbb{R}\supseteq U\to E$ be a curve into a preconvenient vector space, and $\mathscr{S}\subseteq E'$ point separating. Then the following statements are equivalent:*

(1) *c is k-times $\mathscr{S}$-differentiable;*
(2) *There exist curves $c^j\colon U\to E$ $(1\leq j\leq k)$ such that $\ell\circ c$ is k-times differentiable and $(\ell\circ c)^{(j)}=\ell\circ c^j$ for all $\ell\in\mathscr{S}$ and $1\leq j\leq k$.*

Under the equivalent conditions one has $c^j=c^{(j)}$ for all $1\leq j\leq k$.

Proof. The convergence of $(c(t+s)-c(t))/s$ in the topology $\sigma(E,\mathscr{S})$ is equivalent to the convergence of

$$\ell\left(\frac{c(t+s)-c(t)}{s}\right)=\frac{\ell(c(t+s))-\ell(c(t))}{s}$$

and the limits coincide for all $\ell\in\mathscr{S}$. Using this the proposition follows by induction on k. □

4.1.11 Lemma. *Let $I\subseteq\mathbb{R}$ be an open interval, $\mathscr{S}\subseteq E'$ point separating and $c\colon I\to E$ $\mathscr{S}$-differentiable. If $c^{\cdot}=0$ then c is constant.*

Proof. For any $\ell\in\mathscr{S}$ one has $(\ell\circ c)^{\cdot}=\ell\circ c^{\cdot}=0$, hence $\ell\circ c$ is constant and since $\mathscr{S}$ is point separating c is constant. □

Remark. The previous lemma remains true if c is $\sigma(E,\mathscr{S})$-continuous and $\mathscr{S}$-differentiable at all $t\in I\setminus D$ for some countable set D. In fact, from $(\ell\circ c)^{\cdot}(t)=0$ for $t\in I\setminus D$ and the continuity of $\ell\circ c$ it still follows that $\ell\circ c$ is constant, cf. [Dieudonné, 1960, p. 156].

4.1.12 Theorem. *Let $k\in\mathbb{N}$ and $0\leq j\leq k$. For a curve $c\colon\mathbb{R}\supseteq U\to E$ the following statements are equivalent:*

(1) *c is a $\mathscr{Lip}^k$-curve;*
(2) *$\delta^{k+1}c\colon U^{\langle k+1\rangle}\to E$ is bornological;*
(3_j) *$\delta^j c\colon U^{\langle j\rangle}\to E$ has a $\mathscr{Lip}^{k-j}$-extension $\bar\delta^j c\colon U^{j+1}\to E$;*
(4_j) *c is j-times strongly differentiable and $c^{(j)}$ is a $\mathscr{Lip}^{k-j}$-curve;*

(5_j) *c is j-times $\mathcal{S}$-differentiable for some (all) point-separating $\mathcal{S}\subseteq E'$ and $c^{(j)}$ is a $\mathcal{Lip}^{k-j}$-curve.*

Proof. Clearly $1=3_0=4_0=5_0$.

($1\Rightarrow 2$) By composing with $\ell\in E'$ this is immediately deduced from ($1\Rightarrow 3$) of (1.3.22).

($2\Rightarrow 3_j$) Take $\ell\in E'$. Then $\delta^j(\ell\circ c)$ is bornological, hence by ($3\Rightarrow 4_j$) of (1.3.22) a $\mathcal{Lip}^{k-j}$ extension to U^{j+1} exists. Thus by (4.1.8) a $\mathcal{Lip}^{k-j}$ extension of $\delta^j c$ to U^{j+1} exists.

($3_j\Rightarrow 1$) Let $\ell\in E'$. Using that $\delta^j(\ell\circ c)$ has a $\mathcal{Lip}^{k-j}$ extension to U^{j+1} we conclude from ($4_j\Rightarrow 1$) of (1.3.22) that $\ell\circ c$ is k-times Lipschitz differentiable. Thus c is $\mathcal{Lip}^k$.

Thus we have proved: ($1\Leftrightarrow 2\Leftrightarrow 3_j$).

($1+2+3_1\Rightarrow 4_1$) For a compact $A\subseteq U$ take another compact set $K\subseteq U$ such that the interior of K contains A. Since c is $\mathcal{Lip}^k$ it is $\mathcal{Lip}^1$; hence, by (2), $B:=\delta^2 c(K^{\langle 2\rangle})$ is bounded in E. Let now $(t,s)\neq(t',s')\in K^{\langle 2\rangle}$. By (1.3.13) we have

$$\delta^1 c(t,s)-\delta^1 c(t',s')=\tfrac{1}{2}(t-t')\,\delta^2 c(t,t',s')+\tfrac{1}{2}(s-s')\,\delta^2 c(t,s,s')$$

provided $\{t,s\}\neq\{t',s'\}$. Hence

$$\frac{1}{|t-t'|+|s-s'|}(\delta^1 c(t,s)-\delta^1 c(t',s'))$$

lies in the absolutely convex hull of B, i.e. in a bounded subset of E. So we have shown that the Mackey–Cauchy condition for the derivative is satisfied uniformly on A. Thus c is strongly differentiable and $c^{\cdot}(s)=\text{M-lim}_{t\to 0}\delta^1 c(t+s,\,s)$ $=\overline{\delta^1}c(s,s)$. Hence $c^{\cdot}$ is, by (3_1), as restriction of a $\mathcal{Lip}^{k-1}$ map also $\mathcal{Lip}^{k-1}$.

($4_1\Rightarrow 5_1$) This is trivial.

($5_1\Rightarrow 1$) Since $c^{\cdot}$ is at least $\mathcal{Lip}^0$ we may form the Mackey–Riemann integral $e(t):=c(0)+\int_0^t c^{\cdot}$. Let $\ell\in E'$; then $\ell(e(t))=\ell(c(0))+\int_0^t \ell\circ c^{\cdot}$. Since $\ell\circ c^{\cdot}$ is at least continuous one concludes that $\ell\circ e$ is differentiable with derivative $(\ell\circ e)^{\cdot}=\ell\circ c^{\cdot}$ being $(k-1)$-times Lipschitz differentiable. Thus e is $\mathcal{Lip}^k$ and it remains to show that e is equal to c. Since e and c are both $\mathcal{S}$-differentiable and have the same derivative, the previous lemma implies that $e-c$ is constant. Therefore $e=c$, since $e(0)=c(0)$.

Thus the proposition is proved for $j=1$ and for $j=0$.

($4_j\Leftrightarrow 4_{j+1}$) is easily proved by induction using ($1\Leftrightarrow 4_0\Leftrightarrow 4_1$).

($5_j\Leftrightarrow 5_{j+1}$) is easily proved by induction using ($1\Leftrightarrow 5_0\Leftrightarrow 5_1$). □

4.1.13 Theorem. *Let $c\colon \mathbb{R}\supseteq U\to E$ be a curve into a convenient vector space and let $j\in\mathbb{N}_0$. Then the following statements are equivalent:*

(1) *c is $\mathcal{Lip}^\infty$;*

(2) *$\delta^k c\colon U^{\langle k\rangle}\to E$ is bornological for all (infinitely many) $k\in\mathbb{N}$;*

(3_j) *$\delta^j c\colon U^{\langle j\rangle}\to E$ has a $\mathcal{Lip}^\infty$-extension $\overline{\delta}^j c\colon U^{j+1}\to E$;*

(4_j) *c is j-times strongly differentiable and* $c^{(j)}$ *is a* $\mathcal{Lip}^\infty$*-curve;*
(5_j) *c is j-times* $\mathcal{S}$*-differentiable for some (all) point-separating* $\mathcal{S} \subseteq E'$ *and* $c^{(j)}$ *is a* $\mathcal{Lip}^\infty$*-curve.*

Proof. This follows easily from (4.1.12). □

4.1.14 Proposition. *For a* $\mathcal{Lip}^0$*-curve* $c: \mathbb{R} \to E$ *let* $\int c$ *denote the curve* $(\int c)(t) := \int_0^t c$.
(i) *If c is a* $\mathcal{Lip}^k$*-curve then* $\int c$ *is* $\mathcal{Lip}^{k+1}$ *and* $c = (\int c)^\cdot$.
(ii) *If c is a* $\mathcal{Lip}^{k+1}$*-curve then* $c^\cdot$ *is* $\mathcal{Lip}^k$ *and* $c = c(0) + \int(c^\cdot)$.

Proof. (i) Let $\ell \in E'$. Then $\ell \circ \int c = \int(\ell \circ c)$. Since $\ell \circ c$ is k-times locally Lipschitz differentiable one concludes that $\int(\ell \circ c)$ is differentiable with derivative $\ell \circ c$, hence $\ell \circ \int c$ is $\mathcal{Lip}^{k+1}$ with derivative equal to c.

(ii) By ($1 \Rightarrow 5_1$) of (4.1.12) we know that $c^\cdot$ is $\mathcal{Lip}^k$. Thus by (i) $\int c^\cdot$ exists and is $\mathcal{Lip}^{k+1}$ and the derivative of $\int c^\cdot$ is $c^\cdot$. By (4.1.11), $c - \int c^\cdot$ is constant and at 0 its value is obviously $c(0) - 0$. □

Next we give a proposition estimating the increment of a curve c by means of its derivative, the estimation being expressed in terms of a convex set. This proposition generalizes the mean value theorem for curves in Banach spaces where the estimations are expressed in terms of the norm; cf. [Dieudonné, 1960, p. 153].

4.1.15 Proposition. (*General Mean Value Theorem.*) *Let* $c: \mathbb{R} \to E$ *be a* $\mathcal{Lip}^1$*-curve into a convenient vector space;* $A \subseteq E$ *an* M*-closed convex set; and* $h: \mathbb{R} \to \mathbb{R}$ *a monotonic differentiable function. If* $c^\cdot(t) \in h^\cdot(t) A$ *for all* $t \in [a, b]$ *then* $c(b) - c(a) \in (h(b) - h(a)) A$.

Proof. We show first that certain Riemann sums of $h^\cdot$ converge to $h(b) - h(a)$. For a fixed partition $a = t_0 \leq \ldots \leq t_n = b$ one chooses by the classical mean value theorem $\tau_j \in [t_{j-1}, t_j]$ with $h(t_j) - h(t_{j-1}) = (t_j - t_{j-1}) h^\cdot(\tau_j)$. Then for the marked partition $P := (t_0, \ldots, t_n; \tau_1, \ldots, \tau_n)$ one has, using that A is convex and $h^\cdot(t) \geq 0$: $R_{c^\cdot}(P) \in \Sigma_j((t_j - t_{j-1}) h^\cdot(\tau_j) A) \subseteq (\Sigma_j (t_j - t_{j-1}) h^\cdot(\tau_j)) A = (h(b) - h(a)) A$. Taking the M-limit of Riemann sums so chosen and using (ii) of (4.1.14) and the M-closedness of A one obtains

$$c(b) - c(a) = \int_a^b c^\cdot = \lim_P R_{c^\cdot}(P) \in (h(b) - h(a)) A. \qquad \square$$

Remark. The previous proposition remains true if h is monotonic, continuous and for all $t \in [a, b] \setminus D$ differentiable at t with derivative satisfying $c^\cdot(t) \in h^\cdot(t) A$, where D is some countable set.

This can be proved by showing that certain Riemann sums of $h^\cdot$ still converge to $h(b) - h(a)$ (One uses that $\inf\{h^\cdot(t); t \in [a, b] \setminus D\} \leq (h(b) - h(a))/(b - a)$). Then the rest follows easily using the product of a sequence in E which is Mackey convergent to $c(b)-c(a)$ with a sequence of real numbers converging to $1/(h(b)-h(a))$ is Mackey convergent to $(c(b)-c(a))/(h(b)-h(a))$.

Furthermore one may weaken the differentiability properties of c provided one assumes A to be closed in a stronger sense. More precisely:

Assume that for some point separating $\mathscr{S}\subseteq E'$ the curve c is $\sigma(E,\mathscr{S})$-continuous and $\mathscr{S}$-differentiable in all $t\in[a,b]\setminus D$ and satisfies $c^{\cdot}(t)\in h^{\cdot}(t)A$ for a $\sigma(E,\mathscr{S})$-closed convex set $A\subseteq E$ and a countable set $D\subseteq[a,b]$. Then $c(b)-c(a)\in(h(b)-h(a))A$.

Using Hahn–Banach this can be easily reduced to the situation $E=\mathbb{R}$; cf. [Dieudonné, 1960, p. 153].

As will be proved later, many function spaces such as the space $\mathscr{Lip}^k(X,E)$ of $\mathscr{Lip}^k$-maps from a $\mathscr{Lip}^k$-space X into a convenient vector space E are again convenient vector spaces. For curves $c\colon \mathbb{R}\to\mathscr{Lip}^k(X,E)$ into these spaces it is often easy to test whether $\mathrm{ev}_x\circ c\colon \mathbb{R}\to E$ is smooth, where $\mathrm{ev}_x\colon \mathscr{Lip}^k(X,E)\to E$ denotes the evaluation maps. So it is natural to ask under which conditions the smoothness of these composites implies the smoothness of c. If the bornology of the function space is the initial one induced by the evaluations (as for example for the function space $L(E,F)$, cf. (3.6.5)) then no additional condition is necessary. In general this is, however, not the case. But often the evaluation maps form a family of morphisms which has a certain property allowing useful results to be given in the indicated direction. More generally one has in many situations in a natural way a point separating family $\mathscr{S}$ of $\underline{\mathrm{Con}}$-morphisms and a curve $c\colon \mathbb{R}\to E$ for which one can guess the derivatives, i.e. one finds curves $c^k\colon \mathbb{R}\to E$ with the property that $\ell\circ c$ is smooth with derivatives $(\ell\circ c)^{(k)}=\ell\circ c^k$ for all $k\in\mathbb{N}$ and $\ell\in\mathscr{S}$. Under some mild condition on the family $\mathscr{S}$ it is then enough to assume that the c^k are bornological in order to obtain that c is smooth and has c^k as kth derivative.

We will start with discussing this property in the case where the point separating family of $\underline{\mathrm{Con}}$-morphisms consists of real valued functionals only and prove the respective criterion for smoothness (and $\mathscr{Lip}^0$-ness) of a curve. In (4.1.20) we will show that the case of general families can be reduced to this one. We then give several examples; further ones will follow later, when the respective function spaces are available.

At the first reading a shortcut can be made by skipping the rest of this section together with (4.3.10), (5) of (4.3.30) and (4.4.15)–(4.4.35).

4.1.16 Lemma. *Let E be a convenient vector space and $\mathscr{S}\subseteq E'$ a subset. Then the following statements are equivalent:*

(1) *the bornology of E has a basis of $\sigma(E,\mathscr{S})$-closed sets, i.e. for every bounded $B\subseteq E$ there exists a $\sigma(E,\mathscr{S})$-closed bounded $A\subseteq E$ with $B\subseteq A$;*
(2) *the $\sigma(E,\mathscr{S})$-closure of bounded sets is bounded;*
(3) *$\bigcap_{\ell\in\mathscr{S}_0}\ell^{-1}\left(\overline{\ell(B)}\right)$ is bounded for every bounded $B\subseteq E$, where $\mathscr{S}_0$ denotes the vector space generated by $\mathscr{S}$.*

Proof. (1⇒2) trivial.

(2⇒3) Use that by the bipolar theorem [Jarchow, 1981, p. 149] $\bigcap_{\ell \in \mathcal{S}_0} \ell^{-1}\,\overline{(\ell(B))}$ is the $\sigma(E, \mathcal{S}_0)$-closure of B if B is absolutely convex.

(3⇒1) Clearly $\bigcap_{\ell \in \mathcal{S}_0} \ell^{-1}\overline{(\ell(B))}$ is a $\sigma(E, \mathcal{S})$-closed subset containing B. □

Remark. It is obvious that these equivalent conditions on $\mathcal{S}$ imply that $\mathcal{S}$ separates points.

Now some applications of this condition on a subset $\mathcal{S}$, first in order to determine the $\mathcal{Lip}^0$-curves and then the smooth curves.

4.1.17 Lemma. *Let $\mathcal{S} \subseteq E'$ be such that the bornology of E has a basis of $\sigma(E, \mathcal{S})$-closed sets; let $c\colon \mathbb{R} \supseteq U \to E$ be a $\mathcal{Lip}^0$-curve; and A be a bounded subset of $\mathbb{R}$ with closure contained in U. If c is $\mathcal{S}$-differentiable at t for all $t \in A$ then $\{c^{\cdot}(t); t \in A\}$ is bounded.*

Proof. Let $I \subseteq U$ be a compact set such that A is contained in the interior of I. By assumption δc is bounded on $I^{\langle 1 \rangle}$. So let B be a $\sigma(E, \mathcal{S})$-closed subset containing $\delta c(I^{\langle 1 \rangle})$. For all $t \in A$ one has $c^{\cdot}(t) = \lim_{s \to 0} \delta c(t, t+s) \in B$, hence $\{c^{\cdot}(t); t \in A\} \subseteq B$ is bounded. □

4.1.18 Proposition. *Let E be a convenient vector space, $\mathcal{S} \subseteq E'$ a subset such that the bornology of E has a basis of $\sigma(E, \mathcal{S})$-closed sets. Let $c\colon \mathbb{R} \supseteq U \to E$ be a* $\mathcal{S}$-differentiable curve. Then c is $\mathcal{Lip}^0$ *if and only if $c^{\cdot}$ is bornological.*

Proof. If c is $\mathcal{Lip}^0$ then $c^{\cdot}$ is bornological by lemma (4.1.17).

Conversely suppose $c^{\cdot}$ is bornological. Let $I \subseteq U$ be a compact set. Choose a $\sigma(E, \mathcal{S})$-closed absolutely convex bounded set B that contains $c^{\cdot}(I)$. Let $\ell \in \mathcal{S}$, and $t, s \in I$, $t \neq s$. Then by the classical mean value theorem there exists an $r \in [t, s]$ dependent on ℓ such that $\ell(\delta c(t, s)) = \delta(\ell \circ c)(t, s) = (\ell \circ c)^{\cdot}(r) = \ell(c^{\cdot}(r)) \in \ell(B)$. Since B was chosen absolutely convex and $\sigma(E, \mathcal{S})$-closed one concludes that $\delta c(t, s) \in B$, i.e. $\delta c(I^{\langle 1 \rangle}) \subseteq \bigcap_{\ell \in \mathcal{S}} \ell^{-1}\ell(B) \subseteq B$ is bounded, cf. (2⇒3) of (4.1.16). □

4.1.19 Theorem. *Let $c\colon \mathbb{R} \supseteq U \to E$ be a curve into a convenient vector space E, and let $\mathcal{S} \subseteq E'$ be a subset such that the bornology of E has a basis of $\sigma(E, \mathcal{S})$-closed sets. Then the following statements are equivalent:*

(1) *c is $\mathcal{Lip}^\infty$;*
(2) *c is k-times $\mathcal{S}$-differentiable and $c^{(k)}$ is bornological for all $k \in \mathbb{N}_0$.*

Proof. (1⇒2) trivial.

(2⇒1) By assumption $c^{(k)}$ is $\mathcal{S}$-differentiable and its derivative $c^{(k+1)}$ is bornological. Hence it follows from (4.1.18) that $c^{(k)}$ is $\mathcal{Lip}^0$. Using now (5_k⇒1) of (4.1.12) one deduces that c is $\mathcal{Lip}^k$ for all k and hence $\mathcal{Lip}^\infty$ □

The following proposition shows that the general situation mentioned before, where one has a family $f_j: E \to E_j$ of linear morphisms, can be reduced to that of a subset $\mathscr{S} \subseteq E'$.

4.1.20 Proposition. *Let $f_j: E \to E_j$ $(j \in J)$ be a family of* $\underline{\text{Con}}$*-morphisms, $\mathscr{S} := \bigcup_{j \in J} f_j^*(E_j')$, $c: \mathbb{R} \to E$ a curve, $B \subseteq E$ bounded and absolutely convex, and $k \in N_{0,\infty}$.*

(i) *The curve c is k-times $\mathscr{S}$-differentiable if and only if there exist curves c^i $(i < k+1)$ such that for all $j \in J$, $f_j \circ c$ is k-times weakly differentiable and $(f_j \circ c)^{(i)} = f_j \circ c^i$ for all $i < k+1$.*

(ii) *The set B is $\sigma(E, \mathscr{S})$-closed if and only if it is closed for the initial locally convex topology on E induced by the family f_j $(j \in J)$.*

Proof. (i) c is k-times $\mathscr{S}$-differentiable iff c^i (for $i < k+1$) exist with $(\ell_j \circ f_j \circ c)^{(i)} = \ell_j \circ f_j \circ c^i$ for all $j \in J$ and all $\ell_j \in E_j'$. This is equivalent to $f_j \circ c$ being k-times weakly differentiable and $\ell_j \circ (f_j \circ c)^{(i)} = \ell_j \circ (f_j \circ c^i)$.

(ii) Since closed and weakly closed is equivalent for absolutely convex subsets of a locally convex space, cf. [Jarchow, 1981, p. 149] it is enough to show that on E the weak topology associated to the initial locally convex topology induced by f_j $(j \in J)$ is exactly the topology $\sigma(E, \mathscr{S})$. This follows since taking initial locally convex topologies commutes with taking associated weak topologies, cf. [Jarchow, 1981, p. 167]. □

In order to apply the above results (4.1.19) and (4.1.18) we give some

4.1.21 Examples

(i) Let $f_j: E \to E_j$ $(j \in J)$ be a family of $\underline{\text{Con}}$-morphisms which is an initial $\underline{\ell^\infty}$-source. Then the bornology of E has a basis of sets being closed for the initial locally convex topology. It is enough to show that $\bigcap f_j^{-1}(\overline{f_j(B)})$ is bounded for every closed absolutely convex bounded set. Since the bornology on E is by assumption the initial one induced by the f_j it is enough to show that $f_i(\bigcap f_j^{-1}(\overline{f_j(B)}))$ is bounded. But this is trivial since it is contained in $f_i(f_i^{-1}(\overline{f_i(B)})) \subseteq \overline{f_i(B)}$ and this is bounded as closure of a bounded set. For initial $\underline{\ell^\infty}$-sources theorem (4.1.19) is of course not interesting, since in this situation the boundedness assumption on the derivatives is not necessary.

(ii) The bornology of $\ell^\infty(\mathbb{N}, E)$ has a basis of subsets closed in the initial locally convex topology induced by $\text{ev}_n: \ell^\infty(\mathbb{N}, E) \to E$ $(n \in \mathbb{N})$. A basis of the bornology on $\ell^\infty(\mathbb{N}, E)$ is given by the sets $B_0 := \{c; c(\mathbb{N}) \subseteq B\}$ ($B \subseteq E$ bounded and closed in the locally convex topology). It is enough to show that $B_0 = \bigcap_{n \in \mathbb{N}} \text{ev}_n^{-1} \text{ev}_n(B_0)$. Clearly $\text{ev}_n(B_0) = B$, so $c \in \bigcap \text{ev}_n^{-1} \text{ev}_n(B_0)$ iff $c(n) \in B$ for all n, i.e. $c \in B_0$.

(iii) Let c_0 be the closed subspace of ℓ^∞ formed by the sequences converging to 0. Then the bornology of c_0 has a basis of $\sigma(c_0, \{\text{ev}_n; n \in \mathbb{N}\})$-closed sets. To prove this one only has to combine the initiality of the inclusion $c_0 \subseteq \ell^\infty$ with (ii) for $E := \mathbb{R}$.

(iv) Let X be an ℓ^∞-space, E a convenient vector space, and $\mathscr{S} \subseteq E'$ such that the bornology of E has a basis of $\sigma(E, \mathscr{S})$-closed sets. Then the bornology on $\ell^\infty(X, E)$ has a basis of $\sigma(\ell^\infty(X, E), \{\ell \circ \mathrm{ev}_x; \ell \in S, x \in X\})$-closed sets. To prove this one only has to combine the initiality of $c^*: \ell^\infty(X, E) \to \ell^\infty(\mathbb{N}, E)$ $(c \in \ell^\infty(\mathbb{N}, X))$ with example (ii).

(v) Let K be a compact Hausdorff space and $D \subseteq K$ a dense subset. Then the bornology of the Banach space $C(K)$ of continuous real valued functions with the maximum norm has a basis of $\sigma(C(K), \{\mathrm{ev}_x; x \in D\})$-closed subsets. It is enough to show that $B = \bigcap_{x \in D} \mathrm{ev}_x^{-1}(\mathrm{ev}_x(B))$, where B denotes the closed unit ball $B = \{f \in C(K, \mathbb{R}); |f(x)| \leq 1 \text{ for all } x \in K\}$. It is obvious that $\mathrm{ev}_x(B) = [-1, 1]$. Suppose there is an $f \in C(K, \mathbb{R})$ with $f(x) \in [-1, 1]$ for all $x \in D$ but $f \notin B$. Then there is an $x \in K$ with $|f(x)| > 1$ and by continuity of f and denseness of D there is a $y \in D$ with $|f(y)| > 1$. This is a contradiction.

(vi) The bornology of $\coprod_{j \in J} E_j$ has a basis of $\sigma(\coprod E_j, \{\ell \circ \mathrm{pr}_j; j \in J, \ell \in E_j'\})$-closed sets, where $\mathrm{pr}_j: \coprod_{i \in J} E_i \to E_j$ denotes the projection. This is obvious, since a bounded B is contained in a finite subsum of the coproduct.

A general proposition that allows to obtain families $\mathscr{S}$ with the desired condition is the following:

4.1.22 Proposition. *Let $\mathscr{S} \subseteq E \subseteq (E')'$ be a linear subspace that separates points of E'. Then the bornology of E' has a basis of $\sigma(E', \mathscr{S})$-closed sets.*

Proof. It is enough to show that closed bounded subsets B of E' are $\sigma(E', \mathscr{S})$-closed. By the linear uniform boundedness principle (3.6.4) the bounded subsets of E' are the equicontinuous ones. By the Alaoğlu–Bourbaki theorem [Jarchow, 1981, p. 157] these sets are relatively $\sigma(E', E)$-compact and their closures taken in $\prod_E \mathbb{R}$ are contained in E'. So the closed ones are $\sigma(E', E)$-compact and since $\sigma(E', \mathscr{S})$ is a coarser Hausdorff topology they are also $\sigma(E', \mathscr{S})$-compact, hence $\sigma(E', \mathscr{S})$-closed. □

4.1.23 Corollary. *Let E be a reflexive convenient vector space, i.e. $\iota_E: E \to E''$ is bijective. Then the bornology of E has a basis of $\sigma(E, \mathscr{S})$-closed sets for every point-separating set $\mathscr{S} \subseteq E'$.*

Example. Let ℓ^p for $1 \leq p < \infty$ denote the classical Banach space of p-summable sequences. The bornology of ℓ^p has a basis of $\sigma(\ell^p, \{\mathrm{ev}_n; n \in \mathbb{N}\})$-closed sets. This follows since for $p > 1$ the Banach space ℓ^p is the dual of ℓ^q, where q is given by $(1/p) + (1/q) = 1$. The Banach space ℓ^1 is the dual of the Banach space c_0 of sequences converging to 0.

4.1.24 Proposition. *Let E be a convenient vector space, $\mathscr{S} \subseteq E'$ be a subset and $\bar{\mathscr{S}}$ be the closure of $\mathscr{S}$ with respect to the topology of uniform convergence on bounded subsets of E (a 0-neighborhood basis of this topology is given by the polars*

$B^0 := \{\ell \in E'; \ell(B) \subseteq [-1, 1]\}$ *of all bounded* $B \subseteq E$). *If the bornology of E has a basis of* $\sigma(E, \bar{\mathscr{S}})$*-closed sets then it has a basis of* $\sigma(E, \mathscr{S})$*-closed sets.*

Proof. It is enough to show that for any bounded absolutely convex $B \subseteq E$ one has $\bigcap_{\ell \in \mathscr{S}} \ell^{-1}(\overline{\ell(B)}) = \bigcap_{\ell \in \bar{\mathscr{S}}} \ell^{-1}(\overline{\ell(B)})$.

($\supseteq$) is trivial, since $\mathscr{S} \subseteq \bar{\mathscr{S}}$.

($\subseteq$) Admit that this is not true, then there exists an $x \in E$ with $\ell(x) \in \overline{\ell(B)}$ for all $\ell \in \mathscr{S}$ but $\ell_0(x) \notin \overline{\ell_0(B)}$ for some $\ell_0 \in \bar{\mathscr{S}}$. Now choose an absolutely convex 0-neighborhood U in $\mathbb{R}$ with $\ell_0(x) + 3U \subseteq \mathbb{R} \setminus \overline{\ell_0(B)}$. Then $\ell_0 + \{\ell \in E'; \ell(\{x\} \cup B) \subseteq U\}$ is a neighborhood of ℓ_0 with respect to the topology of uniform convergence on bounded sets. Since $\mathscr{S}$ is dense in $\bar{\mathscr{S}}$ with respect to this topology there has to exist an $\ell_1 \in \mathscr{S}$ that is in this neighborhood. In order to reach a contradiction we claim that $\ell_1(x) \notin \overline{\ell_1(B)}$. By construction $(\ell_1 - \ell_0)(x) \in U$ and $(\ell_1 - \ell_0)(B) \subseteq U$ and thus $\ell_1(x) \in \ell_0(x) + U$ and $\overline{\ell_1(B)} \subseteq \overline{\ell_0(B) + U} \subseteq \overline{\ell_0(B)} + 2U$, cf. [Jarchow, 1981, p. 31]. The claim now follows using that $\ell_0(x) + U$ and $\overline{\ell_0(B)} + 2U$ are disjoint. □

An application of this proposition will be given in (4.4.34).

4.2 Curve spaces

In the preceding section we considered $\mathscr{Lip}^k$-curves of a convenient vector space E. They obviously form a vector space $\mathscr{Lip}^k(\mathbb{R}, E)$, and we will show now that this space has in fact a natural convenient vector space structure. These curve spaces will be used later in order to study more general function spaces.

We begin with the case $k < \infty$ and make use of the following linear maps:

$$\delta^p : \mathscr{Lip}^k(\mathbb{R}, E) \to \ell^\infty(\mathbb{R}^{\langle p \rangle}, E) \qquad (0 \le p \le k+1);$$

$$\mathscr{D}^p : \mathscr{Lip}^k(\mathbb{R}, E) \to \mathscr{Lip}^{k-p}(\mathbb{R}, E) \qquad (0 \le p \le k);$$

$$\mathrm{ev}_t : \mathscr{Lip}^k(\mathbb{R}, E) \to E \qquad (t \in \mathbb{R});$$

where $\delta^p c$ is the difference quotient of c of order p; $\mathscr{D}^p c := c^{(p)}$ is the pth derivative of c; and ev_t is the evaluation at t.

4.2.1 Lemma. *Let* $k \in \mathbb{N}_0$; $a_i \in \mathbb{R}$ *all different; and* $b_i \in \mathbb{R}$. *For a subset of* $\mathscr{Lip}^k(\mathbb{R}, E)$ *it is equivalent to be bounded under any of the following five families of maps:*

(1) $\delta^0, \dots, \delta^{k+1}$;

(2) $\mathrm{ev}_{b_0}, \mathrm{ev}_{b_1} \circ \mathscr{D}, \dots, \mathrm{ev}_{b_k} \circ \mathscr{D}^k, \delta^1 \circ \mathscr{D}^k$;

(3) $\mathrm{ev}_{a_0}, \dots, \mathrm{ev}_{a_{k-1}}, \mathrm{ev}_{b_k} \circ \mathscr{D}^k, \delta^1 \circ \mathscr{D}^k$;

(4) $\mathrm{ev}_{a_0}, \dots, \mathrm{ev}_{a_k}, \delta^1 \circ \mathscr{D}^k$;

(5) $\mathrm{ev}_{a_0}, \dots, \mathrm{ev}_{a_k}, \delta^{k+1}$.

Proof. The bornology of E is the initial one induced by the maps $\ell \in E'$, and that of $\ell^\infty(\mathbb{R}^{\langle i\rangle}, E)$ is the initial one induced by the maps $\ell_*: \ell^\infty(\mathbb{R}^{\langle i\rangle}, E) \to \ell^\infty(\mathbb{R}^{\langle i\rangle}, \mathbb{R})$ for $\ell \in E'$, cf. (1.1.8) or (1.2.9). Using this and furthermore that $\mathscr{D}^i$, δ^i and ev_t commute with ℓ_* one reduces the general case to the special case $E=\mathbb{R}$ which we study in what follows. So let $B \subseteq \mathscr{Lip}^k(\mathbb{R}, \mathbb{R})$ be an arbitrary subset.

(1⇒2) Let $0 \le p \le k$; $f \in B$; $b \in \mathbb{R}$. Choose $I :=]b-1, b+1[$. According to the remark following (1.3.15), $f^{(p)}(b)$ is in the closure of $\delta^p f(I^{\langle p\rangle})$; therefore $(\mathrm{ev}_b \circ \mathscr{D}^p)(B)$ is in the closure of the bounded set $\delta^p B(I^{\langle p\rangle})$ and hence is bounded. Finally, for any bounded open interval J, $\delta^{k+1} B(J^{\langle k+1\rangle})$ is bounded by some value M; then by (1.3.18), $\delta^1 \mathscr{D}^k B(J^{\langle 1\rangle})$ is bounded by the same value M. So we conclude that $\delta^1 \mathscr{D}^k B$ is bounded in $\ell^\infty(\mathbb{R}^{\langle 1\rangle}, \mathbb{R})$.

(2⇒3) Let I be a bounded interval containing the points a_i and b_i. From the boundedness of $(\delta^1 \circ \mathscr{D}^k) B(I^{\langle 1\rangle})$ and $\mathscr{D}^k B(b_k)$ we deduce immediately that $\mathscr{D}^k B(I)$ and hence $\mathscr{D}^k B(b_0)$ is bounded. Using that $\mathscr{D}^k B(I)$ is bounded one further deduces that also $\delta \mathscr{D}^{k-1} B(I^{\langle 1\rangle})$ is bounded (by the same value, according to the first mean value theorem (1.3.15) for $k=j=1$). Combining this with the hypothesis that $\mathscr{D}^{k-1} B(b_{k-1})$ is bounded one obtains as before: $\mathscr{D}^{k-1} B(I)$ is bounded. Repeating the same arguments gives: $\mathscr{D}^0 B(I) = B(I)$ is bounded. From this the boundedness of $B(a_i)$ follows.

(3⇒4) Let I be a bounded interval containing the points a_i and b_k. From $\mathscr{D}^k(B)(b_k)$ and $(\delta^1 \circ \mathscr{D}^k) B(I^{\langle 1\rangle})$ both bounded one deduces that $\mathscr{D}^k B(I)$ is bounded, and hence, by (1.3.15) for $j=k$, also $\delta^k B(I^{\langle k\rangle})$. Using (1.3.2), i.e. the identity $\delta^k f(a_0, \ldots, a_k) = k! \sum_{i=0}^k \beta_i \cdot f(a_i)$, one deduces, since the terms of the sum for $i<k$ are bounded for $f \in B$, that also $B(a_k)$ is bounded.

(4⇒5) follows immediately using (1.3.15) for $j=k-1$.

(5⇒1) Using again $\delta^k f(a_0, \ldots, a_k) = k! \sum_{i=0}^k \beta_i \cdot f(a_i)$ one first remarks that $\delta^k B(a_0, \ldots, a_k)$ is bounded. Furthermore, for any bounded interval $I \subseteq \mathbb{R}$, $\delta^{k+1} B(I^{\langle k+1\rangle})$ is bounded. So, using (1.3.14), $\delta^k B(I^{\langle k\rangle})$ is bounded, and this shows that $\delta^k B \subseteq \ell^\infty(\mathbb{R}^{\langle k\rangle}, \mathbb{R})$ is bounded. Repeating the same argument one successively obtains that $\delta^{k-1} B, \ldots, \delta^1 B$ are all bounded. □

4.2.2 Definition. For $k \in \mathbb{N}_0$ we will from now on denote with $\mathscr{Lip}^k(\mathbb{R}, E)$ the vector space formed by the $\mathscr{Lip}^k$-curves of E together with the initial preconvenient vector space structure induced by the maps $\delta^i: \mathscr{Lip}^k(\mathbb{R}, E) \to \ell^\infty(\mathbb{R}^{\langle i\rangle}, E)$ $(i=0, \ldots, k+1)$.

4.2.3 Proposition. *The structure of the preconvenient vector space* $\mathscr{Lip}^k(\mathbb{R}, E)$ *is the initial one induced by any of the five families of maps considered in* (4.2.1).

Proof.. This is a direct consequence of (4.2.1). □

4.2.4 Proposition. *Let E be a convenient vector space and $k \in \mathbb{N}$. Differentiation and integration yield morphisms of preconvenient vector spaces as follows:*

(i) $\mathscr{D}: \mathscr{Lip}^k(\mathbb{R}, E) \to \mathscr{Lip}^{k-1}(\mathbb{R}, E)$, *defined by* $\mathscr{D}(c) := \dot{c}$;

(ii) $\int: \mathscr{Lip}^{k-1}(\mathbb{R}, E) \to \mathscr{Lip}^k(\mathbb{R}, E)$, *defined by* $(\int c)(t) := \int_0^t c$.

Proof. (i) is trivially verified if one uses the characterization of the structure of $\mathcal{Lip}^{k-1}(\mathbb{R}, E)$ as being the initial one induced by the family (4) of (4.2.1), where of course one has to replace k by $k-1$. For the verification of (ii), the family (2) of (4.2.1) is adequate. □

4.2.5 Proposition. *Let E be a convenient vector space and $k \in \mathbb{N}$. One has a* Pre-*isomorphism $\mathcal{Lip}^k(\mathbb{R}, E) \cong E \sqcap \mathcal{Lip}^{k-1}(\mathbb{R}, E)$.*

Proof. One considers the map $(\mathrm{ev}_0, \mathcal{D})$: $\mathcal{Lip}^k(\mathbb{R}, E) \to E \sqcap \mathcal{Lip}^{k-1}(\mathbb{R}, E)$. It has an inverse, by (4.1.14), namely $(a, c) \mapsto a + \int c$. Both are morphisms according to the previous proposition (4.2.4). □

4.2.6 Proposition. *For a convenient vector space E and $k \in \mathbb{N}_0$ the curve space $\mathcal{Lip}^k(\mathbb{R}, E)$ is also convenient.*

Proof. By the above proposition (4.2.5) it is enough to prove that $\mathcal{Lip}^0(\mathbb{R}, E)$ is convenient. The map $(\mathrm{ev}_0, \delta^1)$: $\mathcal{Lip}^0(\mathbb{R}, E) \to E \sqcap \ell^\infty(\mathbb{R}^{\langle 1 \rangle}, E)$ is a Pre-embedding. Hence $\mathcal{Lip}^0(\mathbb{R}, E)$ is isomorphic to its image. Since $E \sqcap \ell^\infty(\mathbb{R}^{\langle 1 \rangle}, E)$ is complete by (ii) in (3.6.1) the image is also complete provided it is closed with respect to the Mackey closure topology, see the closed embedding lemma (2.6.4). We verify this by showing that this image is the intersection of the kernels of the Pre-morphisms:

$$h_{r,s,t}\colon E \sqcap \ell^\infty(\mathbb{R}^{\langle 1 \rangle}, E) \to E \quad \text{for} \quad (r, s, t) \in \mathbb{R}^{\langle 2 \rangle} \quad \text{where}$$

$$h_{r,s,t}(a, g) := (r-s)g(r, s) + (s-t)g(s, t) + (t-r)g(r, t).$$

The verification that $h_{r,s,t} \circ (\mathrm{ev}_0, \delta^1) = 0$ is trivial. So let conversely $h_{r,s,t}(a, g) = 0$ for all $(r, s, t) \in \mathbb{R}^{\langle 2 \rangle}$. The non-symmetric choice of $h_{r,s,t}$ allows to deduce that g is symmetric: $0 = 0 + 0 = h_{r,s,t}(a, g) + h_{s,r,t}(a, g) = (r-s) \cdot (g(r, s) - g(s, r))$, hence $g(r, s) = g(s, r)$ for $r \neq s$. Define now c: $\mathbb{R} \to E$ by $c(0) := a$ and $c(t) := a + t g(t, 0)$ *for* $t \neq 0$. One verifies that $\delta^1 c = g$ and $c(0) = a$, i.e. $(a, g) = (\mathrm{ev}_0, \delta^1)(c)$ as to be shown. □

We now consider the vector space $\mathcal{Lip}^\infty(\mathbb{R}, E)$ formed by the $\mathcal{Lip}^\infty$-curves of a convenient vector space E:

4.2.7 Proposition. *Let $k \in \mathbb{N}_0$; $a_i \in \mathbb{R}$ all different; and $b_i \in \mathbb{R}$. For a subset $B \subseteq \mathcal{Lip}^\infty(\mathbb{R}, E)$ it is equivalent to be bounded under any of the following four families of maps:*

(1) δ^i: $\mathcal{Lip}^\infty(\mathbb{R}, E) \to \ell^\infty(\mathbb{R}^{\langle i \rangle}, E)$ *for all* $i \in \mathbb{N}_0$;
(2) ι_i: $\mathcal{Lip}^\infty(\mathbb{R}, E) \to \mathcal{Lip}^i(\mathbb{R}, E)$ *for infinitely many* i;
(3) $\mathrm{ev}_{b_i} \circ \mathcal{D}^i$: $\mathcal{Lip}^\infty(\mathbb{R}, E) \to E$ *for* $0 \leq i \leq k$ *and*
$\mathcal{D}^i$: $\mathcal{Lip}^\infty(\mathbb{R}, E) \to \ell^\infty(\mathbb{R}, E)$ *for* $i > k$;
(4) ev_{a_i}: $\mathcal{Lip}^\infty(\mathbb{R}, E) \to E$ *for* $0 \leq i \leq k$ *and*
δ^i: $\mathcal{Lip}^\infty(\mathbb{R}, E) \to \ell^\infty(\mathbb{R}^{\langle i \rangle}, E)$ *for* $i > k$.

Proof. Follows directly from (4.2.1) and the remark after (1.3.15). □

4.2.8 Definition. $\mathscr{Lip}^\infty(\mathbb{R}, E)$ will from now on denote the vector space formed by the $\mathscr{Lip}^\infty$-curves of E together with the initial preconvenient vector space structure induced by the maps $\delta^i\colon \mathscr{Lip}^\infty(\mathbb{R}, E) \to \ell^\infty(\mathbb{R}^{\langle i\rangle}, E)$ $(i \in \mathbb{N}_0)$.

4.2.9 Proposition. *The structure of the preconvenient vector space* $\mathscr{Lip}^\infty(\mathbb{R}, E)$ *is the initial one induced by any of the four families of maps considered in* (4.2.7).

Proof. This is a direct consequence of (4.2.7). □

4.2.10 Proposition. *For any convenient vector space E the curve space* $\mathscr{Lip}^\infty(\mathbb{R}, E)$ *is also convenient.*

Proof. The maps of (2) in (4.2.7) yield a Pre-embedding $\iota\colon \mathscr{Lip}^\infty(\mathbb{R}, E) \to \prod_{k\in\mathbb{N}_0} \mathscr{Lip}^k(\mathbb{R}, E)$. As in the proof of (4.2.6) it is enough to show that the image of this morphism is equal to the intersection of the kernels of some morphisms. It is trivial that the maps $h_k\colon \prod_{i\in\mathbb{N}_0} \mathscr{Lip}^i(\mathbb{R}, E) \to \mathscr{Lip}^k(\mathbb{R}, E)$ for $k \in \mathbb{N}_0$ defined by $h_k(c_0, \ldots) := c_{k+1} - c_k$ do the job. □

Remark. $\mathscr{Lip}^\infty(\mathbb{R}, E) = \bigcap_k \mathscr{Lip}^k(\mathbb{R}, E)$ is the projective limit of the spaces $\mathscr{Lip}^k(\mathbb{R}, E)$ in the category Pre. Hence the previous proposition can also be deduced from (4.2.6) using that Con is reflective in Pre, cf. (2.6.5).

The following proposition is a special case of the differentiable uniform boundedness principle (4.4.7).

4.2.11 Proposition. *Let F be a convenient vector space. The structure of* $\mathscr{Lip}^k(\mathbb{R}, F)$ *introduced in* (4.2.2) *respectively* (4.2.8) *is the coarsest convenient vector space structure making all evaluations* ev_t $(t \in \mathbb{R})$ *morphisms.*

Proof. Let E be a convenient vector space and $m\colon E \to \mathscr{Lip}^k(\mathbb{R}, F)$ a linear map such that $\mathrm{ev}_t \circ m \in L(E, F)$ for all $t \in \mathbb{R}$. We have to show that m is a morphism which means that $\delta^i \circ m\colon E \to \ell^\infty(\mathbb{R}^{\langle i\rangle}, F)$ is an ℓ^∞-morphism for $i < k+2$ or, using (3.6.6), that $\mathrm{ev}_{(t_0, \ldots, t_i)} \circ \delta^i \circ m\colon E \to F$ is an ℓ^∞-morphism for $i < k+2$ and $(t_0, \ldots, t_i) \in \mathbb{R}^{\langle i\rangle}$. But this obviously holds, since using (1.3.2) we can write $\mathrm{ev}_{(t_0, \ldots, t_i)} \circ \delta^i \circ m = \sum_{j=0}^{i} \beta_j \cdot (\mathrm{ev}_{t_j} \circ m)$. □

We now consider some functoriality properties of the curve spaces $\mathscr{Lip}^k(\mathbb{R}, E)$.

4.2.12 Lemma. *Let F be a convenient vector space. For any* $\mathscr{Lip}^k$*-function* $f\colon \mathbb{R} \to \mathbb{R}$ *the induced map* $f^*\colon \mathscr{Lip}^k(\mathbb{R}, F) \to \mathscr{Lip}^k(\mathbb{R}, F)$ *is a morphism of convenient vector spaces.*

Proof. Since by (4.2.6) and (4.2.10) we know that $\mathscr{Lip}^k(\mathbb{R}, F)$ is convenient and since f^* is a linear map we can apply the preceding proposition (4.2.11) and hence have only to show that $\mathrm{ev}_t \circ f^*$ is a morphism. But this is trivial, since $\mathrm{ev}_t \circ f^* = \mathrm{ev}_{f(t)}$. □

4.2.13 Lemma. *For any linear morphism $m: E \to F$ of convenient vector spaces the induced map $m_*: \mathcal{L}ip^k(\mathbb{R}, E) \to \mathcal{L}ip^k(\mathbb{R}, F)$ is also a morphism of convenient vector spaces.*

Proof. One either uses (4.2.11) and the commutative diagram

$$\begin{array}{ccc} \mathcal{L}ip^k(\mathbb{R}, E) & \xrightarrow{m_*} & \mathcal{L}ip^k(\mathbb{R}, F) \\ \downarrow{\scriptstyle \mathrm{ev}_t} & & \downarrow{\scriptstyle \mathrm{ev}_t} \\ E & \xrightarrow{m} & F \end{array}$$

or that the bornology of $\mathcal{L}ip^k(\mathbb{R}, F)$ is the initial one induced by the maps δ^i $(i<k+2)$ and the commuting diagram

$$\begin{array}{ccc} \mathcal{L}ip^k(\mathbb{R}, E) & \xrightarrow{m_*} & \mathcal{L}ip^k(\mathbb{R}, F) \\ \downarrow{\scriptstyle \delta^i} & & \downarrow{\scriptstyle \delta^i} \\ \ell^\infty(\mathbb{R}^{\langle i \rangle}, E) & \xrightarrow{m_*} & \ell^\infty(\mathbb{R}^{\langle i \rangle}, F). \end{array}$$

□

In order to study $\mathcal{L}ip^k$-functions in section 4.3 we often have to construct certain curves, and for that purpose the following lemma will be quite useful. We start with a

4.2.14 Definition. A sequence $n \mapsto x_n$ in a convex bornological vector space E is called *fast falling* iff for every real polynomial p the set $\{p(n)x_n; n \in \mathbb{N}\}$ is bounded or equivalently iff $\{n^k x_n; n \in \mathbb{N}\}$ is bounded for every $k \in \mathbb{N}$.

4.2.15 Proposition. (*General Curve Lemma.*) *Let E be a preconvenient vector space, $n \mapsto c_n \in C^\infty(\mathbb{R}, E)$ a fast falling sequence of curves; $\varepsilon_n \geq 0$ with $\Sigma \varepsilon_n < \infty$. Then there exists a smooth curve $c: \mathbb{R} \to E$ and a convergent sequence of reals t_n such that $c(t+t_n) = c_n(t)$ for $|t| \leq \varepsilon_n$ and $c(\lim t_n) = 0$ (so c 'joins' all the pieces $c_n|_{[-\varepsilon_n, \varepsilon_n]}$ within a finite interval).*

Proof. Let

$$r_n := \sum_{k<n} \left(\frac{2}{k^2} + 2\varepsilon_k \right) \quad \text{and} \quad t_n := \frac{r_n + r_{n+1}}{2}.$$

Then $0 = r_1 < \ldots < r_n < r_{n+1} < \ldots < r_\infty := \lim r_n < \infty$ and $r_{n+1} - r_n = 2/n^2 + 2\varepsilon_n$. Using a fixed smooth function $h: \mathbb{R} \to [0, 1]$ with $h(t) = 1$ for $t \geq 0$ and $h(t) = 0$ for $t \leq -1$ one constructs smooth functions $h_n: \mathbb{R} \to [0, 1]$ with the properties $h_n(t) = 0$ for $|t| \geq 1/n^2 + \varepsilon_n$, $h_n(t) = 1$ for $|t| \leq \varepsilon_n$ and $|h_n^{(i)}(t)| \leq (n^2)^i H_i$ for all $t\in\mathbb{R}$ where $H_i:=\max\{|h^{(i)}(t)|;\ t\in\mathbb{R}\}$ (e.g. $h_n(t):=h(n^2(\varepsilon_n+t))\cdot h(n^2(\varepsilon_n-t))$). Now define $c(t):=\sum_{n\in\mathbb{N}}h_n(t-t_n)c_n(t-t_n)$. For every $t\in\mathbb{R}$ there is at most one non-zero summand; hence c is well-defined and $c(t)=c_n(t-t_n)$ for $-\varepsilon_n\leq t-t_n\leq\varepsilon_n$. It remains to show that $\ell\circ c$ is smooth for any $\ell\in E'$. We have $(\ell\circ c)(t)=\sum_{n\in\mathbb{N}}f_n(t)$ for $t\in\mathbb{R}$, where $f_n(t+t_n):=h_n(t)\cdot(\ell\circ c_n)(t)$. Since each f_n is smooth the assertion follows if we show that for any $k\in\mathbb{N}$ the series $\Sigma f_n^{(k)}(t)$ converges uniformly in t. So we estimate:

$$\begin{aligned}\sup\{|f_n^{(k)}(t)|;t\in\mathbb{R}\}&=\sup\left\{|f_n^{(k)}(s+t_n)|;\ |s|\leq\frac{1}{n^2}+\varepsilon_n\right\}\\ &\leq\sum_{i=0}^{k}\binom{k}{i}n^{2i}H_i\cdot\sup\left\{|(\ell\circ c_n)^{(k-i)}(s)|;|s|\leq\frac{1}{n^2}+\varepsilon_n\right\},\end{aligned}$$

hence

$$\begin{aligned}n^2\cdot\sup\{|f_n^{(k)}(t)|;t\in\mathbb{R}\}\leq\left(\sum_{i=1}^{k}\binom{k}{i}n^{2i+2}H_i\right)\cdot\sup\{&|(\ell\circ c_n)^{(j)}(s)|;\\ &|s|\leq 1+\max\{\varepsilon_n;n\in\mathbb{N}\}\quad\text{and}\quad 0\leq j\leq k\}.\end{aligned}$$

The hypothesis that c_n is fast falling implies that the right side of the above inequality is bounded with respect to $n\in\mathbb{N}$. This shows that $\Sigma f_n^{(k)}(t)$ converges uniformly in t. □

If the curves are polynomials of bounded degree, the application of the general curve lemma is simplified to a large extent by the following

4.2.16 Proposition. *Let $c_n\colon\mathbb{R}\to E$ be polynomials of bounded degree. If for every $\ell\in E'$ the sequence $n\mapsto\sup\{|(\ell\circ c_n)(t)|;\ |t|\leq 1\}$ is fast falling in $\mathbb{R}$ then the sequence c_n is fast falling in $C^\infty(\mathbb{R},E)$.*

Proof. The bornology of $C^\infty(\mathbb{R},E)$ is the initial one induced by the maps $\ell_*\colon C^\infty(\mathbb{R},E)\to C^\infty(\mathbb{R},\mathbb{R})$, where ℓ varies in E'. Thus we only have to show the result for $E=\mathbb{R}$. Let d be an upper bound of the degrees and let Poly^d denote the $(d+1)$-dimensional vector space of polynomials of degree at most d. Since $c\mapsto\sup\{|c(t)|;\ |t|\leq 1\}$ is a norm on Poly^d, it defines a vector bornology that is separated. Thus the inclusion $\mathrm{Poly}^d\to C^\infty(\mathbb{R},\mathbb{R})$ is bornological. By assumption c_n is fast falling in Poly^d; hence also in $C^\infty(\mathbb{R},\mathbb{R})$. □

Explicit constants for the boundedness of the inclusion $\mathrm{Poly}^d\to C^\infty(\mathbb{R},\mathbb{R})$ are given by [Rivlin, 1974, p. 93, p. 119]: For $k\in\mathbb{N}$, $r>0$, and polynomial c of degree at most d one has

$$\max\{|c^{(k)}(t)|;|t|\leq r\}\leq\max\{|T_d^{(k)}(t)|;|t|\leq r\}\cdot\max\{|c(t)|;|t|\leq 1\},$$

where T_d denotes the Tschebyschef polynomial of degree d. The maximum on the left side is a typical seminorm on $C^\infty(\mathbb{R},\mathbb{R})$.

4.3 Differentiable maps

Now we take up the discussion of general $\mathcal{Lip}^k$-maps $f\colon E\supseteq U\to F$. For this notion we refer to the beginning of section 4.1.

4.3.1 Lemma. *Let $f\colon E\supseteq U\to F$ be a map; and $0\le k\le j\le\infty$. Then the following statements are equivalent:*

(1) *$f\circ c\colon \mathbb{R}\to F$ is $\mathcal{Lip}^k$ for all $\mathcal{Lip}^j$-curves $c\colon \mathbb{R}\to U$;*
(2) *$f\circ c\colon c^{-1}(U)\to F$ is $\mathcal{Lip}^k$ for all $\mathcal{Lip}^j$-curves $c\colon \mathbb{R}\to E$.*

Proof. ($1\Leftarrow 2$) is obvious.

($1\Rightarrow 2$) By composing with $\ell\in F'$ one reduces the considerations to $F=\mathbb{R}$. Let $c\colon \mathbb{R}\to E$ be a $\mathcal{Lip}^j$-curve and $t_0\in c^{-1}(U)$. Since U is open in the Mackey closure topology which is the final one induced by the $\mathcal{Lip}^j$-curves there exists a $\delta>0$ such that $[t_0-2\delta,\ t_0+2\delta]\subseteq c^{-1}(U)$. Choose a smooth $h\colon \mathbb{R}\to t_0+[-2\delta,2\delta]$ with $h(t)=t$ for $|t-t_0|\le\delta$. Then $c_1:=c\circ h$ is $\mathcal{Lip}^j$ and has its image in $c(t_0+[-2\delta,2\delta])\subseteq c(c^{-1}(U))\subseteq U$, hence $f\circ c_1$ is k-times Lipschitz differentiable and equals $f\circ c$ on $t_0+[-\delta,\delta]$ showing that $f\circ c$ is k-times Lipschitz differentiable around t_0. Since $t_0\in c^{-1}(U)$ was arbitrary one concludes that $f\circ c$ is k-times Lipschitz differentiable. □

4.3.2 Proposition. ($\mathcal{Lip}^k$*-ness is a local concept*). *Let $f\colon E\supseteq U\to F$ be a map.*

(i) *If $f\colon U\to F$ is $\mathcal{Lip}^k$ and $U_1\subseteq U$ open then $f|_{U_1}\colon U_1\to F$ is $\mathcal{Lip}^k$.*
(ii) *If $U=\bigcup_{i\in I}U_i$ is an open covering and $f|_{U_i}$ is $\mathcal{Lip}^k$ for all $i\in I$ then $f\colon U\to E$ is $\mathcal{Lip}^k$.*

Proof. (i) is trivial since the inclusion $U_1\to U$ is a $\mathcal{Lip}^k$-morphism.

(ii) By composing with $\ell\in F'$ we may again assume $F=\mathbb{R}$. Let $c\colon \mathbb{R}\to U$ be a $\mathcal{Lip}^k$-curve. Then by assumption and the previous lemma (4.3.1) $(f\circ c)|_{c^{-1}(U_i)}=f|_{U_i}\circ c\colon c^{-1}(U_i)\to\mathbb{R}$ is k-times Lipschitz differentiable. Since the $c^{-1}(U_i)$ cover $c^{-1}(U)$, $f\circ c\colon c^{-1}(U)\to F$ is k-times Lipschitz differentiable and therefore f is $\mathcal{Lip}^k$ by the previous lemma (4.3.1). □

Of central importance is the following

4.3.3 Proposition. (*Smooth curves suffice.*) *Let $f\colon E\supseteq U\to F$ be a map and $k\in\mathbb{N}_{0,\infty}$. If $f\circ c$ is $\mathcal{Lip}^k$ for all smooth curves c then the same is true for all $\mathcal{Lip}^k$-curves c.*

Remark. Lemma (4.3.1) shows that it does not matter whether one admits curves $c\colon \mathbb{R}\to E$ or only curves $c\colon \mathbb{R}\to U$.

Proof. To simplify notation we prove this theorem for $k-1$ with $0\le k-1\le\infty$. By composing with $\ell\in F'$ the general situation can be immediately reduced to that with $F=\mathbb{R}$, which we shall consider now. Let us assume indirectly that there exists a $\mathscr{Lip}^{k-1}$-curve e such that $f\circ e\notin\mathscr{Lip}^{k-1}$. Because of lemma (4.3.1) we can assume that e has values in U. According to (1.3.22) there exists a $t_0\in\mathbb{R}$ such that the difference quotient $\delta^k(f\circ e)$ is unbounded on every neighborhood of t_0. By means of translations we reduce the consideration to the case $t_0=0$ and $e(t_0)=0$. Choose

$$a^n=(a_0^n,\ldots,a_k^n)\in\left[-\frac{1}{4^n},\frac{1}{4^n}\right]^{\langle k\rangle}$$

such that $|\delta^k(f\circ e)(a^n)|\geq n2^{kn}$ and define e_n to be the interpolation polynomial with $e_n(a_0^n):=e(a_0^n),\ldots,e_n(a_k^n):=e(a_k^n)$. Explicitly one has by (i) in (1.3.7):

$$e_n(t):=e(a_0^n)+\frac{1}{1!}(t-a_0^n)\delta^1 e(a_0^n,a_1^n)+\cdots$$
$$+\frac{1}{k!}(t-a_0^n)\cdot\ldots\cdot(t-a_{k-1}^n)\delta^k e(a_0^n,\ldots,a_k^n).$$

Let $c_n(t):=e_n(t/2^n)$. Thus, with $\alpha^n:=2^n\cdot a^n$, one has

$$|\delta^k(f\circ c_n)(\alpha^n)|=\frac{1}{2^{kn}}|\delta^k(f\circ e)(a^n)|\geq\frac{n2^{kn}}{2^{kn}}=n,$$

and we remark that $|\alpha_i^n|=2^n|a_i^n|\leq 1/2^n$. Let us verify that $\{c_n;\,n\in\mathbb{N}\}$ is fast falling in $C^\infty(\mathbb{R},E)$, using (4.2.16). So let $\ell\in E'$ and $|t|\leq 1$. Since e is a $\mathscr{Lip}^{k-1}$-curve and $|a_i^n|\leq 1$ for all n and i there exists a $K>0$ such that $|\ell(\delta^i e(a^n))|\leq K$ for all $n\in\mathbb{N}$ and $0\leq i\leq k$. Using that $|t-\alpha_i^n|\leq 2$ and that $e(a_0^n)=a_0^n\cdot\delta^1 e(a_0^n,0)$ we obtain:

$$(\ell\circ c_n)(t)=(\ell\circ e)(a_0^n)+\ldots$$
$$+\frac{1}{k!}(2^{-n}t-a_0^n)\cdot\ldots\cdot(2^{-n}t-a_{k-1}^n)\delta^k(\ell\circ e)(a_0^n,\ldots,a_k^n);$$
$$|(\ell\circ c_n)(t)|\leq|a_0^n|K+\frac{1}{1!2^n}|t-\alpha_0^n|K+\ldots+\frac{1}{k!2^{kn}}|t-\alpha_0^n|\cdot\ldots\cdot|t-\alpha_{k-1}^n|K$$
$$\leq K\left[\frac{1}{4^n}+\frac{2}{2^n}+\ldots+\frac{2^k}{k!2^{kn}}\right]$$

and this is fast falling in $\mathbb{R}$ for $n\to\infty$.

Applying now the general curve lemma (4.2.15) with $\varepsilon_n:=1/2^n$ we get a smooth curve c with $c(t+t_n)=c_n(t)$ for $|t|\leq\varepsilon_n$. Hence $|\delta^k(f\circ c)(a_0^n+t_n,\ldots,a_k^n+t_n)|\geq n$ which shows that $\delta^k(f\circ c)(V^{\langle k\rangle})$ is unbounded for any neighborhood V of $t_\infty:=\lim_{n\to\infty}t_n$. Since $c(t_n)=c_n(0)=e_n(0)=e(a_0^n)\to e(0)$ we have that $f\circ c$ is not k-times Lipschitz differentiable in a neighborhood of t_∞ in $c^{-1}(U)$. □

4.3.4 Corollary. *Let $f\colon E\supseteq U\to F$ be a map and $0\leq j\leq k\leq\infty$. If f is a $\mathscr{Lip}^k$-map then it is a $\mathscr{Lip}^j$-map.*

4.3.5 Theorem. *Let E, F be convenient vector spaces; $k \in \mathbb{N}_{0,\infty}$; X be a $\mathcal{Lip}^k$-space; and $f: X \times E \to F$ be linear in the second factor E. Then the following statements are equivalent:*

(1) *$f: X \sqcap E \to F$ is $\mathcal{Lip}^k$, where $X \sqcap E$ denotes the product of X with E considered as $\mathcal{Lip}^k$-space;*

(2) *f is partially $\mathcal{Lip}^k$, i.e. $f(x, \_)$ and $f(\_, y)$ are $\mathcal{Lip}^k$ for all $x \in X$ and $y \in E$;*

(3) *for all $x \in X$ one has $f^\vee(x) = f(x, \_) \in L(E, F)$ and $f^\vee: X \to L(E, F)$ is $\mathcal{Lip}^k$.*

Proof. $(1 \Rightarrow 2)$ is obvious, since thc partial maps are obtained by composing f with the smooth maps $x \mapsto (x, y)$ and $y \mapsto (x, y)$.

$(2 \Rightarrow 3)$ $f^\vee(x) \in L(E, F)$ since it is linear and $\mathcal{Lip}^k$. The map $f^\vee$ is $\mathcal{Lip}^k$, since by (3.6.5) it is enough to verify that all composites $\mathrm{ev}_y \circ f^\vee$ are $\mathcal{Lip}^k$, which holds by assumption because $\mathrm{ev}_y \circ f^\vee = f(\_, y)$.

$(3 \Rightarrow 1)$ Since $f^\vee$ is $\mathcal{Lip}^k$ so is $f^\vee \sqcap \mathrm{id}: X \sqcap E \to L(E, F) \sqcap E$. Furthermore $\mathrm{ev}: L(E, F) \sqcap E \to F$ is bilinear and $\mathcal{Lip}^k$ by (3.7.1), hence $f = \mathrm{ev} \circ (f^\vee \sqcap \mathrm{id})$ is $\mathcal{Lip}^k$. □

In order to obtain a recursive characterization of $\mathcal{Lip}^k$-maps we begin with $k=0$ and for this we need the following

4.3.6 Definition. (i) Let E be a convenient vector space. A subset $K \subseteq E$ is called *bornologically compact* or shortly *b-compact* iff there exists a bounded absolutely convex $B \subseteq E$ such that K is compact in the normed space E_B, cf. (2.1.15).

For $U \subseteq E$ the family of b-compact sets contained in U is a basis for a bornology. It will be called the *b-compact* bornology of U. (It is induced by the b-compact bornology of E iff U is M-closed in E.) The b-compact bornology of E is a convex vector bornology since the closed convex hull of a compact subset of a Banach space is compact.

Let $f: E \supseteq U \to F$ be a map.

(ii) f is called $\mathcal{Lip}^{-1}$ iff $f(K)$ is bounded for every b-compact set $K \subseteq U$ or, equivalently, iff f is bounded on M-convergent sequences in U having also their limit point in U.

(iii) The *directional difference quotient* ϑf: $E^2 \sqcap \mathbb{R}^2 \supseteq U_\vartheta \to F$ of f is defined by $\vartheta f(x, y; t, s) := (f(x+ty) - f(x+sy))/(t-s)$ where $U_\vartheta := \{(x, y; t, s) \in E^2 \sqcap \mathbb{R}^2;\ x+ty \in U,\ x+sy \in U,\ t \neq s\}$.

Remark. The assumption that U is M-open in E implies that U_ϑ is M-open in $E^2 \sqcap \mathbb{R}^2$.

4.3.7 Lemma. (*Characterization of $\mathcal{Lip}^{-1}$-maps.*) *Let $f: E \supseteq U \to F$ be a map and $k \in \mathbb{N}_{0,\infty}$. Then the following statements are equivalent:*

(1) *f is a $\mathcal{Lip}^{-1}$-map;*

(2_k) *$f \circ c$ is an ℓ^∞-map for every $\mathcal{Lip}^k$-curve $c: \mathbb{R} \to U \subseteq E$;*

(3) *$f \circ x$ is an ℓ^∞-map for every sequence $x: \mathbb{N} \to U$ which is M-convergent to some $x_\infty \in U$.*

Proof. ($1\Rightarrow 2_0$) is trivial, since $c(K)$ is b-compact in U for every compact $K\subseteq\mathbb{R}$ and any $\mathcal{Lip}^0$-curve c.

($2_0\Rightarrow 2_k\Rightarrow 2_\infty$). Trivial since less and less composites have to be ℓ^∞-curves.

($2_\infty\Rightarrow 3$) Suppose $f(x_n)$ is unbounded for some sequence (x_n) converging Mackey to $x_\infty\in U$. Using the special curve lemma (2.3.4) we conclude that there is a smooth curve c passing in finite time through infinitely many x_n. Thus we have a contradiction to the assumption that $f\circ c$ is an ℓ^∞-map.

($3\Rightarrow 1$) Let $K\subseteq U$ be b-compact, i.e. compact in some Banach space E_B. Suppose $f(K)$ is unbounded. Then there exists an $x\colon\mathbb{N}\to K$ with $(f\circ x)(\mathbb{N})$ unbounded. Since K is compact in the metrizable space E_B we may assume that x converges in E_B to some $x_\infty\in K\subseteq U$. Thus x is M-convergent to x_∞, which gives a contradiction to (3). □

4.3.8 Theorem. (*Characterization of $\mathcal{Lip}^0$-maps.*) *Let E and F be convenient vector spaces, $U\subseteq E$ be* M*-open and $f\colon U\to F$ be a map. Then the following statements are equivalent:*

(1) *f is $\mathcal{Lip}^0$;*

(2) *$\ell\circ f|_{E_B}\colon E_B\to\mathbb{R}$ is locally Lipschitzian for every absolutely convex bounded $B\subseteq E$ and every $\ell\in F'$;*

(3) $\left\{\dfrac{f(x)-f(y)}{\|x-y\|_B};\ x,y\in K,\ x\neq y\right\}$; *is bounded for all compact $K\subseteq U\cap E_B$ with $B\subseteq E$ bounded and absolutely convex;*

(4) *$\{r(f(x)-f(y));\ r\in\mathbb{R},\ x,y\in K,\ r(x-y)\in B\}$ is bounded for all bounded $B\subseteq E$ and b-compact $K\subseteq U$;*

(5) *ϑf is bounded on $K_1\times K_2\times[-1,1]^{\langle 1\rangle}$ for all b-compact K_1,K_2 with $K_1+[-1,1]K_2\subseteq U$; for ϑf cf.* (4.3.6).

Proof. ($1\Rightarrow 2$) Obviously $\ell\circ f|_{E_B}\colon E_B\to E\to F\to\mathbb{R}$ is $\mathcal{Lip}^0$, hence by the characterization of $\mathcal{Lip}^0$-maps between normed spaces (1.4.2) one concludes that it is locally Lipschitzian.

($2\Rightarrow 3$) By composing with $\ell\in F'$ we may assume that $F=\mathbb{R}$. Let $K\subseteq U\cap E_B$ be compact. Now (3) follows by applying (1.4.2) to the map $f|_{E_B}$.

($3\Rightarrow 4$) By enlarging the bounded absolutely convex set B if necessary we may assume that K is compact in E_B. For $r(x-y)\in B$ one has $|r|\cdot\|x-y\|_B=\|r(x-y)\|_B\le 1$, i.e. $|r|\ge\|x-y\|_B$. Consequently $\{r(f(x)-f(y));\ x,\ y\in K,\ r(x-y)\in B\}$ is a subset of the absolutely convex hull of the bounded set

$$\left\{\frac{f(x)-f(y)}{\|x-y\|_B};\ x,y\in K,\ x\neq y\right\}.$$

($4\Rightarrow 5$) Let K_1, K_2 be b-compact with $K_1+[-1,\ 1]K_2\subseteq U$. By setting $x:=a+tv$, $y:=a+sv$, $r:=1/(t-s)$, one verifies that

$$\left\{\frac{f(a+tv)-f(a+sv)}{t-s};\ t,s\in[-1,1],\ t\neq s,\ a\in K_1,\ v\in K_2\right\}$$

$$\subseteq\{r(f(x)-f(y));\ r(x-y)\in K_2,\ x,y\in K\},$$

where $K:=K_1+[-1,1]\cdot K_2$. The assertion now follows since with K_1 and K_2 also K is b-compact.

(5⇒1) Let $c\colon \mathbb{R}\to U$ be smooth. We have to show that $f\circ c$ is $\mathscr{Lip}^0$ in a neighborhood of any $t_0\in\mathbb{R}$. Without loss of generality let $t_0=0$. Since the extension $h:=\bar{\delta}c\colon \mathbb{R}^2\to E$ of the difference quotient is also smooth, cf. (2) in (4.1.13), $K_2:=\bar{\delta}c([-1,1]^2)$ and $K_1:=c([-1,1])$ are b-compact.

Now

$$\left\{\frac{f(c(t))-f(c(s))}{t-s}=\frac{f(c(s)+(t-s)h(t,s))-f(c(s))}{t-s};\ t,s\in[-1,1],\ t\neq s\right\}$$

is a subset of the bounded set

$$\left\{\frac{f(x+t'v)-f(x+s'v)}{t'-s'};\ t',s'\in[-1,1],\ t'\neq s',\ x\in K_1,\ v\in K_2\right\},$$

as verified by setting $x:=c(s)$, $v:=h(t,s)$, $t':=t-s$ and $s':=0$. □

4.3.9 Definition. Let $f\colon E\supseteq U\to F$ be a map and $\mathscr{S}\subseteq F'$ be point separating.

(i) f is called *$\mathscr{S}$-differentiable at x in direction v* iff the curve $t\mapsto f(x+tv)$ is $\mathscr{S}$-differentiable at 0. The derivative of this curve at 0 is called the *differential of f at x in direction v* and is denoted by $df(x,v)$; cf. (i) in (4.1.9).

f is called *$\mathscr{S}$-differentiable at x* iff f is $\mathscr{S}$-differentiable at x in direction v for all $v\in E$.

f is called *$\mathscr{S}$-differentiable* iff f is $\mathscr{S}$-differentiable at x for all $x\in U$. The map $df\colon U\times E\to F$ is then called the *differential* of f.

For curves this definition of $\mathscr{S}$-differentiability is equivalent with that given in (4.1.9).

(ii) f is called *strongly differentiable* iff for all b-compact $K_1\subseteq U$ and $K_2\subseteq E$ the limit

$$f'(x)(v):=\text{M-lim}_{t,s\to 0,t\neq s}\frac{f(x+tv)-f(x+sv)}{t-s}$$

exists uniformly for $x\in K_1$ and $v\in K_2$; and $f'(x)\in L(E,F)$ for all $x\in U$ (The difference quotient makes sense since there exists an $\varepsilon>0$ with $K_1+[-\varepsilon,\varepsilon]\cdot K_2\subseteq U$.) The map $f'\colon U\to L(E,F)$ will be called the *derivative* of f.

For curves this definition of strong differentiability is equivalent with that given in (4.1.9). The derivative $c^{\cdot}$ of c defined in (4.1.9) corresponds to the derivative c' as defined now via the canonical isomorphism $E\cong L(\mathbb{R},E)$, i.e. $c'(t)(s)=s\cdot c^{\cdot}(t)$.

The contrast between strong differentiability and $\mathscr{S}$-differentiability consists not only in a stronger form of convergence but also in the assumption that the derivative at each point is a linear morphism. We shall show, however, in (4.3.12) that combined with the condition that the differentials, respectively the derivatives, are $\mathcal{Lip}^k$-maps they both become equivalent.

We first give a generalization of (4.1.18).

4.3.10 Proposition. *Let E and F be convenient vector spaces, $\mathcal{S} \subseteq F'$ a subset such that the bornology of F has a basis of $\sigma(F, \mathcal{S})$-closed sets. Let $f: E \supseteq U \to F$ be a $\mathcal{S}$-differentiable map. Then f is $\mathcal{Lip}^0$ iff* d*f is $\mathcal{Lip}^{-1}$.*

Proof. ($\Leftarrow$) By (4.3.8) we have to show that

$$\vartheta f(x, v; t, s) := \frac{f(x+tv) - f(x+sv)}{t-s}$$

is bounded for x, v varying in b-compact subsets K_1, K_2 satisfying $K_1 + [-1, 1]K_2 \subseteq U$ and $t, s \in [-1, 1]$ with $t \neq s$. For $\ell \in \mathcal{S}$ we have by the classical mean value theorem that $\ell(\vartheta f(x, v; t, s)) = \ell(\mathrm{d}f(x+rv, v))$ for some $r \in [-1, 1]$. Thus $\vartheta f(x, v; t, s) \in \bigcap_{\ell \in \mathcal{S}} \ell^{-1} \ell(B)$; where B is the bounded image under df of $(K_1 + [-1, 1]K_2) \times K_2$ (Use that a $\mathcal{Lip}^{-1}$-map is bounded on b-compact sets.) By the condition on the bornology we conclude that this intersection is bounded; cf. (4.1.16).

($\Rightarrow$) Assume df is not bounded on some M-converging sequence (x_n, v_n). Using homogeneity of d$f(x, _)$ one can assume that the v_n converge Mackey to 0. And by passing to a subsequence we may assume that the polynomials $t \mapsto x_n + tv_n$ are fast falling. Hence by the general curve lemma (4.2.15) there exist a smooth curve c and a bounded sequence of reals t_n such that for every n one has $c(t+t_n) = x_n + tv_n$ for small t. Hence d$f(x_n, v_n) = (f \circ c)^{\cdot}(t_n)$ is unbounded, which is a contradiction to the corresponding proposition (4.1.18) for curves. □

Let us now generalize the result (4.1.5) on integrals with respect to a parameter.

4.3.11 Proposition. *Let $f: E \sqcap \mathbb{R} \supseteq U \to F$ be a $\mathcal{Lip}^0$-map. Then the domain of definition $W := \{x \in E;\ (x, t) \in U$ for all $t \in [0, 1]\}$ of the map $g: x \mapsto \int_0^1 f(x, t)\mathrm{d}t$ is M-open in E and $g: E \supseteq W \to F$ is $\mathcal{Lip}^0$.*

Proof. We first show that W is M-open in E. Let $c: \mathbb{R} \to E$ be a smooth curve with $c(0) \in W$, i.e. $\{c(0)\} \times [0, 1] \subseteq U$. Since $(s, t) \mapsto (c(s), t)$ is smooth there is a neighborhood W_0 of $\{0\} \times [0, 1]$ in $\mathbb{R}^2$ with $(c(s), t) \in U$ for all $(s, t) \in W_0$. Thus there exists an $\varepsilon > 0$ with $[-\varepsilon, \varepsilon] \times [0, 1] \subseteq W_0$, i.e. $c(s) \in W$ for $|s| \leq \varepsilon$.

Now let $c: \mathbb{R} \to W$ be a smooth curve. Then $f_0: \mathbb{R}^2 \supseteq U_0 := \{(s, t);\ (c(s), t) \in U\} \to F$ defined by $f_0(s, t) := f(c(s), t)$ is a $\mathcal{Lip}^0$-map; hence by (4.1.5) the map $g_0: \mathbb{R} \to F$ defined by $g_0(s) := \int_0^1 f_0\,(s, t)\mathrm{d}t = g(c(s))$ is $\mathcal{Lip}^0$, i.e. g is $\mathcal{Lip}^0$. □

4.3.12 Proposition. *Let $f: E \supseteq U \to F$ be a map and $\mathcal{S} \subseteq F'$ be point separating. Then the following statements are equivalent:*

(1) *f is $\mathcal{S}$-differentiable and* d*f $\mathcal{Lip}^0$;*
(2) *f is strongly differentiable and f' is $\mathcal{Lip}^0$;*

(3) *$\vartheta f: U_\vartheta \to F$ has a $\mathcal{L}ip^0$-extension $\bar\vartheta f: U_{\bar\vartheta} \to F$, where $U_{\bar\vartheta} := \{(x, y;\ t, s) \in E^2 \sqcap \mathbb{R}^2;\ x + tv \in U,\ x + sv \in U\}$; cf. (iii) in (4.3.6).*

Proof. (1⇒3) Define

$$\bar\vartheta f(x, y; t, s) := \begin{cases} \vartheta f(x, y; t, s) & \text{for } t \neq s \\ \mathrm{d}f(x + ty, y) & \text{for } t = s. \end{cases}$$

Then $\bar\vartheta f|_{U_\vartheta} = \vartheta f$ is $\mathcal{L}ip^0$, since f is $\mathcal{L}ip^0$. Furthermore $\tilde U := \{(x, y; t, s);\ x + ry \in U$ for all r in the interval from t to $s\}$ is M-open and $(x, y; t, s) \mapsto \bar\vartheta f(x, y; t, s) = \int_0^1 \mathrm{d}f(x + ty + r(s-t)y, y)\mathrm{d}r$ is $\mathcal{L}ip^0$ on $\tilde U$ by (4.3.11). Some $U_{\bar\vartheta} = U_\vartheta \cup \tilde U$, the map $\bar\vartheta f$ is $\mathcal{L}ip^0$ on $U_{\bar\vartheta}$ by (ii) of (4.3.2).

(3⇒2) We show first that the directional difference quotients are uniformly M-convergent to the corresponding differential. So let $K_1 \subseteq U$, $K_2 \subseteq E$ be b-compact. Choose an $\varepsilon > 0$ such that $K_1 + [-\varepsilon, \varepsilon] K_2 \subseteq U$. It is enough to prove that

$$B := \left\{ \frac{1}{|t| + |s|} \left(\frac{f(x + tv) - f(x + sv)}{t - s} - \mathrm{d}f(x, v) \right);\ x \in K_1,\ v \in K_2,\ t, s \in [-\varepsilon, \varepsilon],\ t \neq s \right\}$$

is bounded, where $\mathrm{d}f(x, v)$ is defined as $\bar\vartheta f(x, v; 0, 0)$.

By assumption $\bar\vartheta f: U_{\bar\vartheta} \cap (E_{B_1} \times E_{B_2} \times \mathbb{R}^2) \to F$ is locally Lipschitzian, where B_1 (resp. B_2) is an absolutely convex bounded subset of E in which K_1 (resp. K_2) is compact. Thus, with $\|(t, s)\|_1 = |t| + |s|$:

$$\left\{ (\bar\vartheta f(x, y; t, s) - \bar\vartheta f(x, y; 0, 0)) \frac{1}{\|(t, s)\|_1};\ (x, y, t, s) \in K_1 \times K_2 \times [-\varepsilon, \varepsilon]^2,\ t \neq s \right\}$$

is bounded. Since

$$(\bar\vartheta f(x, y; t, s) - \bar\vartheta f(x, y; 0, 0)) \frac{1}{\|(t, s)\|_1} = \left(\frac{f(x + ty) - f(x + sy)}{t - s} - \mathrm{d}f(x, y) \right) \frac{1}{|t| + |s|}$$

the claim is proved.

Let us show next that $\mathrm{d}f(x, \_)$ is linear. By composing f with $\ell \in F'$ we may assume that $F = \mathbb{R}$. Obviously $\mathrm{d}f(x, \_)$ is homogeneous (consider for $\lambda \in \mathbb{R}$ the smooth curve $t \mapsto x + t\lambda v$). To prove the additivity consider

$$\frac{f(x + tv + tw) - f(x + tv)}{t} = \frac{f(x + tv + tw) - f(x)}{t} - \frac{f(x + tv) - f(x)}{t}$$

which converges to $\mathrm{d}f(x, v + w) - \mathrm{d}f(x, v)$ for $t \to 0$.

By property (3) we know that $\bar\vartheta f$ is $\mathcal{L}ip^0$. Thus

$$\frac{f(x + tv + tw) - f(x + tv)}{t} = \bar\vartheta f(x + tv, w; t, 0) \to \bar\vartheta f(x, w; 0, 0) = \mathrm{d}f(x, w).$$

Hence $\mathrm{d}f(x, w) = \mathrm{d}f(x, v + w) - \mathrm{d}f(x, v)$.

Since $\mathrm{d}f(x, \_)$ is $\mathcal{L}ip^0$ and linear it is in $L(E, F)$. Thus we have shown that f is strongly differentiable and that, by (4.3.5) for $k = 0$, the derivative f' is $\mathcal{L}ip^0$.

(1⇐2) Strongly differentiable implies $\mathscr{S}$-differentiable with the differential $\mathrm{d}f$ given by $\mathrm{d}f(x, v) = f'(x)(v)$. Using (4.3.5) for $k=0$ it follows that $\mathrm{d}f$ is $\mathscr{Lip}^0$. □

4.3.13 Definition. A map $f: E \supseteq U \to F$ is called *$\mathscr{Lip}$-differentiable* iff it satisfies the equivalent conditions of (4.3.12).

This definition is consistent with (ii) of (1.3.19).

4.3.14 Lemma. (*Special Chain Rule.*) *Let $f: E \supseteq U \to F$ be a $\mathscr{Lip}$-differentiable map; $c: \mathbb{R} \to U$ a $\mathscr{Lip}^1$-curve. Then $f \circ c$ is $\mathscr{Lip}^1$ and $(f \circ c)^{\cdot}(s) = \mathrm{d}f(c(s), c^{\cdot}(s))$ for all $s \in \mathbb{R}$.*

Proof. Since $\mathrm{d}f \circ (c, c^{\cdot})$ is $\mathscr{Lip}^0$ we only have to show that $\ell \circ f \circ c$ is differentiable with derivative $\ell \circ \mathrm{d}f \circ (c, c^{\cdot}) = \mathrm{d}(\ell \circ f) \circ (c, c^{\cdot})$ for all $\ell \in F'$. This is only a statement about $\ell \circ f$, so we may assume that $F = \mathbb{R}$.

$$\frac{f(c(t+s)) - f(c(s))}{t} = \frac{1}{t} \cdot \left(f\left(c(s) + \frac{c(t+s) - c(s)}{t} \cdot t \right) - f(c(s)) \right)$$

converges to $\mathrm{d}f(c(s), c^{\cdot}(s))$ as $t \to 0$ since on one hand $(1/t) \cdot (f(x + \bar{\delta}c(r) \cdot t) - f(x)) \to \mathrm{d}f(x, \bar{\delta}c(r))$ as $t \to 0$ uniformly for r in any compact interval (f is assumed to be strongly differentiable), where $x := c(s)$ and $\bar{\delta}c$ is the $\mathscr{Lip}^0$-extension of $r \mapsto (c(s+r) - c(s))/r$; and on the other hand $\mathrm{d}f(x, \bar{\delta}c(r)) \to \mathrm{d}f(x, c^{\cdot}(s))$ for $r \to 0$. □

In order to compare $\mathscr{Lip}$-differentiability of a map between Banach spaces with the standard concept of Fréchet-differentiability we recall the following classical

4.3.15 Definition. A map $f: E \supseteq U \to F$ between Banach spaces is called *Fréchet differentiable* (resp. *strictly Fréchet differentiable*) iff it is continuous and for every $x \in U$ there exists a (necessarily unique) linear map $m_x: E \to F$ such that

$$\lim_{v \to 0} \frac{\| f(x+v) - f(x) - m_x(v) \|}{\| v \|} = 0$$

$$\left(\text{resp. } \lim_{v, w \to 0, v \neq w} \frac{\| f(x+v) - f(x+w) - m_x(v-w) \|}{\| v - w \|} = 0 \right).$$

The continuity of f implies that m_x is continuous. Any Fréchet differentiable map is obviously weakly differentiable and $m_x(v) = \mathrm{d}f(x, v)$. In classical calculus the derivative f' is defined by $f'(x) := m_x$.

4.3.16 Proposition. *Let $f: E \supseteq U \to F$ be a map between Banach spaces. Then the following statements are equivalent:*

(1) *f is $\mathscr{Lip}$-differentiable;*
(2) *f is strictly Fréchet differentiable and f' is locally Lipschitzian;*
(3) *f is Fréchet differentiable and f' is locally Lipschitzian.*

Proof. (1⇒2) Since f is $\mathscr{Lip}$-differentiable there exists by (4.3.12) a $\mathscr{Lip}^0$-extension $\bar{\vartheta}f: U_{\bar{\vartheta}} \to F$ of the directional difference quotient ϑf. Thus for a given $p \in U$ there exist $\delta > 0$ and $N \in \mathbb{R}$ such that $\|\bar{\vartheta}f(x, y; t, s) - \bar{\vartheta}f(x', y'; t', s')\| \leq (\|x - x'\| + \|y - y'\| + |t - t'| + |s - s'|) \cdot N$ provided $\|x - p\| \leq \delta$, $\|x' - p\| \leq \delta$, $\|y\| \leq \delta$, $\|y'\| \leq \delta$, $|t| \leq \delta$, $|t'| \leq \delta$, $|s| \leq \delta$, $|s'| \leq \delta$. In particular for $t \neq 0$, $s = -t$, $x' = p$, $y' = y$, $t' = s' = 0$ one has

$$\left\| \frac{f(x+ty) - f(x-ty)}{2t} - \mathrm{d}f(p, y) \right\| \leq (\|x - p\| + 2|t|) \cdot N$$

provided $\|x - p\| \leq \delta$, $\|y\| \leq \delta$, $0 \neq |t| \leq \delta$. We have to show that

$$\frac{\|f(p+v) - f(p+w) - \mathrm{d}f(p, v-w)\|}{\|v - w\|} \to 0 \quad \text{for } v, w \to 0,\ v \neq w.$$

Let $\|v\|, \|w\| \leq \delta$ and $\|v - w\| \leq 2\delta^2$ and set

$$x := p + \frac{v+w}{2}, \quad y := \delta \frac{v - w}{\|v - w\|}, \quad t := \frac{\|v - w\|}{2\delta}.$$

Then

$$ty = \frac{v - w}{2}, \ \|x - p\| \leq \delta, \ \|y\| \leq \delta, \ 0 \neq |t| \leq \delta.$$

Thus we obtain

$$\frac{\|f(p+v) - f(p+w) - \mathrm{d}f(p, v-w)\|}{\|v - w\|} = \frac{\|f(x+ty) - f(x-ty) - \mathrm{d}f(p, 2ty)\|}{\|v - w\|}$$

$$\leq N \cdot 2|t| \cdot \frac{\|x - p\| + 2|t|}{\|v - w\|} = N \cdot \frac{1}{\delta} \cdot \left(\frac{\|v + w\|}{2} + \frac{\|v - w\|}{\delta} \right) \to 0 \text{ for } v, w \to 0.$$

(2⇒3) is trivial.

(3⇒2) Since f is Fréchet differentiable it is obviously weakly differentiable. Since f' is assumed to be locally Lipschitzian it is $\mathscr{Lip}^0$ by (1.4.2), and thus $\mathrm{d}f$ is $\mathscr{Lip}^0$ by (4.3.5). Therefore f is $\mathscr{Lip}$-differentiable. □

We start now the investigation of differentiability properties of $\mathscr{Lip}^k$-maps. For a $\mathscr{Lip}^{k+1}$-map f the existence of the differential $\mathrm{d}f$ is an immediate consequence of (4.1.12). We want to prove that $\mathrm{d}f$ is in fact $\mathscr{Lip}^k$ and shall deduce this easily from (4.3.23), which says that for any $\mathscr{Lip}^{k+1}$-map $g: \mathbb{R}^2 \to \mathbb{R}$ the function $\partial_2 g(\_, 0): \mathbb{R} \to \mathbb{R}$ is $\mathscr{Lip}^k$. This is a trivial consequence of Boman's characterization of $\mathscr{Lip}^k$-functions on $\mathbb{R}^2$ [Boman, 1967]. We will, however, prove (4.3.23) directly by means of some lemmas on functions of two variables. We thus not only obtain the desired generalization of Boman's result from finite-dimensional vector spaces to arbitrary convenient vector spaces, but also a new proof of Boman's original result.

4.3.17 Notation. Let $k \in \mathbb{N}$ be fixed. Then $r_0, \ldots, r_k$ denote the unique rational numbers satisfying the equations:

$$\begin{aligned} r_0+r_1+\ r_2+\cdots+\ r_k&=0\\ r_1+\ 2r_2+\cdots+kr_k&=1\\ r_1+2^2r_2+\cdots+k^2r_k&=0\\ &\ \vdots\\ r_1+2^kr_2+\cdots+k^kr_k&=0 \end{aligned}$$

(They exist since the determinant of Vandermonde is non-zero.)

4.3.18 Lemma. (*Approximation of a Derivative.*) *For any $f\in \mathscr{Lip}^k(\mathbb{R},\mathbb{R})$ and any $A>0$ there exists N such that $|f^{\cdot}(a)\cdot h-\sum_{i=0}^k r_i\cdot f(a+ih)|\le N\cdot|h|^{k+1}$ for all $a,h\in[-A,A]$.*

Proof. Consider $f_a:\mathbb{R}\to\mathbb{R}$ defined by $f_a(h):=f^{\cdot}(a)h-\sum_{i=0}^k r_i\cdot f(a+ih)$. According to the choice of the coefficients r_i one has $f_a(0)=f_a^{\cdot}(0)=\cdots=f_a^{(k)}(0)=0$.

Since $f^{(k)}$ is Lipschitzian on $[-(k+1)A,(k+1)A]$ there exists an N_1 such that $|f_a^{(k)}(h)|\le N_1\cdot|h|$ for all $a,h\in[-A,A]$ which implies successively

$$|f_a^{(k-1)}(h)|\le N_1\frac{|h|^2}{2},\ldots,|f_a(h)|\le N_1\frac{|h|^{k+1}}{(k+1)!}$$

for all $a,h\in[-A,A]$. Hence $N:=N_1/(k+1)!$ suffices. □

4.3.19 Lemma. *For $f\in\mathscr{Lip}^1(\mathbb{R}^2,\mathbb{R})$, $\partial_2 f(\_,0):\mathbb{R}\to\mathbb{R}$ is bornological.*

Proof. Suppose $\partial_2 f(\_,0)$ is unbounded in every neighborhood of some point a. Without loss of generality $a=0$. Then there exist

$$a_n\in\left[-\frac{1}{2^n},\frac{1}{2^n}\right]$$

with $|\partial_2 f(a_n,0)|\ge n2^n$. For the curve $c_n:\mathbb{R}\to\mathbb{R}^2$ defined by

$$c_n(t):=\left(a_n,t\frac{1}{2^n}\right)$$

one has

$$|(f\circ c_n)^{\cdot}(0)|=\frac{1}{2^n}|\partial_2 f(a_n,0)|\ge n.$$

Since the sequence (c_n) is fast falling we can apply the general curve lemma (4.2.15) with $\varepsilon_n=1/2^n$. It yields a smooth curve $c:\mathbb{R}\to\mathbb{R}^2$ for which $(f\circ c)^{\cdot}$ would be unbounded on a bounded interval which is a contradiction with $f\circ c\in\mathscr{Lip}^1$. □

4.3.20 Proposition. (*Approximation of a Partial Derivative.*) *Let $k\in\mathbb{N}$ and $r_0,\ldots,r_k$ as in (4.3.17). For any $f\in\mathscr{Lip}^k(\mathbb{R}^2,\mathbb{R})$ and any $A>0$ there exists an $N\in\mathbb{R}$ such that $|\partial_2 f(a,0)\cdot h-\sum_{i=0}^k r_i f(a,ih)|\le N\cdot|h|^{k+1}$ for all $a,h\in[-A,A]$.*

Proof. By lemma (4.3.19) there certainly exists for any $\delta \geq 0$ an N such that the above inequality is fulfilled for $|h| \geq \delta$. Proceed now indirectly. Then there exist $a_n \in [-A, A]$ and

$$h_n \in \left[-\frac{1}{2^n}, \frac{1}{2^n}\right]$$

with $|\partial_2 f(a_n, 0) \cdot h_n - \sum_{i=0}^{k} r_i f(a_n, ih_n)| \geq n2^{n(k+1)}|h_n|^{k+1}$, and by passing to a subsequence we may assume that the a_n converge to some point a, say $a = 0$, and that

$$a_n \in \left[-\frac{1}{2^n}, \frac{1}{2^n}\right].$$

Let $c_n\colon \mathbb{R} \to \mathbb{R}^2$ be the curve

$$t \mapsto \left(a_n, \frac{t}{2^n}\right).$$

Let $f_n := f \circ c_n$ and $v_n := 2^n h_n$. One obtains

$$|f_n^{\cdot}(0) \cdot v_n - \textstyle\sum_{i=0}^{k} r_i f_n(iv_n)| = \left|\frac{1}{2^n} \partial_2 f(a_n, 0) \cdot 2^n h_n - \sum_{i=0}^{k} r_i f(a_n, ih_n)\right| \geq$$

$$n2^{n(k+1)}|h_n|^{k+1} = n|v_n|^{k+} \quad .$$

Since the sequence formed by the curves c_n is fast falling we can apply once more the general curve lemma (4.2.15) with $\varepsilon_n := 1/2^n$. It yields a smooth curve $c\colon \mathbb{R} \to \mathbb{R}^2$ such that for $g := f \circ c$ one would have $|g^{\cdot}(t_n)v_n - \sum_{i=0}^{k} r_i g(t_n + iv_n)| \geq n|v_n|^{k+1}_{.}$ and this is a contradiction with lemma (4.3.18) since t_n and v_n are bounded. □

4.3.21 Lemma. *Let $g\colon \mathbb{R} \to \mathbb{R}$ and consider $G\colon \mathbb{R}^2 \to \mathbb{R}$ defined by $G(t, s) := g(t)s$. Then for any $a \in \mathbb{R}$ and $h \neq 0$ one has $(k+1)\delta_{\text{eq}}^{k} g(a; h) = \delta_{\text{eq}}^{k+1}(G \circ c)(a; h)$, where $c\colon \mathbb{R} \to \mathbb{R}^2$ is defined by $c(t) := (t-h, t-a)$.*

Proof.

$$\frac{1}{k+1}\delta_{\text{eq}}^{k+1}(G \circ c)(a; h)$$

$$= k! \textstyle\sum_{i=0}^{k+1} (G \circ c)(a + ih) \prod_{k+1 \geq r \neq i} (ih - rh)^{-1}$$

$$= k! \textstyle\sum_{i=1}^{k+1} g(a + (i-1)h)(ih - 0h) \prod_{k+1 \geq r \neq i} (ih - rh)^{-1}$$

$$= k! \textstyle\sum_{j=0}^{k} g(a + jh) \prod_{k \geq s \neq j} (jh - sh)^{-1} = \delta_{\text{eq}}^{k} g(a; h) \qquad (j := i-1,\ s := r-1).$$ □

4.3.22 Lemma. *For each $k \in \mathbb{N}_0$ there exists a constant C_k with the property: for any function $G\colon \mathbb{R}^2 \to \mathbb{R}$ with $|G(t, s)| \leq N|s|^{k+1}$ for $t, s \in [-(k+1), k+1]$ and any curve $c\colon \mathbb{R} \to \mathbb{R}^2$ of the form $c(t) = (t-h, t-a)$ with $a, h \in [-1, 1]$, $h \neq 0$ one has $|\delta_{\text{eq}}^{k+1}(G \circ c)(a; h)| \leq NC_k$.*

Proof. In fact,

$$|\delta_{\text{eq}}^{k+1}(G\circ c)(a;h)|\le(k+1)!\sum_{i=1}^{k+1}\frac{|G((i-1)h+a,ih)|}{|ih|^{k+1}}i^{k+1}\Pi_{r\neq i}|i-r|^{-1},$$

hence $C_k:=(k+1)!\sum_{i=1}^{k+1}i^{k+1}\Pi_{r\neq i}|i-r|^{-1}$ suffices. □

4.3.23 Lemma. *For $k\in\mathbb{N}$ and $G\in\mathcal{Lip}^k(\mathbb{R}^2,\mathbb{R})$ one has $\partial_2 G(\_,0)\in\mathcal{Lip}^{k-1}(\mathbb{R},\mathbb{R})$.*

Proof. Let $G_0(t,s):=\sum_{i=0}^k r_i\cdot G(t,is)$ with r_i defined as in (4.3.17); $g(t):=\partial_2 G(t,0)$; $G_1(t,s):=g(t)\cdot s$; $G_2(t,s):=G_1(t,s)-G_0(t,s)$. Then G_0 is $\mathcal{Lip}^k$ and $\partial_2 G_0(t,0)=\partial_2 G(t,0)$.

Suppose g is not $\mathcal{Lip}^{k-1}$. Then there exists an $a\in\mathbb{R}$ such that $\delta_{\text{eq}}^k g$ is unbounded in every neighborhood of $(a,0)$. Without loss of generality $a=0$. Choose a_n, h_n with $|a_n|\le 1/4^n$ and $0<h_n\le 1/4^n$ such that $|\delta_{\text{eq}}^k g(a_n;h_n)|>n\cdot 2^{n(k+1)}$.

Consider the curves $e_n:\mathbb{R}\to\mathbb{R}^2$ defined by $e_n(t):=(t-h_n,t-a_n)$ and $c_n:\mathbb{R}\to\mathbb{R}^2$ defined by $c_n(t):=e_n(t/2^n+a_n)$. Let $v_n:=2^n h_n$.

Since the curves c_n form a fast-falling sequence we can apply the general curve lemma (4.2.15) with $\varepsilon_n:=(k+1)/2^n$ to obtain a smooth curve $c:\mathbb{R}\to\mathbb{R}^2$ such that $c(t+t_n)=c_n(t)$ for $0\le t\le(k+1)/2^n$.

The set $\{|\delta_{\text{eq}}^{k+1}(G_0\circ c)(t_n;v_n)|;\ n\in\mathbb{N}\}$ is bounded since the t_n and v_n form bounded sequences.

By (4.3.20) there exists an N such that $|G_2(t,s)|\le N|s|^{k+1}$ for all $t,s\in[-(k+1),k+1]$. Using (4.3.22) one obtains

$$|\delta_{\text{eq}}^{k+1}(G_2\circ c)(t_n;v_n)|=|\delta_{\text{eq}}^{k+1}(G_2\circ c_n)(0;v_n)|$$

$$=\frac{1}{2^{n(k+1)}}|\delta_{\text{eq}}^{k+1}(G_2\circ e_n)(a_n;h_n)|\le\frac{1}{2^{n(k+1)}}NC_k$$

which is also bounded. Together this gives for $G_1=G_0+G_2$: $\delta_{\text{eq}}^{k+1}(G_1\circ c)(t_n,v_n)$ is bounded.

But using (4.3.21) one obtains a contradiction with this:

$$|\delta_{\text{eq}}^{k+1}(G_1\circ c)(t_n;v_n)|=|\delta_{\text{eq}}^{k+1}(G_1\circ c_n)(0;v_n)|$$

$$=\frac{1}{2^{n(k+1)}}|\delta_{\text{eq}}^{k+1}(G_1\circ e_n)(a_n;h_n)|=\frac{k+1}{2^{n(k+1)}}|\delta_{\text{eq}}^k g(a_n;h_n)|\ge n(k+1)$$

which is unbounded. □

Now we are able to prove that for a map between convenient vector spaces $\mathcal{Lip}^k$-ness can be characterized recursively.

4.3.24 Theorem. (*Recursiveness of $\mathcal{Lip}^k$.*) *Let E and F be convenient vector spaces, $U\subseteq E$ be* M-*open, $f:U\to F$ be a map and $k\in\mathbb{N}_0$. Then the following statements are equivalent:*

(1) *f is $\mathscr{Lip}^{k+1}$;*
(2) *f is $\mathscr{Lip}$-differentiable and f' is $\mathscr{Lip}^k$;*
(3) *f is $\mathscr{Lip}$-differentiable and $\mathrm{d}f$ is $\mathscr{Lip}^k$.*

Proof. (1⇒3) Let us first show that the differential $\mathrm{d}f$ exists: Since for $x \in U$ and $v \in E$ the map $c: t \mapsto x + tv$ is a $\mathscr{Lip}^k$-curve in E, the composite $e := f \circ c$ is defined and $\mathscr{Lip}^k$ in a neighbourhood of 0. So by (4.1.12)

$$e^{\cdot}(0) := \frac{\mathrm{d}}{\mathrm{d}t}\bigg|_{t=0} f(x+tv) = (\text{weak-}) \lim_{t \to 0} \frac{f(x+tv) - f(x)}{t}$$

exists.

In order to show that $\mathrm{d}f$ is $\mathscr{Lip}^k$, let $(c_1, c_2): \mathbb{R} \to U \sqcap E$ be any $\mathscr{Lip}^\infty$-curve of $U \sqcap E$; $\ell \in F'$. Since $g: \mathbb{R}^2 \to E$ defined by $g(t, s) := c_1(t) + s c_2(t)$ is a $\mathscr{Lip}^\infty$-map, $\ell \circ f \circ g$ is $\mathscr{Lip}^k$ as composite of $\mathscr{Lip}^k$-maps. One has

$$(\ell \circ \mathrm{d}f \circ (c_1, c_2))(t) = \ell\left(\text{M-}\lim_{s \to 0} \frac{f(c_1(t) + s c_2(t)) - f(c_1(t))}{s}\right)$$

$$= \lim_{s \to 0} \frac{(\ell \circ f \circ g)(t, s) - (\ell \circ f \circ g)(t, 0)}{s} = \partial_2(\ell \circ f \circ g)(t, 0).$$

So by (4.3.23) the composite $\ell \circ \mathrm{d}f \circ (c_1, c_2) \in \mathscr{Lip}^{k-1}$ as to be shown.

(3⇒1) Let $c: \mathbb{R} \to U$ be a smooth curve. By the special chain rule (4.3.14) we know that $f \circ c$ is $\mathscr{Lip}^1$ and $(f \circ c)^{\cdot} = \mathrm{d}f \circ (c, c^{\cdot})$, which is by assumption $\mathscr{Lip}^k$. Hence $f \circ c$ is $\mathscr{Lip}^{k+1}$.

(2⇔3) holds by (4.3.5). □

4.3.25 Proposition. (*Chain Rule.*) *Let $f: E \supseteq U \to W \subseteq F$ and $g: F \supseteq W \to G$ be two $\mathscr{Lip}^1$-maps. Then $g \circ f: E \supseteq U \to G$ is $\mathscr{Lip}^1$ and $(g \circ f)'(x) = g'(f(x)) \circ f'(x)$ for all $x \in U$.*

Proof. Clearly $g \circ f$ is $\mathscr{Lip}^1$. Since $(g \circ f)'(x)(v) = (g \circ c)^{\cdot}(0)$ for $c(t) := f(x + tv)$ the formula follows directly from the special chain rule (4.3.14). □

4.3.26 Definition. Let $f: E \supseteq U \to F$ be a map.

(i) f is called $(k+1)$-*times $\mathscr{S}$-differentiable* iff it is $\mathscr{S}$-differentiable and $\mathrm{d}f(\_, v): U \to F$ is k-times $\mathscr{S}$-differentiable for all $v \in E$. The *differential* $\mathrm{d}^{k+1}f: U \sqcap E^{k+1} \to F$ of order $k+1$ is then defined by $\mathrm{d}^{k+1}f\,(x; v_0, \ldots, v_k) := \mathrm{d}^k[\mathrm{d}f(\_, v_0)](x; v_1, \ldots, v_k)$.

This recursive definition starts with the convention that every function is 0-times $\mathscr{S}$-differentiable and $\mathrm{d}^0 f := f$. We remark that 1-times $\mathscr{S}$-differentiable is the same as $\mathscr{S}$-differentiable (4.3.9) and $\mathrm{d}^1 f = \mathrm{d}f$. The function f is called ∞-times $\mathscr{S}$-differentiable iff f is k-times $\mathscr{S}$-differentiable for all $k \in \mathbb{N}$.

(ii) f is called $(k+1)$-*times strongly differentiable* iff it is strongly differentiable and $f': U \to L(E, F)$ is k-times strongly differentiable. The *derivative* $f^{(k+1)}: U \to L(E, \ldots, E; F)$ of order $k+1$ is then defined by $f^{(k+1)}(x)(v_0, \ldots, v_k)$

$:=[f']^{(k)}(x)(v_1,\dots,v_k)(v_0)$; i.e. $f^{(k+1)}(x)$ corresponds to $[f']^{(k)}(x)$ via the canonical isomorphism $L(E,\dots,E,E;F)\cong L(E,\dots,E;L(E;F))$.

This recursive definition starts with the convention that every function f is 0-times strongly differentiable and $f^{(0)}:=f$. We remark that 1-times strongly differentiable is the same as strongly differentiable (4.3.9) and $f^{(1)}=f'$.

f is called ∞-times strongly differentiable iff f is k-times strongly differentiable for all $k\in\mathbb{N}$.

(iii) f is called $(k+1)$*-times* $\mathcal{Lip}$*-differentiable* iff it is $\mathcal{Lip}$-differentiable and $\mathrm{d}f(\_,v)\colon U\to F$ is k-times $\mathcal{Lip}$-differentiable for all $v\in E$. This is exactly the case if f is $(k+1)$-times strongly differentiable and $f^{(i)}$ is $\mathcal{Lip}^0$ for all $i=0\dots k+1$.

f is called 0-times $\mathcal{Lip}$-differentiable iff f is $\mathcal{Lip}^0$.

f is called ∞-times $\mathcal{Lip}$-differentiable iff f is k-times $\mathcal{Lip}$-differentiable for all $k\in\mathbb{N}$.

This definition is consistent with (ii) of (1.3.19).

Now we can prove that a map between convenient vector spaces is $\mathcal{Lip}^k$ iff it is k-times $\mathcal{Lip}$-differentiable, a result established by [Boman, 1967] for maps $\mathbb{R}^m\to\mathbb{R}$.

4.3.27 Theorem. *($\mathcal{Lip}^k$-maps as k-times differentiable maps.) Let E and F be convenient vector spaces, $U\subseteq E$ be M-open, $f\colon U\to F$ be a map, $\mathcal{S}\subseteq F'$ point separating and $k\in\mathbb{N}_0$. Then the following statements are equivalent:*

(1) *f is $\mathcal{Lip}^k$;*
(2) *f is k-times $\mathcal{S}$-differentiable and $d^jf\colon U\sqcap E^j\to F$ is $\mathcal{Lip}^0$ for all $j\le k$;*
(3) *f is k-times $\mathcal{Lip}$-differentiable;*
(4) *f is k-times strongly differentiable and $f^{(j)}\colon U\to L(E,\dots,E;F)$ is $\mathcal{Lip}^{k-j}$ for all $j\le k$.*

Proof. We prove this by induction on k. For $k=0$ this is trivial. Now for $(k+1)$:

($1\Rightarrow4$) By (4.3.24), f is strongly differentiable and f' is $\mathcal{Lip}^k$. Thus by induction hypothesis f' is k-times strongly differentiable and $[f']^{(j)}$ is $\mathcal{Lip}^{k-j}$ for all $j=0\dots k$. Hence f is $(k+1)$-times strongly differentiable, f is $\mathcal{Lip}^k$, f' is $\mathcal{Lip}^{k-1}$ and $f^{(j)}$ is $\mathcal{Lip}^{k-(j-1)}$ for $1\le j\le k+1$.

($4\Rightarrow3$) One uses that $\mathcal{Lip}^{k-j}$ implies $\mathcal{Lip}^0$ by (4.3.4).

($3\Rightarrow2$) We show that $\mathrm{d}^kf(x;v_0,\dots,v_k)=f^{(k)}(x)(v_0,\dots,v_k)$:

$f^{(k+1)}(x)(v_0,\dots,v_k)=$	(by definition)
$=([f']^{(k)}(x)(v_1,\dots,v_k))(v_0)=$	(by induction hypothesis)
$=(\mathrm{d}^k[f'](x;v_1,\dots,v_k))(v_0)=$	(ev_0 is a linear morphism)
$=\mathrm{d}^k(f'(\_)(v_0))(x;v_1,\dots,v_k)=$	(by (4.3.12))
$=\mathrm{d}^k(\mathrm{d}f(\_,v_0))(x;v_1,\dots,v_k)=$	(by definition)
$=\mathrm{d}^{k+1}f(x;v_0,\dots,v_k)$.	

($2\Rightarrow1$) By assumption f is $\mathcal{S}$-differentiable and $\mathrm{d}f(\_,v_0)$ is k-times $\mathcal{S}$-differentiable with differential $\mathrm{d}^j[\mathrm{d}f(\_,v_0)](x;v_1,\dots,v_j)=$

$d^{j+1}f(x; v_0, \dots, v_j)$. By induction hypothesis $df(\_, v_0)$ is $\mathscr{Lip}^k$. Since df exist and is $\mathscr{Lip}^0$ one concludes that $df(x, \_)$ is linear by (4.3.12). Thus df is $\mathscr{Lip}^k$, by (4.3.5), and hence f is $\mathscr{Lip}^{k+1}$ by (4.3.24). □

Remark. In (2) it is not enough to assume that the highest derivative $d^k f$ is $\mathscr{Lip}^0$. In fact, consider a linear non-bornological map f. Then $df(x, v) = f(v)$ and $d^2 f(x; v, w) = 0$. Thus $d^2 f$ is smooth but f is not $\mathscr{Lip}^1$.

4.3.28 Corollary. *Let $f: E \supseteq U \to F$ be a $\mathscr{Lip}^k$-map. Then $d^k f(x; v_1, \dots, v_k) = f^{(k)}(x)(v_1, \dots, v_k)$ is symmetric in $v_1, \dots, v_k$.*

Proof. By composing with $\ell \in F'$ we may assume that $F = \mathbb{R}$. Consider the smooth map $h: (t_1, \dots, t_k) \mapsto x + t_1 v_1 + \cdots + t_k v_k$. Then $g := \ell \circ f \circ h$ is $\mathscr{Lip}^k$. Since $d^k f(x; v_{\sigma(1)}, \dots, v_{\sigma(k)}) = (\partial_{\sigma(1)} \dots \partial_{\sigma(k)} g)(0, \dots, 0)$, the result follows from the classical theorem according to which $\partial_{\sigma(1)} \dots \partial_{\sigma(k)} g = \partial_1 \dots \partial_k g$ for any k-times continuously differentiable function $g: \mathbb{R}^k \to \mathbb{R}$. □

Having shown that the derivative of order k of a map $f: E \supseteq U \to F$ is a map $f^{(k)}: U \to L(E, \dots, E; F)$ whose values are symmetric it is natural to ask which maps $g: U \to L(E, \dots, E; F)$ appear as derivatives. We give a first characterization for $k = 1$ here. Under additional differentiability conditions on g we will again take up this question in (4.5.5).

4.3.29 Lemma. *Let $g: U \to F$ be a $\mathscr{Lip}^0$-map where $U \subseteq E$ is a convex subset containing 0. Then there exists a $\mathscr{Lip}^1$-map $f: U \to F$ (which is given by $f(x) := \int_0^1 g(tx)(x)dt$) with $f' = g$ if and only if $\int_0^1 g(t(x+v))(x+v)dt - \int_0^1 g(tx)(x)dt = \int_0^1 g(x+tv)(v)dt$ for all $x, x+v \in U$.*

Proof. If $g = f'$ for some $\mathscr{Lip}^1$-map f the equation obviously holds, since $\int_0^1 f'(y+tw)(w)dt = g(y+w) - g(y)$ for every $y, y+w \in U$.

Assume, conversely, that the equation holds. Then f is weakly differentiable with differential $df(x, v) = g(x)(v)$ since

$$s \mapsto \frac{f(x+sv) - f(x)}{s} = \frac{1}{s} \int_0^1 g(x+tsv)(sv)dt = \int_0^1 g(x+tsv)(v)dt$$

is by (4.1.5) $\mathscr{Lip}^0$, hence converges to $g(x)(v)$. Using that g is $\mathscr{Lip}^0$ we have: f is $\mathscr{Lip}^1$ and $f' = g$. □

4.3.30 Proposition. ($\mathscr{Lip}^\infty$*-maps as* ∞*-times differentiable maps.*) *Let E and F be convenient vector spaces, $U \subseteq E$ be M-*open, *$f: U \to F$ be a map, and let $\mathscr{S} \subseteq F'$ be such that the bornology of F has a basis of $\sigma(F, \mathscr{S})$-closed sets. Then the following statements are equivalent:*

(1) *f is $\mathscr{Lip}^\infty$;*
(2) *f is ∞-times $\mathscr{S}$-differentiable and $d^j f: U \sqcap E^j \to F$ is $\mathscr{Lip}^0$ for all j;*
(3) *f is ∞-times $\mathscr{Lip}$-differentiable;*

(4) *f is ∞-times strongly differentiable and $f^{(j)}: U \to L(E, \ldots, E; F)$ is $\mathscr{Lip}^{\infty}$ for all j.*

(5) *f is ∞-times $\mathscr{S}$-differentiable and $\mathrm{d}^j f: U \sqcap E^j \to F$ is $\mathscr{Lip}^{-1}$ for all j.*

Proof. The equivalences ($1 \Leftrightarrow 2 \Leftrightarrow 3 \Leftrightarrow 4$) follow immediately from (4.3.27) without using the assumption on the bornology of F.

($3 \Rightarrow 5$) is trivial.

($5 \Rightarrow 1$) By (4.3.10) f is $\mathscr{Lip}^0$.

We next show that f is $\mathscr{Lip}^1$. By assumption f and $\mathrm{d}f(\_, v)$ are $\mathscr{S}$-differentiable for any $v \in E$. It is enough to show that $\mathrm{d}f$ is $\mathscr{Lip}^0$ since this implies by (4.3.27) that f is $\mathscr{Lip}^1$. Applying the case $k=0$ to $\mathrm{d}f(\_, v)$ we conclude that $\mathrm{d}f(\_, v)$ is $\mathscr{Lip}^0$. We next show that $\mathrm{d}f(\_, v)$ is linear. Only the additivity $\mathrm{d}f(x, v+w) = \mathrm{d}f(x, v) + \mathrm{d}f(x, w)$ is not trivial. For this we consider

$$\frac{f(x+tv+sw)-f(x+tv)}{s} - \mathrm{d}f(x+tv, w) = \int_0^1 \mathrm{d}f(x+tv+s\sigma w, w) - \mathrm{d}f(x+tv, w)\,\mathrm{d}\sigma$$

$$= s\int_0^1\int_0^1 \mathrm{d}^2 f(x+tv+s\sigma\tau w, w, \sigma w)\,\mathrm{d}\tau\,\mathrm{d}\sigma.$$

This M-converges uniformly to 0 for $s \to 0$ and $t \in [-1, 1]$. Now divide the equation $f(x+sv+sw)-f(x) = (f(x+sv+sw)-f(x+sv)) + (f(x+sv)-f(x))$ by s and take the limit for $s \to 0$. This gives $\mathrm{d}f(x, v+w) = \mathrm{d}f(x, v) + \mathrm{d}f(x, w)$. By assumption $\mathrm{d}f(x, \_)$ is $\mathscr{Lip}^{-1}$ and hence even $\mathscr{Lip}^0$, using for example (4.3.10). Now (4.3.5) implies that $\mathrm{d}f$ is $\mathscr{Lip}^0$.

Finally we prove by induction that f is $\mathscr{Lip}^k$ for all $k \in \mathbb{N}$. Since f is $\mathscr{Lip}^1$ one has $\mathrm{d}f(x, v) = f'(x)(v)$ and by assumption $\mathrm{ev}_v \circ f' = \mathrm{d}f(\_, v)$ satisfies (5). Thus by induction hypothesis $\mathrm{ev}_v \circ f'$ is $\mathscr{Lip}^k$ and using the corollary (3.6.5) of the linear uniform boundedness principle we conclude that f' is $\mathscr{Lip}^k$. Thus f is $\mathscr{Lip}^{k+1}$. □

Remark. For a comparison of $\mathscr{Lip}^{\infty}$ with more classically considered differentiability concepts (cf. [Keller, 1974]) and, in particular, for such maps between Fréchet spaces see [Kriegl, 1983].

4.4 Function spaces and exponential laws

We shall first describe explicitly how the function space $\mathscr{Lip}^k(X, E)$ can be equipped with a convenient vector space structure, X being an arbitrary $\mathscr{Lip}^k$-space and E a convenient vector space. In particular X can be any subset of a convenient vector space, or any classical differentiable manifold whose coordinate transformations are k-times Lipschitz differentiable. We then show that this structure is natural by verifying that it satisfies a universal property by which it is characterized. We shall also compare these function spaces with classical ones. Finally we will examine the differentiability of natural maps between function spaces. Since function spaces are often written in exponential form, some of these results are also called exponential laws, cf. (8.6.4).

4.4.1 Definition. Let $k \in \mathbb{N}_{0,\infty}$, X a $\mathscr{Lip}^k$-space and E a convenient vector space. By $\mathscr{Lip}^k(X, E)$ we denote from now on the vector space formed by the $\mathscr{Lip}^k$-maps $X \to E$ together with the initial Pre-structure induced by the linear maps $c^*\colon \mathscr{Lip}^k(X, E) \to \mathscr{Lip}^k(\mathbb{R}, E)$ for $c \in C$, where C is the set $\mathscr{Lip}^k(\mathbb{R}, X)$ of structure curves of X.

Remark. We already saw in (4.2.6) and (4.2.10) that $\mathscr{Lip}^k(\mathbb{R}, E)$ is a convenient vector space. For $X = \mathbb{R}$ the structure described above in (4.4.1) coincides with the one of (4.2.2) and of (4.2.8) since for a $\mathscr{Lip}^k$-function $f\colon \mathbb{R} \to \mathbb{R}$ the map $f^*\colon \mathscr{Lip}^k(\mathbb{R}, E) \to \mathscr{Lip}^k(\mathbb{R}, E)$ is a morphism of convenient vector space; cf. (4.2.12).

4.4.2 Theorem. *For $k \in \mathbb{N}_{0,\infty}$, any $\mathscr{Lip}^k$-space X and any convenient vector space E, the function space $\mathscr{Lip}^k(X, E)$ is also a convenient vector space.*

Proof. Let $m\colon \mathscr{Lip}^k(X, E) \to \prod_C \mathscr{Lip}^k(\mathbb{R}, E)$ be the Pre-morphism characterized by $\mathrm{pr}_c \circ m = c^*$ for $c \in C$, where C denotes again the set of $\mathscr{Lip}^k$-curves $\mathbb{R} \to X$. This morphism m is trivially injective; and by the definition of the structure of $\mathscr{Lip}^k(X, E)$ it is also initial. Hence $\mathscr{Lip}^k(X, E)$ is isomorphic to its image under m, and by the closed embedding lemma (2.6.4) it is enough to show that this image is M-closed in $\prod_C \mathscr{Lip}^k(\mathbb{R}, E)$, since we know by (4.2.6) and (4.2.10) and (3.3.1) that this product is complete. But the M-closedness follows because the image can be described by the equations $\mathrm{ev}_{t_1} \circ \mathrm{pr}_{c_1} = \mathrm{ev}_{t_2} \circ \mathrm{pr}_{c_2}$ for $t_1, t_2 \in \mathbb{R}$ and $c_1, c_2 \in C$ satisfying $c_1(t_1) = c_2(t_2)$, hence is an intersection of kernels of Pre-morphisms. □

An alternate more direct but less elegant proof of the completeness of $\mathscr{Lip}^k(X, E)$ can be given along the following lines: If (g_n) is a Mackey–Cauchy sequence in $\mathscr{Lip}^k(X, E)$, then $(g_n(x))$ is a Mackey–Cauchy sequence in E (since $\mathrm{ev}_x\colon \mathscr{Lip}^k(X, E) \to E$ is a Pre-morphism) and hence $g(x) := \text{M-lim}_{n\to\infty} g_n(x)$ exists. One then verifies that $g \in \mathscr{Lip}^k(X, E)$ and $g = \text{M-lim}_{n\to\infty} g_n$.

4.4.3 Proposition. *One has, for $k \in \mathbb{N}_{0,\infty}$, a functor $\mathscr{Lip}^k\colon (\mathscr{Lip}^k)^{\mathrm{op}} \times$ Con $\to$ Con for which $\mathscr{Lip}^k(X, E)$ is the function space defined in* (4.4.1) *and* $\mathscr{Lip}^k(g, m) := g^* \circ m_*$.

Proof. Obviously g^* is linear. That it is bornological follows from (4.2.12) since $c^*\colon \mathscr{Lip}^k(X, E) \to \mathscr{Lip}^k(\mathbb{R}, E)$ $(c \in \mathscr{Lip}^k(\mathbb{R}, X))$ is an initial source and $c^* \circ g^* = (g \circ c)^*$.

That the same holds for m_* is shown similarly, using (4.2.13) instead of (4.2.12). □

4.4.4 Proposition. *Let $k \in \mathbb{N}_{0,\infty}$, X be a $\mathscr{Lip}^k$-space, E a convenient vector space whose Pre-structure is the initial one induced by $\mathscr{S} \subseteq E'$. Then the Pre-structure of $\mathscr{Lip}^k(X, E)$ is the initial one induced by any of the following families:*

(1) $c^*: \mathscr{Lip}^k(X, E) \to \mathscr{Lip}^k(\mathbb{R}, E) \quad (c \in \mathscr{Lip}^k(\mathbb{R}, X))$;
(2) $\ell_*: \mathscr{Lip}^k(X, E) \to \mathscr{Lip}^k(X, \mathbb{R}) \quad (\ell \in \mathscr{S})$;
(3) $\mathscr{Lip}^k(c, \ell): \mathscr{Lip}^k(X, E) \to \mathscr{Lip}^k(\mathbb{R}, \mathbb{R}) \quad (c \in \mathscr{Lip}^k(\mathbb{R}, X), \ell \in \mathscr{S})$.

Proof. (1) is the definiton of the structure of $\mathscr{Lip}^k(X, E)$.

(2) We show this first for the case $X = \mathbb{R}$. One has a commutative diagram:

$$\begin{array}{ccc} \mathscr{Lip}^k(\mathbb{R}, E) & \xrightarrow{\ \ell_*\ } & \mathscr{Lip}^k(\mathbb{R}, \mathbb{R}) \\ \downarrow{\scriptstyle \delta^j} & & \downarrow{\scriptstyle \delta^j} \\ \ell^\infty(\mathbb{R}^{\langle j \rangle}, E) & \xrightarrow{\ \ell_*\ } & \ell^\infty(\mathbb{R}^{\langle j \rangle}, \mathbb{R}) \end{array}$$

Using now that the family $\delta^j (j=0, \ldots, k+1)$ is initial by definition and the family formed by the ℓ_* in the lower row of the diagram for $\ell \in \mathscr{S}$ is initial by (ii) in (1.2.9) one concludes that the family formed by the ℓ_* in the upper row of the diagram is initial by (8.7.2). The general case follows now from (1).

(3) is a combination of (1) and (2), since $\mathscr{Lip}^k(c, \ell) = c^* \circ \ell_*$. □

4.4.5 Proposition. *Let $k \in \mathbb{N}_{0,\infty}$; X be a $\mathscr{Lip}^k$-space, E a preconvenient and F a convenient vector space. The map $f \mapsto \tilde{f}$ where $\tilde{f}(y)(x) := f(x)(y)$ constitutes an isomorphism of convenient vector spaces: $\mathscr{Lip}^k(X, L(E, F)) \cong L(E, \mathscr{Lip}^k(X, F))$.*

Proof. By using evaluations at points of X and E, which are linear morphisms on the function spaces $\mathscr{Lip}^k(X, F)$ and $L(E, F)$, one concludes that for a map of one of the two iterated function spaces the associated map has values in the appropriate function space.

We next show that $f \mapsto \tilde{f}$ defines a bijection in case where $X = \mathbb{R}$:

$f: \mathbb{R} \to L(E, F)$ is $\mathscr{Lip}^k$;
$\Leftrightarrow \delta^j f: \mathbb{R}^{\langle j \rangle} \to L(E, F)$ is bornological for all $j < k+2$;
$\Leftrightarrow \delta^j f(K^{\langle j \rangle}) \subseteq L(E, F)$ is bounded for $j < k+2$ and all compact $K \subseteq \mathbb{R}$;
$\Leftrightarrow (\delta^j \circ \tilde{f})(B)(K^{\langle j \rangle}) = \delta^j f(K^{\langle j \rangle})(B) \subseteq F$ is bounded for $j < k+2$, compact $K \subseteq \mathbb{R}$ and all bounded $B \subseteq E$;
$\Leftrightarrow \tilde{f}(B) \subseteq \mathscr{Lip}^k(\mathbb{R}, F)$ is bounded for all bounded $B \subseteq E$;
$\Leftrightarrow \tilde{f}: E \to \mathscr{Lip}^k(\mathbb{R}, F)$ is bornological.

That $f \mapsto \tilde{f}$ is even a bornological isomorphism follows by the same argument applied to a set of maps instead of a single map f.

Now the result for a general $\mathscr{Lip}^k$-space X follows from that for $X = \mathbb{R}$ by using the finality of the family of structure curves $c: \mathbb{R} \to X$, the initiality of $c^*: \mathscr{Lip}^k(X, F) \to \mathscr{Lip}^k(\mathbb{R}, F)$, i.e. (i) in (4.4.4), and the equation $(f \circ c)^\sim = c^* \circ \tilde{f}$.

□

Using the previous proposition the point (1) of (4.4.4) can be generalized as follows:

4.4.6 Proposition. *Let E be a convenient vector space, and $g_j: X_j \to X$ $(j \in J)$ be a final family of $\mathscr{Lip}^k$-maps whose images cover X. Then $(g_j)^*: \mathscr{Lip}^k(X, E) \to \mathscr{Lip}^k(X_j, E)$ $(j \in J)$ is a Pre-initial source.*

Proof. Let F be an arbitrary preconvenient vector space and $g: F \to \mathscr{Lip}^k(X, E)$ a linear map for which all composites $(g_j)^* \circ g$ are $\mathscr{Lip}^k$. We have to show that g is a morphism. Since every $x \in X$ can be written as $g_j(x_j)$ for some $j \in J$ and some $x_j \in X_j$ we conclude that $(\mathrm{ev}_x \circ g)(y) = g(y)(x) = g(y)(g_j(x_j)) = ((g_j)^* \circ g)(y)(x_j) = (\mathrm{ev}_{x_j} \circ (g_j)^* \circ g)(y)$ and hence $(\mathrm{ev}_x \circ g)$ is a linear morphism. Thus we get a map $\tilde{g}: X \to L(F, E)$, $x \mapsto (\mathrm{ev}_x \circ g)$. We want to show that it is $\mathscr{Lip}^k$. Since the family $g_j: X_j \to X$ is final it is enough to show that $\tilde{g} \circ g_j: X_j \to L(F, E)$ is $\mathscr{Lip}^k$, but this is clear by (4.4.5) since it equals $((g_j)^* \circ g)^{\sim}: x \mapsto (\mathrm{ev}_x \circ ((g_j)^* \circ g)) = \mathrm{ev}_{g_j(x)} \circ g$. Thus the associated map $g: F \to \mathscr{Lip}^k(X, E)$ is a linear morphism by (4.4.5). □

4.4.7 Theorem. (*Differentiable Uniform Boundedness Principle.*) *Let $k \in \mathbb{N}_{0,\infty}$; X be a $\mathscr{Lip}^k$-space; and E a convenient vector space. The structure of $\mathscr{Lip}^k(X, E)$ introduced in* (4.4.1) *is the coarsest convenient vector space structure making all evaluations* ev_x $(x \in X)$ *morphisms. In categorical language this means:* $\mathrm{ev}_x: \mathscr{Lip}^k(X, E) \to E$ $(x \in X)$ *is an initial source with respect to the forgetful functor Con→VS.*

Proof. This follows immediately from the special case $X = \mathbb{R}$ treated in (4.2.11) using the definition (4.4.1) of the structure of $\mathscr{Lip}^k(X, F)$ and the identity $\mathrm{ev}_t \circ c^* = \mathrm{ev}_{c(t)}$. □

4.4.8 Corollary. *Let E_j, F be convenient vector spaces. Then $L(E_1, \ldots, E_m; F)$ is a Pre-subspace of $\mathscr{Lip}^k(E_1 \sqcap \ldots \sqcap E_m, F)$.*

Remark. The much weaker result that it is also a Con-subspace is a trivial consequence, cf. (3.2.2).

Proof. That the inclusion $L(E_1, \ldots, E_m; F) \to \mathscr{Lip}^k(E_1 \sqcap \ldots \sqcap E_m, F)$ is a morphism is an immediate consequence of (4.4.7). Initiality follows since the inclusion composed with the map $\delta^0: \mathscr{Lip}^k(E_1 \sqcap \ldots \sqcap E_m, F) \to \ell^\infty(E_1 \sqcap \ldots \sqcap E_m, F)$ is the inclusion of $L(E_1, \ldots, E_m; F)$ in $\ell^\infty(E_1 \sqcap \ldots \sqcap E_m, F)$ which is by definition initial. □

4.4.9 Proposition. *Let E and F be convenient vector spaces and $U \subseteq E$ M-open. Then the following maps are Con-morphism:*

(*i*) $\mathrm{d}: \mathscr{Lip}^{k+1}(U, F) \to \mathscr{Lip}^k(U \sqcap E, F)$
(*ii*) $(\_)': \mathscr{Lip}^{k+1}(U, F) \to \mathscr{Lip}^k(U, L(E, F))$
(*iii*) $\bar{\mathfrak{d}}: \mathscr{Lip}^{k+1}(U, F) \to \mathscr{Lip}^k(U_{\bar{\mathfrak{d}}}, F)$

Proof. (i) Since by the differentiable uniform bounded principle (4.4.7) the Con-structure of $\mathscr{Lip}^k(U\sqcap E, F)$ is the initial one induced by the evaluations $\mathrm{ev}_{(x,v)}$: $\mathscr{Lip}^k(U\sqcap E, F)\to F$ $(x\in U, v\in E)$ it is enough to show that $f\mapsto df(x,v)$, $\mathscr{Lip}^{k+1}(U,F)\to F$ is a Con-morphism for all $(x,v)\in U\times E$. This map can be factorized as

$$\mathrm{ev}_0\circ\mathscr{D}\circ c^*: \mathscr{Lip}^{k+1}(U,F)\to\mathscr{Lip}^{k+1}(\mathbb{R},F)\to\mathscr{Lip}^k(\mathbb{R},F)\to F,$$

where $c\colon\mathbb{R}\to U$ is any smooth curve which equals $t\mapsto x+tv$ locally around 0. Now (i) follows, since c^* is a Con-morphism by (4.4.4) and $\mathrm{ev}_0\circ\mathscr{D}$ is a Con-morphism by (4.2.3).

(ii) holds provided $\mathrm{ev}_x\circ(\_)'$: $\mathscr{Lip}^k(U,F)\to L(E,F)$ is a Con-morphism for all $x\in U$. It is a Con-morphism provided $\mathrm{ev}_v\circ\mathrm{ev}_x\circ(\_)'$ is one for all $v\in E$. And this now follows from (i) since $f'(x)(v)=df(x,v)$ implies $\mathrm{ev}_v\circ\mathrm{ev}_x\circ(\_)'=\mathrm{ev}_{(x,v)}\circ d$.

(iii) Recall that $U_{\bar\vartheta}\cap\{(x,y;t,s); t\neq s\}=U_\vartheta$ and $\bar\vartheta f(x,y;t,s):=\vartheta f(x,y;t,s)$ for $t\neq s$ and $\bar\vartheta f(x,y;t,t)=df(x+ty,y)$. Since the Con-structure on $\mathscr{Lip}^k(U_{\bar\vartheta},F)$ is the initial one induced by the evaluation maps $\mathrm{ev}_{(x,y;t,s)}$ it is enough to show that the composites $\mathrm{ev}_{(x,y;t,s)}\circ\bar\vartheta$ are morphisms. In case $t=s$ this follows from (i) and for $t\neq s$ this is true since $\delta f(x,y;t,s)=(f(x+ty)-f(x+sy))/(t-s)$ and ev_{x+ty} and ev_{x+sy} are morphisms. □

4.4.10 Proposition. *Let E, F be convenient vector spaces and $U\subseteq E$ be* M*-open. Then the maps $f\mapsto f^{(j)}$: $\mathscr{Lip}^k(U,F)\to\mathscr{Lip}^{k-j}(U,L(E,\ldots,E;F))$ are Con-morphisms for $0\leq j\leq k$.*

Proof. We show this by induction on j. For $j=0$ it is trivial since $f^{(0)}=f$. For $j=1$ it follows from (4.4.9).

If $j>1$, then $f\mapsto f^{(j)}$ is the composite of the maps $\mathscr{Lip}^k(U,F)\xrightarrow{(\_)'}\mathscr{Lip}^{k-1}(U,L(E,F))\xrightarrow{(\_)^{(j-1)}}\mathscr{Lip}^{k-j}(U,L(E,\ldots,E;L(E,F)))\cong$ $\mathscr{Lip}^{k-j}(U,L(E,\ldots,E,E;F))$, hence by the case $(j=1)$ and the induction hypothesis it is a morphism. □

4.4.11 Corollary. *Let $k\in\mathbb{N}_{0,\infty}$, E and F be two convenient vector spaces and $U\subseteq E$ be* M*-open. Then the Pre-structure of $\mathscr{Lip}^k(U,F)$ is the initial one induced by the family $f\mapsto f^{(j)}$ $(j=0\ldots k)$, $\mathscr{Lip}^k(U,F)\to\mathscr{Lip}^0(U,L(E,\ldots,E;F))$.*

Proof. It is enough to show that the Pre-structure of $\mathscr{Lip}^k(U,F)$ is the initial one induced by the two maps $f\mapsto f'$, $\mathscr{Lip}^k(U,F)\to\mathscr{Lip}^{k-1}(U,L(E,F))$ and the inclusion into $\mathscr{Lip}^0(U,F)$. For a $c\in\mathscr{Lip}^k(\mathbb{R},U)$ the map c^*: $\mathscr{Lip}^k(U,F)\to\mathscr{Lip}^k(\mathbb{R},F)$ composed with the isomorphism $(\mathrm{ev}_0,\mathscr{D})$: $\mathscr{Lip}^k(\mathbb{R},F)\to F\sqcap\mathscr{Lip}^{k-1}(\mathbb{R},F)$ can be written by (4.3.14) as composite

$$\mathscr{Lip}^k(U,F)\xrightarrow{(\mathrm{id},(\_)')}\mathscr{Lip}^0(U,F)\sqcap\mathscr{Lip}^{k-1}(U,L(E;F))\xrightarrow{\mathrm{ev}_{c(0)}\sqcap(\_)^\wedge}$$

$$\to F\sqcap\mathscr{Lip}^{k-1}(U\sqcap E,F)\xrightarrow{\mathrm{id}\sqcap(c,c')^*}F\sqcap\mathscr{Lip}^{k-1}(\mathbb{R},F).$$

The map $(\_)^\wedge: \mathscr{Lip}^{k-1}(U, L(E,F)) \to \mathscr{Lip}^{k-1}(U \sqcap E, F)$ is a morphism by (4.4.7). Since the $\underline{\text{Pre}}$-structure of $\mathscr{Lip}^k(U, F)$ is the initial one induced by the family c^* $(c \in \mathscr{Lip}^k(\mathbb{R}, U))$ it is also the initial one induced by the map $\mathscr{Lip}^k(U, F) \xrightarrow{(\mathrm{id}, (\_)')} \mathscr{Lip}^0(U, F) \sqcap \mathscr{Lip}^{k-1}(U, L(E; F))$. Thus the proof is completed in case $k < \infty$. The structure of $\mathscr{Lip}^\infty(U, F)$ is the initial one induced by the inclusions into $\mathscr{Lip}^k(U, F)$ for all $k \in \mathbb{N}$ and so the statement for $k = \infty$ follows immediately. □

4.4.12 Proposition. *Let X be a smooth space and E a convenient vector space. Then $\mathscr{Lip}^\infty(X, E)$ considered as smooth vector space, cf.* (3) *of* (2.4.4), *is identical with $C^\infty(X, E)$ considered in* (1.4.4), *whose smooth structure is the one according to cartesian closedness of the category $\underline{C^\infty}$ of smooth spaces.*

Proof. The underlying vector spaces being the same we only have to compare the smooth structures. By (1.4.4) that of $C^\infty(X, E)$ is the initial one induced by the maps $C^\infty(c, \ell): C^\infty(X, E) \to C^\infty(\mathbb{R}, \mathbb{R})$ for $c \in C^\infty(\mathbb{R}, X)$ and $\ell \in E'$. The $\underline{\text{Pre}}$-structure and hence by (3.1.2) also the smooth structure of $\mathscr{Lip}^\infty(X, E)$ is (using (4.4.1) and (1.2.9)) the initial one induced by the maps $\mathscr{Lip}^\infty(c, \ell): \mathscr{Lip}^\infty(X, E) \to \mathscr{Lip}^\infty(\mathbb{R}, \mathbb{R})$ for the same c and ℓ. Hence it is enough to show that $C^\infty(\mathbb{R}, \mathbb{R})$ and $\mathscr{Lip}^\infty(\mathbb{R}, \mathbb{R})$ have the same smooth structure. By (1.4.3) the one of $C^\infty(\mathbb{R}, \mathbb{R})$ is generated by (i.e. is the initial one induced by) all functions of the form $\psi \circ \delta^k: C^\infty(\mathbb{R}, \mathbb{R}) \to \mathbb{R}$ with $\delta^k: C^\infty(\mathbb{R}, \mathbb{R}) \to \ell^\infty(\mathbb{R}^{\langle k \rangle}, \mathbb{R})$ for $0 \le k < \infty$ and $\psi: \ell^\infty(\mathbb{R}^{\langle k \rangle}, \mathbb{R}) \to \mathbb{R}$ linear and bornological. The $\underline{\text{Pre}}$-structure and hence by (3.1.2) also the smooth structure of $\mathscr{Lip}^\infty(\mathbb{R}, \mathbb{R})$ is by definition the initial one induced by the maps $\delta^k: \mathscr{Lip}^\infty(\mathbb{R}, \mathbb{R}) \to \ell^\infty(\mathbb{R}^{\langle k \rangle}, \mathbb{R})$ and hence also induced by the maps $\psi \circ \delta^k: \mathscr{Lip}^\infty(\mathbb{R}, \mathbb{R}) \to \mathbb{R}$, with k and ψ as before. □

4.4.13 Corollary. *Let X and Y be smooth spaces, E a convenient vector space. The map $f \mapsto f^\vee$ constitutes an isomorphism of convenient vector spaces:*

$$\mathscr{Lip}^\infty(X \sqcap Y, E) \cong \mathscr{Lip}^\infty(X, \mathscr{Lip}^\infty(Y, E)).$$

Proof. One uses cartesian closedness of $\underline{C^\infty}$, cf. (1.4.3), and the remark (8.6.4).

□

We now determine the differentiability class of maps between function spaces and begin with the evaluation map:

4.4.14 Lemma. *Let X be a $\mathscr{Lip}^k$-space, E a convenient vector space. Then the evaluation map* $\mathrm{ev}: \mathscr{Lip}^k(X, E) \sqcap X \to E$ *is* $\mathscr{Lip}^k$.

Proof. Since this map is linear in the first factor and the partial maps $\mathrm{ev}(f, \_) = f$ and $\mathrm{ev}(\_, x) = \mathrm{ev}_x$ are $\mathscr{Lip}^k$ the assertion follows using (4.3.5). □

4.4.15 Proposition. *Let X be a $\mathscr{L}ip^k$-space; E and F convenient vector spaces; and $f \in \mathscr{L}ip^{k+n}(E, F)$. Then $f_*: \mathscr{L}ip^k(X, E) \to \mathscr{L}ip^k(X, F)$ is $\mathscr{L}ip^{n-1}$.*

Proof. First we prove the statement by induction on n, for the case $X = \mathbb{R}$:

We begin with $n=0$ and recall that a map is called $\mathscr{L}ip^{-1}$ iff it transforms smooth curves into ℓ^∞-curves. We have the following implications, where we shortly write 'is ℓ^∞' for 'is an ℓ^∞-map':

$$\begin{aligned}
&c: \mathbb{R} \to \mathscr{L}ip^k(\mathbb{R}, E) \text{ is a smooth curve;}\\
\Rightarrow\ &\delta^i c: \mathbb{R}^{\langle i\rangle} \to \mathscr{L}ip^k(\mathbb{R}, E) \text{ is } \ell^\infty \text{ for all } i;\\
\Rightarrow\ &(\delta_1^i \delta_2^j \hat{c})^\vee = (\delta^j)_* \circ \delta^i c: \mathbb{R}^{\langle i\rangle} \to \ell^\infty(\mathbb{R}^{\langle j\rangle}, E) \text{ is } \ell^\infty \text{ for all } i \text{ and all } j<k+2;\\
\Rightarrow\ &\delta_1^i \delta_2^j \hat{c}: \mathbb{R}^{\langle i\rangle} \sqcap \mathbb{R}^{\langle j\rangle} \to E \text{ is } \ell^\infty \text{ for all } i, j \text{ with } i+j<k+2;\\
\Rightarrow\ &\hat{c}: \mathbb{R} \sqcap \mathbb{R} \to E \text{ is } \mathscr{L}ip^k;\\
\Rightarrow\ &f \circ \hat{c}: \mathbb{R} \sqcap \mathbb{R} \to F \text{ is } \mathscr{L}ip^k;\\
\Rightarrow\ &((\delta^j)_* \circ f_* \circ c)^\wedge = \delta_2^j (f \circ \hat{c}): \mathbb{R} \sqcap \mathbb{R}^{\langle j\rangle} \to F \text{ is } \ell^\infty \text{ for all } j \text{ with } j<k+2;\\
\Rightarrow\ &(\delta^j)_* \circ f_* \circ c: \mathbb{R} \to \ell^\infty(\mathbb{R}^{\langle j\rangle}, F) \text{ is } \ell^\infty \text{ for all } j<k+2;\\
\Rightarrow\ &f_* \circ c: \mathbb{R} \to \mathscr{L}ip^k(\mathbb{R}, F) \text{ is } \ell^\infty.
\end{aligned}$$

Let now $n>0$. We take as $\mathscr{S}$ the set $\{\ell \circ \mathrm{ev}_s; \ell \in F', s \in \mathbb{R}\}$, where $\mathrm{ev}_s: \mathscr{L}ip^k(\mathbb{R}, F) \to F$ is the evaluation at s. The set $\mathscr{S}$ is a point separating subset of $\mathscr{L}ip^k(\mathbb{R}, F)'$. We want to prove that f_* is $\mathscr{S}$-differentiable and that its derivative is the map $(\mathrm{d}f)_*: \mathscr{L}ip^k(\mathbb{R}, E \sqcap E) \to \mathscr{L}ip^k(\mathbb{R}, F)$ composed with the natural isomorphism $\mathscr{L}ip^k(\mathbb{R}, E) \sqcap \mathscr{L}ip^k(\mathbb{R}, E) \cong \mathscr{L}ip^k(\mathbb{R}, E \sqcap E)$. This follows by calculating $\mathrm{d}(\mathrm{ev}_s \circ f_*)$. With $c(t) := g + th$ we have:

$$\begin{aligned}
\mathrm{d}(\mathrm{ev}_s \circ f_*)(g, h) &= (\mathrm{ev}_s \circ f_* \circ c)^\cdot(0) = \lim_{t \to 0} \frac{1}{t}(f(g(s) + th(s)) - f(g(s))) = \mathrm{d}f(g(s), h(s))\\
&= \mathrm{ev}_s(\mathrm{d}f \circ (g, h)) = \mathrm{ev}_s((\mathrm{d}f)_*(g, h)) = (\mathrm{ev}_s \circ (\mathrm{d}f)_*)(g, h).
\end{aligned}$$

Since $\mathrm{d}f: E \sqcap E \to F$ is $\mathscr{L}ip^{k+n-1}$ the induction hypothesis implies that $(\mathrm{d}f)_*$ is $\mathscr{L}ip^{n-2}$. So by (4.3.24) it follows that f_* is $\mathscr{L}ip^{n-1}$.

Let now X be arbitrary. By definition (4.4.1) it is enough to show that $c^* \circ f_*: \mathscr{L}ip^k(X, E) \to \mathscr{L}ip^k(\mathbb{R}, F)$ is $\mathscr{L}ip^n$ for all $\mathscr{L}ip^k$-curves $c: \mathbb{R} \to X$. Using the commutative diagram:

$$\begin{array}{ccc}
\mathscr{L}ip^k(X, E) & \xrightarrow{f_*} & \mathscr{L}ip^k(X, F)\\
\downarrow{\scriptstyle c^*} & & \downarrow{\scriptstyle c^*}\\
\mathscr{L}ip^k(\mathbb{R}, E) & \xrightarrow{f_*} & \mathscr{L}ip^k(\mathbb{R}, E)
\end{array}$$

one reduces this immediately to the case $X = \mathbb{R}$. □

Remark. To obtain the conclusion of the proposition above it is not enough to have merely $f \in \mathscr{L}ip^{k+n-1}(E, F)$.

We first give an example in case $n=1$:

Let $f: \mathbb{R} \to \mathbb{R}$ be a map that is $\mathscr{L}ip^k$ but not $\mathscr{L}ip^{k+1}$. We claim that $f_*: C^\infty(\mathbb{R}, \mathbb{R})$

$\to \mathscr{Lip}^k(\mathbb{R}, \mathbb{R})$ is not $\mathscr{Lip}^0$. Otherwise $\delta^{k+1} \circ f_* \circ c: \mathbb{R} \to C^\infty(\mathbb{R}, \mathbb{R}) \to \mathscr{Lip}^k(\mathbb{R}, \mathbb{R}) \to \ell^\infty(\mathbb{R}^{\langle k+1 \rangle}, \mathbb{R})$ is $\mathscr{Lip}^0$ where $c: \mathbb{R} \to C^\infty(\mathbb{R}, \mathbb{R})$ denotes the curve $t \mapsto (s \mapsto t+s)$. Thus $\delta(\delta^{k+1} \circ f_* \circ c): \mathbb{R}^{\langle 1 \rangle} \to \ell^\infty(\mathbb{R}^{\langle k+1 \rangle}, \mathbb{R})$ is ℓ^∞. An easy calculation shows that $\delta(\delta^{k+1} \circ f_* \circ c)$ $(0, d; t, t+d, \ldots, t+(k+1)d) = \delta^{k+1} f(t, t+d, \ldots, t+(k+2)d) = \delta_{\text{eq}}^{k+1}(t, d)$ which is not locally bounded, since f is not $\mathscr{Lip}^{k+1}$.

In order to obtain an example for arbitrary n, take as f a function $\mathbb{R} \to \mathbb{R}$ whose nth derivative is $\mathscr{Lip}^k$ but not $\mathscr{Lip}^{k+1}$. For $\mathscr{S} := \{\mathrm{ev}_t;\, t \in \mathbb{R}\}$ the map f_* is n-times $\mathscr{S}$-differentiable and the differential $\mathrm{d}^n(f_*)$ of order n equals $(\mathrm{d}^n f)_* = (f^{(k)})_*$, hence is not $\mathscr{Lip}^0$. Thus $f_*: \mathscr{Lip}^k(\mathbb{R}, \mathbb{R}) \to \mathscr{Lip}^k(\mathbb{R}, \mathbb{R})$ is not $\mathscr{Lip}^n$.

4.4.16 Corollary. *Let X be a $\mathscr{Lip}^k$-space and E and F be convenient vector spaces. Then the composition map* $\mathrm{comp}: \mathscr{Lip}^{k+n+1}(E, F) \sqcap \mathscr{Lip}^k(X, E) \to \mathscr{Lip}^k(X, F)$ *is* $\mathscr{Lip}^n$.

Proof. The composition map comp is linear in the second variable, $\mathrm{comp}(f, \_) = f_*$ is $\mathscr{Lip}^n$ and $\mathrm{comp}(\_, g) = g^*$ is smooth. Thus by (4.3.5) the map comp is $\mathscr{Lip}^n$. □

Now we want to prove a Taylor formula for $\mathscr{Lip}^k$-functions f, cf. [Kock, 1984]. Such a formula involves terms of the form $f^{(j)}(x)(v, \ldots, v)$, which can be calculated as follows.

4.4.17 Lemma. *Let $f: E \supseteq U \to F$ be a $\mathscr{Lip}^k$-map, $x \in U$ and $v \in E$. Then $f^{(k)}(x)(v, \ldots, v) = (f \circ c)^{(k)}(0)$, where the curve c is defined locally around 0 by $c(t) := x + tv$.*

Proof. By (i) in (4.3.9) $\mathrm{d}f(x+tv, v)$ was defined as the derivative $(f \circ c)^\cdot(t)$. Thus the lemma is true for $k \leq 1$. The general statement follows now by induction using the recursive definition (ii) of (4.3.26) and the special chain rule:

$$f^{(k+1)}(x)(v, \ldots, v) := [f']^{(k)}(x)(v, \ldots, v)(v) = [f' \circ c]^{(k)}(0)(v) = [\mathrm{ev}_v \circ f' \circ c]^{(k)}(0)$$
$$= [(f \circ c)^\cdot]^{(k)}(0) = (f \circ c)^{(k+1)}(0). \qquad \square$$

4.4.18 Proposition. (*Taylor Expansion.*) *Let $f: E \supseteq U \to F$ be a $\mathscr{Lip}^{k+1}$-map; $x \in U$ and $v \in E$ with $x + [0, 1]v \subseteq U$. Then one has the following expansion of f:*

$$f(x+v) = \sum_{i=0}^{k} \frac{1}{i!} f^{(i)}(x)(v, \ldots, v) + \frac{1}{k!} \int_0^1 (1-s)^k f^{(k+1)}(x+sv)(v, \ldots, v)\mathrm{d}s.$$

Proof. We consider the $\mathscr{Lip}^{k+1}$-curve $c: t \mapsto f(x+tv)$. By (ii) in (4.1.14) we have $c(1) = c(0) + \int_0^1 c^\cdot(s)\mathrm{d}s$. By k-fold partial integration one obtains

$$c(1) = \sum_{i=0}^{k} \frac{1}{i!} c^{(i)}(0) + \frac{1}{k!} \int_0^1 (1-s)^k c^{(k+1)}(s)\mathrm{d}s.$$

The previous lemma (4.4.17), i.e. $c^{(i)}(s)=f^{(i)}(x+sv)(v,\ldots,v)$, yields the desired result. □

Since the derivatives $f^{(j)}(0)$ are symmetric and multilinear, cf. (4.3.28), they can be regained from their restriction to the diagonal and hence from the Taylor polynomial using the following lemma.

4.4.19 Lemma. (*Polarization Formula.*) *Let $m\colon E\times\ldots\times E\to F$ be a symmetric k-linear map between vector spaces and $\Delta\colon E\to E\times\ldots\times E$ the diagonal map, i.e. $\Delta(x):=(x,\ldots,x)$. Then one has for any $x_j\in E$:*

$$k!\cdot m(x_1,\ldots,x_k)=\sum_{j=1}^{k}(-1)^{k-j}\sum_{i_1<\ldots<i_j}(m\circ\Delta)(x_{i_1}+\cdots+x_{i_j}).$$

Proof. We compute $\delta^\kappa g(0,1;\ldots;0,1)$ in two ways, where $g\colon\mathbb{R}^k\to F$ is defined by $g(t_1,\ldots,t_k):=(m\circ\Delta)(t_1x_1+\cdots+t_kx_k)$ and κ is the multi-index $(1,\ldots,1)$. First we use the defining formula (1.3.4) for $\delta^\kappa g$. Since for the respective coefficients β_i one has $\beta_0(0,1)=-1$ and $\beta_1(0,1)=1$ one obtains exactly the right-hand side of the stated equation. For the second computation we use that by the multilinearity of m one has $g(t_1,\ldots,t_k)=\sum_{i_1,\ldots,i_k\in\{1,\ldots,k\}}t_{i_1}\ldots t_{i_k}m(x_{i_1},\ldots,x_{i_k})$, and δ^κ of the term with index $(i_1,\ldots,i_k)$ is 0 if $(i_1,\ldots,i_k)$ is not a permutation of $(1,\ldots,k)$, since this term is then independent of some variable t_i. The remaining terms are, by symmetry of m, all equal to $t_1\ldots t_k m(x_1,\ldots,x_k)$ and using (*v*) of (1.3.5) one finds that $\delta^\kappa g(0,1;\ldots;0,1)$ has the constant value $k!\cdot m(x_1,\ldots,x_k)$. □

4.4.20 Definition. For $k\in\mathbb{N}_0$ a map $f\colon E\to F$ between vector spaces is called *k-homogeneous* if $f(tv)=t^kf(v)$ for all $t\in\mathbb{R}$ and $v\in E$.

Remark. In a case where f is a $\mathcal{Lip}^{k-1}$-map between convenient vector spaces it is enough for k-homogeneity to assume that $f(tv)=t^kf(v)$ for all $t>0$ and $v\in E$. In fact, taking the kth derivative of this equation and then the limit for $t\to 0$ gives by (4.4.17) $f^{(k)}(0)(v,\ldots,v)=k!f(v)$. Since the k-linear map $f^{(k)}(0)$ restricted to the diagonal is obviously k-homogeneous, the same is true for f.

4.4.21 Lemma. *Let $f\colon E\to F$ be $\mathcal{Lip}^1$. Then the following statements are equivalent:*

(1) *f is $(k+1)$-homogeneous;*
(2) *$f(0)=0$ and f' is k-homogeneous.*

Proof. (1⇒2) We differentiate the equation $t^{k+1}f(x)=f(tx)$ in direction of v and obtain by (4.4.17) $t^{k+1}\cdot f'(x)(v)=f'(tx)(tv)=t\cdot f'(tx)(v)$.

(2⇒1) By (4.4.18) we have $f(tx)=f(0)+\int_0^1 f'(stx)(tx)ds=0+\int_0^1 t^kt\cdot f'(sx)(x)ds=t^{k+1}f(x)$. □

4.4.22 Proposition. (*Characterization of Homogeneous Maps.*) *Let $f: E \to F$ be a map between convenient vector spaces, $j \in \mathbb{N}_0$, $k \in \mathbb{N}_{0,\infty}$ and $j \le k$. Then the following statements are equivalent:*

(1_k) *f is $\mathcal{Lip}^k$ and j-homogeneous;*
(2_k) *f is $\mathcal{Lip}^{k+1}$, $f^{(j+1)}=0$ and $f^{(i)}(0)=0$ for all $i<j$;*
(3) *$f = m \circ \Delta$ for some symmetric j-linear $m \in L(E, \dots, E; F)$; where $\Delta(x) := (x, \dots, x)$;*
(4_k) *f is $\mathcal{Lip}^k$ and $df(v, v) = j \cdot f(v)$ (Euler's equation) for all $v \in E$.*

Proof. ($1_\infty \Rightarrow 1_k$) is trivial.

($1_k \Rightarrow 2_\infty$) Since f is j-homogeneous we conclude from (4.4.21) that $f^{(i)}$ is $(j-i)$-homogeneous for all $i \le j$, i.e. $f^{(i)}(tx) = t^{j-i} f^i(x)$. By substituting $t=0$ we conclude that $f^{(i)}(0)=0$ for $i<j$ and $f^{(j)}$ is constant and hence $\mathcal{Lip}^\infty$, and $f^{(j+1)}=0$.

($2_\infty \Rightarrow 2_k \Rightarrow 2_j$) is trivial.

($2_j \Rightarrow 3$) By (4.4.18) we have

$$f(v) = \sum_{i=1}^{j} \frac{1}{i!} f^{(i)}(0)(v, \dots, v) + \int_0^1 (1-s)^j f^{(j+1)}(sv)(v) ds = \frac{1}{j!} f^{(j)}(0)(v, \dots, v),$$

and $f^{(j)}(0)$ is j-linear and symmetric by (4.3.28).

($3 \Rightarrow 1_\infty$) $f(tv) = m(tv, \dots, tv) = t^j \cdot m(v, \dots, v) = t^j \cdot f(v)$, where m is a j-linear symmetric function with $f = m \circ \Delta$.

($1_k \Rightarrow 4_k$) We differentiate the equation $f(tv) = t^j \cdot f(v)$ with respect to t at $t=1$ and obtain: $df(v, v) = f'(v)(v) = j \cdot 1^{j-1} f(v) = j \cdot f(v)$.

($4_k \Rightarrow 1_k$) For fixed v we consider the curve $c(t) = t^{-j} f(tv)$ for $t>0$. We want to show that c is constant. So we take the derivative $\dot{c}(t) = t^{-j} f'(tv)(v) - j \cdot t^{-j-1} f(tv) = t^{-j-1}(j \cdot f(tv) - j \cdot f(tv)) = 0$ (by Euler's equation). Thus $t^{-j} f(tv) = c(t) = c(1) = f(v)$, but this is enough for j-homogeneity. □

4.4.23 Definition. For $j \in \mathbb{N}_0$ and convenient vector spaces E and F we denote with *$Homog_j(E, F)$* the vector space of all j-homogeneous maps from E to F which satisfy the equivalent conditions of (4.4.22).

4.4.24 Proposition. *Let $j \in \mathbb{N}_0$, $k \in \mathbb{N}_{0,\infty}$ with $j \le k$; and let E and F be convenient vector spaces. The initial Pre-structures on* $\text{Homog}_j(E, F)$ *induced by the following maps coincide:*

(1_k) *the inclusion map* $\text{Homog}_j(E, F) \to \mathcal{Lip}^k(E, F)$;
(2) *the map* $\text{ev}_0 \circ f^{(j)}$: $\text{Homog}_j(E, F) \to L(E, \dots, E; F)$.

For both maps there exist natural left inverse morphisms.

Proof. By (ii) in (4.4.9) we know that $f \mapsto (1/j!) f^{(j)}$, $\mathcal{Lip}^k(E, F) \to \mathcal{Lip}^{k-j}(E, L(E, \dots, E; F))$ is a morphism. Composing with the morphisms ev_0 and Δ^* gives a map which restricted to $\text{Homog}_j(E, F)$ is the identity. This shows the existence of the claimed left inverse morphisms and that the initial structures induced by the maps in (1_k) and (2) coincide. □

4.4.25 Definition. For $j \in \mathbb{N}_0$ and convenient vector spaces E and F we will from now on denote with *$Homog_j(E, F)$* the preconvenient vector space whose structure is described in the previous proposition. This makes $\mathrm{Homog}_j(E, F)$ into a convenient vector space since the initial embeddings of (4.4.24) have left inverse morphisms.

4.4.26 Proposition. (*Characterization of Polynomial Maps.*) *Let $j \in \mathbb{N}_0$, $k \in \mathbb{N}_{0,\infty}$, $j \leq k$; let $f: E \to F$ be a map between convenient vector spaces, and $\mathcal{S} \subseteq F'$ be point separating. Then the following statements are equivalent:*

- (1_k) *f is $\mathcal{Lip}^k$ and for every $\ell \in \mathcal{S}$ and $v \in E$ the real function $t \mapsto \ell(f(tv))$ is polynomial of degree at most j.*
- (2_k) *f is $\mathcal{Lip}^{k+1}$ and $f^{(j+1)} = 0$;*
- (3_k) *f is $\mathcal{Lip}^k$ and for one (or equivalently every) x and every v one has $f(x+v) = \sum_{i=0}^{j} \frac{1}{i!} f^{(i)}(x)(v, \ldots, v)$;*
- (4) *$f = \sum_{i=0}^{j} f_i$ for some $f_i \in \mathrm{Homog}_i(E, F)$;*

Proof. ($1_k \Rightarrow 2_k$) By (4.4.17) $f^{(i)}(x)(v, \ldots, v) = c^{(i)}(0)$ for the curve c defined by $c(t) := f(x+tv)$ which composed with an $\ell \in F'$ is polynomial of at most degree j. Thus $f^{(j)}(tx)(v, \ldots, v)$ is independent of t and substituting $t=0$ and $t=1$ gives that it is independent of x. Using the polarization formula (4.4.19) we conclude that $f^{(j+1)}(\_)(v_1, \ldots, v_k)$ is constant and hence f is $\mathcal{Lip}^\infty$ and $f^{(j+1)} = 0$.

($2_k \Rightarrow 3_k$) follows immediately for arbitrary $x \in E$ from the Taylor expansion (4.4.18).

($3_k \Rightarrow 4$) If (3_k) holds for $x=0$, this is obvious, since $f^{(i)}(0)$ is i-linear and symmetric by (4.3.28). If x is arbitrary then $g(v) := f(x+v)$ satisfies (3_k) hence (4), i.e. $f(x+v) = \sum_{i=0}^{j} g_i(v, \ldots, v)$. So the claim follows by substituting $w := x+v$ and developing $g_i(w-x, \ldots, w-x)$ into a finite sum of terms $g_i(w, \ldots, w, -x, \ldots, -x)$, which are homogeneous in w.

($4 \Rightarrow 1_k$) is trivial since $t \mapsto \ell(f_i(tv))$ *is* i-homogeneous. □

4.4.27 Definition. For $j \in \mathbb{N}_0$ a map $f: E \to F$ between convenient vector spaces is called *polynomial of at most degree j* if it satisfies the equivalent conditions of the previous proposition (4.4.26). With *$Poly_j(E, F)$* we denote the vector space of all polynomials from E to F of at most degree j.

4.4.28 Proposition. *Let $j \in \mathbb{N}_0$, $k \in \mathbb{N}_{0,\infty}$, $j \leq k$; let E and F be convenient vector spaces and $U \subseteq E$ be* M*-open with $0 \in U$. Then the initial Pre-structures on* $\mathrm{Poly}_j(E, F)$ *induced by the following maps coincide:*

- (1_k) *the inclusion map* $\mathrm{Poly}_j(E, F) \to \mathcal{Lip}^k(U, F)$;
- (2) *the map* $\mathrm{Poly}_j(E, F) \to \bigoplus_{i=0}^{j} \mathrm{Homog}_i(E, F)$, $f \mapsto ((1/i!) f^{(i)}(0))_{i=0}^{j}$.

The first map admits as left inverse the morphism T^j which associates to a map $f \in \mathcal{Lip}^k(U, F)$ its Taylor polynomial $T^j f: v \mapsto \sum_{i=0}^{j} (1/i!) f^{(i)}(0)(v, \ldots, v)$. The second map is even a bijection with left inverse $(f_i)_{i=0}^{j} \mapsto \sum_{i=0}^{j} f_i$.

Proof. The map $\mathcal{Lip}^k(U, F) \to \bigoplus_{i=0}^{j} \mathrm{Homog}_i(E, F)$, $f \mapsto ((1/i!) f^{(i)}(0))_{i=0}^{j}$ composed with $\Sigma: \bigoplus_{i=0}^{j} \mathrm{Homog}_i(E, F) \to \mathrm{Poly}_j(E, F) \subseteq \mathcal{Lip}^\infty(E, F)$ yields the Taylor expansion T^j. Restricted to $\mathrm{Poly}_j(E, F)$ the Taylor expansion is the identity. Thus we have proved that the claimed maps are left inverse and that the initial structures induced by the maps in (1_k) and (2) coincide. □

4.4.29 Proposition. (*Characterization of Flat Maps.*) *Let $j \in \mathbb{N}_{0,\infty}$, let $f: E \supseteq U \to F$ be a $\mathcal{Lip}^j$-map between convenient vector spaces with $0 \in U$, and let $\mathcal{S} \subseteq F'$ be point separating. Then the following statements are equivalent:*

(1) *$f^{(i)}(0) = 0$ for all $0 \le i < j+1$;*
(2) *For all $\mathcal{Lip}^j$-curves c: $\mathbb{R} \to U$ with $c(0) = 0$ the derivatives of order less than $j+1$ of the composites $f \circ c$ vanish at 0;*
(3) *For all $v \in E$ and $\ell \in \mathcal{S}$ the derivatives of order less than $j+1$ of the real function $t \mapsto \ell(f(tv))$ vanish at 0.*

Proof. ($1 \Rightarrow 2$) this follows by applying the chain rule (4.3.14) inductively.

($2 \Rightarrow 3$) is trivial.

($3 \Rightarrow 1$) is trivial by the polarization formula (4.4.19) since $\ell(f^{(i)}(0)(v, \ldots, v)) = c^{(i)}(0)$ by (4.4.17), where $c(t) := \ell(f(tv))$. □

4.4.30 Definition. For $j \in \mathbb{N}_{0,\infty}$ a $\mathcal{Lip}^j$-map $f: E \supseteq U \to F$ is called *j-flat* (*at* 0) if the equivalent conditions of the previous proposition are satisfied. The vector space of all j-flat $\mathcal{Lip}^k$-maps (for $k \in \mathbb{N}_{0,\infty}$, $j \le k$) with the initial Pre-structure induced from the inclusion in $\mathcal{Lip}^k(U, F)$ will be denoted by $\mathcal{Lip}^k_{j\text{-}flat}(U, F)$. It is in fact a convenient vector space by the closed embedding lemma (2.6.4), since it is given by the equations $f^{(i)}(0) = 0$ $(i < j+1)$.

4.4.31 Theorem. *Let E and F be convenient vector spaces, $U \subseteq E$ be* M*-open with $0 \in U$; $j \in \mathbb{N}_0$, $k \in \mathbb{N}_{0,\infty}$, $j \le k$. Then one has the decomposition*

$$\mathcal{Lip}^k(U, F) = \mathrm{Poly}_j(E, F) \oplus \mathcal{Lip}^k_{j\text{-}flat}(U, F)$$

given by $f = T^j f + (f - T^j f)$ where $T^j f$ is the Taylor polynomial of order j of f.

Proof. Since T^j is a left inverse to the inclusion map $\mathrm{Poly}_j(E, F) \to \mathcal{Lip}^k(U, F)$ by (4.4.28) and the kernel of T^j is $\mathcal{Lip}^k_{j\text{-}flat}(U, F)$ by (4.4.29) the stated decomposition follows. □

4.4.32 Remark. We shall see in (7.1.3) that the analogous result for $j = \infty$ fails to be true: i.e. $\mathcal{Lip}^\infty_{\infty\text{-}flat}(E, F)$ is not a direct summand of $\mathcal{Lip}^\infty(E, F)$, not even for $E = F = \mathbb{R}$.

Now we take up the discussion about point separating families of linear functionals initiated in section 4.1:

4.4.33 Proposition. *Let X be a $\mathscr{Lip}^k$-space and F a convenient vector space. Then the bornology of $\mathscr{Lip}^k(X, F)$ has a basis of $\sigma(\mathscr{Lip}^k(X, F), \{\ell \circ \mathrm{ev}_x;\ x \in X, \ell \in F'\})$-closed sets, where ev_x: $\mathscr{Lip}^k(X, F) \to F$ denotes the evaluation map.*

Proof. Since c^*: $\mathscr{Lip}^k(X, F) \to \mathscr{Lip}^k(\mathbb{R}, F)$ ($c \in \mathscr{Lip}^k(\mathbb{R}, X)$) is an initial family of Pre-morphisms, and since $\{\mathrm{ev}_t \circ c^*;\ c \in \mathscr{Lip}^k(\mathbb{R}, X),\ t \in \mathbb{R}\} = \{\mathrm{ev}_x;\ x \in X\}$, it is enough to show the statement for $X = \mathbb{R}$. In this case the difference quotients δ^j: $\mathscr{Lip}^k(\mathbb{R}, F) \to \ell^\infty_? (\mathbb{R}^{\langle j \rangle}, F)$ $(0 \leq j < k+2)$ form an initial family of Pre-morphisms and the sets $\{\mathrm{ev}_t;\ t \in \mathbb{R}\}$ and $\{\mathrm{ev}_x \circ \delta^j;\ x \in \mathbb{R}^{\langle j \rangle},\ 0 \leq j < k+2\}$ generate the same linear subspace of $L(\mathscr{Lip}^k(\mathbb{R}, F), F)$ the result follows from the corresponding one in (iv) of (4.1.21) for $\ell^\infty(X, F)$. □

The above proposition can be improved by taking only the evaluations ev_x with x in some dense subset:

4.4.34 Corollary. *Let X be a $\mathscr{Lip}^k$-space and $D \subseteq X$ a subset of X which is dense for the final topology induced by the $\mathscr{Lip}^k$-curves into X; let F be a convenient vector space. Then the bornology on $\mathscr{Lip}^k(X, F)$ has a basis of $\sigma(\mathscr{Lip}^k(X, F), \{\ell \circ \mathrm{ev}_x;\ \ell \in F', x \in D\})$-closed sets.*

Proof. Because of (4.1.24) and (4.4.33) it is enough to prove that $\{\mathrm{ev}_x;\ x \in D\}$ is dense in $\{\mathrm{ev}_x;\ x \in X\}$ with respect to the topology of uniform convergence on bounded subsets of $\mathscr{Lip}^k(X, F)$. In (4.4.14) we proved that ev: $\mathscr{Lip}^k(X, F) \sqcap X \to F$ is $\mathscr{Lip}^k$. Hence the associated map ι: $X \to L(\mathscr{Lip}^k(X, F), F)$ is $\mathscr{Lip}^k$ and in particular continuous for the final topologies induced by the $\mathscr{Lip}^k$-curves. Thus for the closures in these topologies one has the inclusions: $\overline{\iota D = \{\mathrm{ev}_x;\ x \in D\}} \supseteq \iota(\bar{D}) = \iota X = \{\mathrm{ev}_x;\ x \in X\}$. Since the Mackey-closure topology on $L(\mathscr{Lip}^k(X; F), F)$, i.e. the final topology induced by the $\mathscr{Lip}^k$-curves, is finer than the topology of uniform convergence on bounded sets, cf. (3.6.8), the result follows. □

In order to show that this result is interesting even in the special case where $X = F = \mathbb{R}$ and $k = \infty$ we give an

4.4.35 Example. Suppose bornological curves c^k: $\mathbb{R} \to \mathscr{Lip}^\infty(\mathbb{R}, \mathbb{R})$ are given for all $k \in \mathbb{N}_0$. If $\mathrm{ev}_s \circ c^0$ is smooth with derivatives $(\mathrm{ev}_s \circ c^0)^{(k)} = \mathrm{ev}_s \circ c^k$ for all k and all $s \in D$ for some dense set $D \subseteq \mathbb{R}$, then c^0 is smooth and c^k is its kth derivative (one combines (4.1.19) and (4.4.34)).

Another formulation of the same result is the following: let f: $\mathbb{R}^2 \to \mathbb{R}$ be a function. Suppose that for some dense set $D \subseteq \mathbb{R}$ the first partial derivatives $\partial_1^k f(t, s)$ exist for all $k \in \mathbb{N}_0$, $t \in \mathbb{R}$ and $s \in D$; that furthermore for all $t \in \mathbb{R}$ the function $\partial_1^k f(t, \_)$ has a smooth extension; and that $\partial_2^i \partial_1^k f$ is bornological for all

$i, k \in \mathbb{N}_0$. Then $f\colon \mathbb{R}^2 \to \mathbb{R}$ is smooth (put $c^k := (\partial_1^k f)^\vee$ to reduce this to the result above).

In particular (for $D = \mathbb{R}$) any function $f\colon \mathbb{R}^2 \to \mathbb{R}$ for which all partial derivatives $\partial_2^i \partial_1^k f$ exist and are locally bounded is smooth.

We shall show that quite often the locally convex topology of our convenient function spaces coincides with some classical function space topology. For this we shall first describe their bornology as the von Neumann bornology of a locally convex topology $\mathscr{T}$. Then $\mathscr{T}$ is the locally convex topology of the convenient vector space exactly if it is bornological, and this is in particular the case if $\mathscr{T}$ admits a countable basis for the 0-neighborhoods.

We will use the following kind of difference quotients, cf. (3) in (4.3.8):

4.4.36 Definition. For $f\colon E \supseteq U \to F$ and $B \subseteq E$ absolutely convex and bounded, $\delta_B^1 f\colon U_B^{\langle 1 \rangle} \to F$ is defined by

$$\delta_B^1 f(x, y) := \frac{f(x) - f(y)}{\|x - y\|_B}$$

where $U_B := U \cap E_B$.

We consider first the $\mathscr{Lip}^0$-function spaces.

4.4.37 Lemma. *Let E and F be convenient vector spaces, U an* M-*open subset of E, $\mathscr{K}$ a basis of the* b-*compact bornology of U formed by* b-*compact sets and $\mathscr{B}$ a basis of the bornology of E formed by absolutely convex sets. Then a set $A \subseteq \mathscr{Lip}^0(U, F)$ is bounded iff the sets $A(K)$ and $\delta_B^1 A(K^{\langle 1 \rangle})$ are bounded in F for every $K \in \mathscr{K}$ and $B \in \mathscr{B}$ with $K \subseteq E_B$ compact..*

Proof. ($\Leftarrow$) Let $c\colon \mathbb{R} \to U$ be $\mathscr{Lip}^0$ and $I \subseteq \mathbb{R}$ be a compact interval. Then we choose an absolutely convex bounded set B which contains $c(I)$ and $\delta^1 c(I^{\langle 1 \rangle})$. Since $c\colon I \to U \cap E_B \subseteq E_B$ is a $\mathscr{Lip}^0$-curve of the normed space E_B we can choose a $K \in \mathscr{K}$ which contains $c(I)$. Finally we enlarge B such that it is an element of $\mathscr{B}$ and K is compact in E_B. By assumption the sets $A(K)$ and $\delta_B^1 A(K^{\langle 1 \rangle})$ are bounded. Writing

$$\delta(f \circ c)(t, s) = \delta_B^1 f(c(t), c(s)) \frac{\|c(t) - c(s)\|_B}{t - s}$$

in case $c(t) \neq c(s)$ one deduces that $\delta(f \circ c)(I^{\langle 1 \rangle})$ and $(f \circ c)(I)$ are bounded for $f \in A$. Thus $c^*(A) \subseteq \mathscr{Lip}^0(\mathbb{R}, F)$ is bounded. Since c was an arbitrary $\mathscr{Lip}^0$-curve in U we conclude that $A \subseteq \mathscr{Lip}^0(U, F)$ is bounded.

($\Rightarrow$) Admit first that $A(K)$ is unbounded for some $K \in \mathscr{K}$. Since every sequence in K has an M-converging subsequence, one concludes, using the special curve lemma (2.3.4), that $A(c(I))$ is unbounded for some smooth curve $c\colon \mathbb{R} \to U$ and some compact interval $I \subseteq \mathbb{R}$. Since $c^*(A)$ is bounded in $\mathscr{Lip}^0(\mathbb{R}, F)$ and therefore in $\ell^\infty(\mathbb{R}, F)$ this is a contradiction.

Admit now that $\delta_B^1 A(K^{\langle 1 \rangle})$ is unbounded for some $K \in \mathscr{K}$ and $B \in \mathscr{B}$ with $K \subseteq E_B$ compact. Then there exists a functional $\ell \in F'$ and sequences $a_n, b_n \in K$

and $f_n \in A$ such that $|\ell(\delta_B^1 f_n(a_n, b_n))| \geq n^{n+1}$. By going if necessary to subsequences and using that we already showed that $A(K)$ is bounded we may assume that a_n and b_n converge in E_B to a common limit, say 0, and that this convergence is so fast that $n^{2n}\|a_n\|_B$ and $n^{2n}\|b_n\|_B$ remain bounded. Let now $2d_n := \|a_n - b_n\|_B \cdot n^n$ and consider the line p_n: $\mathbb{R} \to E_B$ characterized by $p_n(-d_n) = a_n$ and $p_n(d_n) = b_n$. Since the sequence p_n is obviously fast falling in $C^\infty(\mathbb{R}, E_B)$ we can apply the general curve lemma (4.2.15) to obtain a smooth curve c: $\mathbb{R} \to E_B$ joining the pieces $p_n|_{[-d_n, d_n]}$ and one has $c(t_n) = a_n$ and $c(s_n) = b_n$ for some t_n and s_n in a bounded interval and satisfying $t_n - s_n = 2d_n$. Since

$$\frac{\ell(f_n(c(t_n))) - \ell(f_n(c(s_n)))|}{t_n - s_n} = |\ell(\delta_B^1 f_n(a_n, b_n))| \cdot \frac{\|a_n - b_n\|_B}{t_n - s_n} \geq n^{n+1} \cdot \frac{1}{n^n} = n$$

the set $\{\ell \circ f_n \circ c;\, n \in \mathbb{N}\}$ is not bounded in $\mathscr{Lip}^0(\mathbb{R}, \mathbb{R})$, hence A is not bounded in $\mathscr{Lip}^0(U, F)$. □

In view of the last lemma it is natural to try to characterize those convenient vector spaces for which a given M-open subset has a countable basis of the b-compact bornology. We start by considering convenient vector spaces that have a countable basis of the compact bornology associated to the locally convex topology.

4.4.38 Proposition. *For any convenient vector space E the following statements are equivalent:*

(1) *The compact bornology of the locally convex topology of E has a countable basis;*
(2) *The bornology of E has a countable basis and the locally convex topology of E is Montel;*
(3) *E is the dual of a Fréchet Montel space (for which one can choose E′).*

Proof. (1⇒2) Let $\{K_n;\, n \in \mathbb{N}\}$ be a countable basis of the bornology generated by the compact subsets of E.

It is enough to show that every bounded $B \subseteq E$ is contained in K_n for some n, since this implies that E is Montel and $\{K_n;\, n \in \mathbb{N}\}$ is a basis of the bornology of E. Admit that $B \nsubseteq K_n$ for every $n \in \mathbb{N}$. Then $B \nsubseteq n \cdot K_n$ for every n, since $n \cdot K_n$ is compact and is thus contained in some K_m. Choose $b_n \in B$ with $b_n \notin n \cdot K_n$. Since $(1/n)b_n$ converges (Mackey) to 0, the set $K := \{0\} \cup \{(1/j)b_j;\, j \in \mathbb{N}\}$ is compact and thus contained in some K_n. In particular $(1/n)b_n \in K_n$, which is a contradiction.

(2⇒3) Since the bornology of E has a countable basis we conclude that the locally convex topology on $F := E'$ is the strong topology and is metrizable. The locally convex topology of F is even Montel, since every bounded subset of E' is equicontinuous, hence by the Alaoğlu-Bourbaki theorem [Jarchow, 1981, p. 157] relatively compact in the topology of uniform convergence on precompact subsets. But this is the strong topology, since E is Montel.

It remains to show that $F' \cong E$. It is enough to show that ι: $E \to E'' = F'$ is onto. So let $x \in F' = E''$. Then $x \in U^0$ for some 0-neighborhood U in F ($x \in F'$ implies that $x^{-1}([-1,1])$ is the desired 0-neighborhood in F). By the definition of the strong topology there has to exist an absolutely convex bounded $B \subseteq E$ with $B^0 \subseteq U$. Thus x is contained in the bipolar of B with respect to (F', F). Since B is relatively compact in E, the image by ι of its closure is compact in E'' and thus contains the bipolar of B. Hence $x \in \iota(E)$.

(2⇐3) Let F be a Fréchet Montel space. The bornology of $E := F'$ is formed by the equicontinuous subsets, hence has a countable basis $\{U^0; U \in \mathcal{U}\}$, where $\mathcal{U}$ is a countable 0-neighborhood basis of the locally convex topology of F and U^0 denotes the polar $\{\ell \in F'; |\ell(x)| \leq 1 \text{ for all } x \in U\}$ of U. By the Alaoğlu-Bourbaki theorem [Jarchow, 1981, p. 157] these sets U^0 are compact in the topology of uniform convergence on precompact subsets of F. Since F is Montel this topology is the strong topology on E and it remains to show that it is the locally convex topology of E. For this it is enough to show that it is bornological. Since F as locally convex space is semi-reflexive [Jarchow, 1981, p. 230], the strong topology on E is barrelled [Jarchow, 1981, p. 227] and since E is metrizable it thus is bornological [Jarchow, 1981, p. 280].

(2⇒1) since for Montel spaces the bornology and the compact bornology coincide by definition. □

Now we are able to characterize the convenient vector spaces for which an M-open subset has a countable basis of the b-compact bornology.

4.4.39 Proposition. *Let E be a convenient vector space and $U \neq \emptyset$ an* M*-open subset of E. Then the following statements are equivalent:*

(1) *The* b*-compact bornology of U has a countable basis;*
(2) *The* b*-compact bornology of E has a countable basis;*
(3) *The bornology of E has a countable basis and every bounded subset is contained in a* b*-compact subset;*
(4) *E is the dual of a Fréchet Schwartz space (for which one can choose E').*

Proof. (1⇒2) We may assume that one has a countable basis $\{K_n; n \in \mathbb{N}\}$ of the b-compact bornology of U with $K_n \subseteq K_{n+1}$ for all n and that $0 \in U$. We show that $\{nK_n; n \in \mathbb{N}\}$ is a basis of the b-compact bornology of E. So let $K \subseteq E$ be bornologically compact. Then it gets absorbed by U, i.e. $K \subseteq n \cdot U$ for some $n \in \mathbb{N}$, otherwise $k_n \in K$ exist with $k_n \notin n \cdot U$, but $(1/n)k_n$ converges Mackey to 0, contradiction. Since $(1/n)K$ is bornologically compact there exists an $m \in \mathbb{N}$ with $(1/n)K \subseteq K_m$. For the maximum N of n and m we obtain

$$\frac{1}{N}K \subseteq \frac{1}{n}K \subseteq K_m \subseteq K_N,$$

i.e. $K \subseteq N \cdot K_N$.

(2⇒3) Similarly to the proof of (2⇒3) of the previous proposition (4.4.38) one shows that every bounded B is contained in K_n for some $n \in \mathbb{N}$. Moreover, K_n is by definition compact in E_B for some bounded B, which itself has to be contained in K_m for some $m \in \mathbb{N}$. Thus the inclusion $E_B \to E_{K_m}$ is continuous and K_n is compact in E_{K_m}.

($3 \Rightarrow 4$) By the previous proposition (4.4.38) $F := E'$ is a Fréchet Montel space and $E \cong F'$. It remains to show that the locally convex topology of F is Schwartz, i.e. for every absolutely convex 0-neighborhood U of F there is another one W containing U and such that the image of U is precompact in the normed space $F_{(W)} := F_W / \{x; \|x\|_W = 0\}$, or equivalently [Jarchow, 1981, p. 201] such that U^0 is compact in $(F')_{W^0}$. Since $F' \cong E$ and the sets U^0 form a basis of the bornology of E this amounts exactly to saying that every bounded set is contained in a bornologically compact set.

($1 \Leftarrow 4$) Let F be a Fréchet Schwartz space, $\mathscr{U} := \{U_n; n \in \mathbb{N}\}$ be a countable 0-neighborhood basis of the locally convex topology of F and $E := F'$. The family $\{U^0;\ U \in \mathscr{U}\}$ forms a countable basis of the bornology of E and, since F is Schwartz, we may assume that $K_n := (U_n)^0$ is compact in $E_{n+1} := E_{K_{n+1}}$. Let $\|\text{–}\|_n := \|\text{–}\|_{K_n}$ be the norm on E_n and W_n the open unit ball of E_n. For every $m \in \mathbb{N}$ the family $\{(1/m) W_{n+1} + x;\ x \in K_n\}$ forms an open covering of the compact set $K_n \subseteq E_{n+1}$, hence there exists a finite set $D_{m,n} \subseteq K_n$ such that $\{(1/m) W_{n+1} + x;\ x \in D_{m,n}\}$ still covers K_n. We claim that $\{(1/m) K_{n+1} + x;\ m, n \in \mathbb{N}, x \in D_{m,n}, (1/m) K_{n+1} + x \subseteq U\}$ is a subbasis of the b-compact bornology of U. In order to show this let $K \subseteq U$ be a b-compact set, i.e. $K \subseteq K_n$ for some $n \in \mathbb{N}$. Then $\{(1/m) W_{n+1} + x;\ n \in \mathbb{N},\ x \in D_{m,n},\ (1/m) K_{n+1} + x \subseteq U\}$ is an open covering of K in E_{n+1} and hence has a finite subcovering, i.e. K is contained in a finite union of sets $(1/m) K_{n+1} + x \subseteq U$. In fact, for any $y \in K$ there exists an $m \in \mathbb{N}$ with $\{x; \|y - x\|_{n+1} \leq 2/m\} \subseteq U$. By the definition of $D_{m,n}$ there exists an $x \in D_{m,n}$ with $y \in (1/m) U_{n+1} + x$, i.e. $\|x - y\|_{n+1} < 1/m$. Let $z \in ((1/m) K_{n+1} + x)$. Since $K_{n+1} \subseteq \{z;\ \|z\|_{n+1} \leq 1\}$ one concludes that $\|z - x\|_{n+1} \leq 1/m$ and hence $\|z - y\|_{n+1} \leq \|z - x\|_{n+1} + \|x - y\|_{n+1} < 2 \cdot (1/m)$, which shows that $z \in U$. □

4.4.40 Corollary. *Let E and F be convenient vector spaces, U an* M*-open subset of E, $\mathscr{K}$ a basis of the* b*-compact bornology of U formed by* b*-compact sets and $\mathscr{B}$ a basis of the bornology of E formed by absolutely convex sets. Then we have:*

(i) *The bornology of $\mathscr{Lip}^0(U, F)$ is the von Neumann bornology of the topology of uniform convergence (in the locally convex topology of F) of f on K and $\delta_B^1 f$ on $K^{\langle 1 \rangle}$ for $K \in \mathscr{K}$, $B \in \mathscr{B}$ and $K \subseteq E_B$ compact.*

  A 0-neighborhood basis of this topology is given by the sets $\{f \in \mathscr{Lip}^0(U, F); f(K) \subseteq V$ and $\delta_B^1 f(K^{\langle 1 \rangle}) \subseteq V\}$ with $K \in \mathscr{K}$, $B \in \mathscr{B}$, $K \subseteq E_B$ compact and $V \in \mathscr{V}_0$ for some 0-neighborhood basis $\mathscr{V}_0$ of the locally convex topology of E.

(ii) *The locally convex topology of $\mathscr{Lip}^0(U, F)$ is the topology described in* (i) *if the* b*-compact bornology of U has a countable basis and F is metrizable.*

Proof. (i) We have to show that $A \subseteq \mathscr{Lip}^0(U, F)$ is bounded iff it gets absorbed by the sets $\{f \in \mathscr{Lip}^0(U, F); f(K) \subseteq V\}$ and $\{f \in \mathscr{Lip}^0(U, F);\ \delta_B^1 f(K^{\langle 1 \rangle}) \subseteq V\}$ with $K \in \mathscr{K}$, $B \in \mathscr{B}$ and $K \subseteq E_B$ compact. By lemma (4.4.37), A is bounded iff each of the sets $A(K)$ and $\delta_B^1 A(K^{\langle 1\rangle})$ is bounded and hence absorbed by V, i.e. is contained in $N\cdot V$ for some $N\in\mathbb{N}$. This is exactly the case when A is contained in $N\cdot\{f\in \mathscr{Lip}^0(U,F);\, f(K)\subseteq V\}$ and in $N\cdot\{f\in\mathscr{Lip}^0(U,F);\, \delta_B^1 f(K^{\langle 1\rangle})\subseteq V\}$.

(ii) Let $\mathscr{K}$ be a countable basis of the b-compact bornology of U. Then by (4.4.39) there is a countable basis $\mathscr{B}$ of the bornology of E (formed by absolutely convex sets). Since F is metrizable its locally convex topology has a countable 0-neighborhood basis $\mathscr{V}_0$. Hence the topology on $\mathscr{Lip}^0(U,F)$ which was described in (i) has a countable 0-neighborhood basis and thus is bornological. Since by (i) its von Neumann bornology is the bornology of $\mathscr{Lip}^0(U,F)$, the assertion follows. □

We next consider the spaces $\mathscr{Lip}^k(U,F)$ with $k\in\mathbb{N}_\infty$.

4.4.41 Corollary. *The locally convex topology of $\mathscr{Lip}^k(U,F)$ is the bornologification of the topology of uniform convergence (in the locally convex topology of $L(E,\ldots,E;F)$) of $f^{(i)}$ on K for $i<k+1$ and (in case where $k<\infty$) of $\delta_B^1 f^{(k)}$ on $K^{\langle 1\rangle}$ with compact $K\subseteq E_B$. Thus the locally convex topology of $\mathscr{Lip}^k(U,F)$ is metrizable if the* b-*compact bornology of U has a countable basis and F is metrizable.*

Proof. One uses (4.4.11), i.e. the initiality of the family of maps $f\mapsto f^{(j)}$, $\mathscr{Lip}^k(U,F)\to\mathscr{Lip}^0(U,L(E,\ldots,E;F))$ for $j<k+1$. It remains to show that the condition on $\delta_B^1 f^{(j)}$ is implied by that on $f^{(j+1)}$. First remark that it is enough to show uniform boundedness of $K^{\langle 1\rangle}$ for those K having the additional property that $\{tx+(1-t)y;\, x,y\in K,\, t\in[0,1]\}$ is a subset of U, since these K form a subbasis of the b-compact bornology of U. In fact if $K\subseteq E_B$ is an arbitrary b-compact subset of U then we consider the continuous map $(t,x_1,x_2)\mapsto(tx_1+(1-t)x_2)$, $[0,1]\sqcap E_B\sqcap E_B\to E$ which maps $[0,1]\times\{x\}\times\{x\}$ to x for all $x\in K$. Thus there exists a neighborhood U_x of x in E_B such that $[0,1]\times U_x\times U_x$ is mapped into U. These sets U_x $(x\in K)$ form a covering of K with M-open subsets, hence admit a finite subcovering. A closed refinement of this finite subcovering then consists of b-compact sets with the desired additional property.

Now let $K\subseteq E_B$ be such a set and let $g:=f^{(j)}$. Then

$$\delta_B^1 g(x,y):=\frac{g(x)-g(y)}{\|x-y\|_B}=\int_0^1 g'(x+t(y-x))\left(\frac{y-x}{\|y-x\|_B}\right)dt$$

and since the maps g' are by assumption bounded on the b-compact set $\{tx+(1-t)y;\, t\in[0,1],\, x,y\in K\}$ and since $(y-x)/\|y-x\|_B$ is bounded, the difference quotient $\delta_B^1 g$ is uniformly bounded on $K^{\langle 1\rangle}$.

In order to obtain the metrizability by applying (4.4.40) we only need to know that $L(E,F)$ is metrizable if F is metrizable and E has a countable basis of the b-compact bornology. But also this follows from (4.4.40) since $L(E,F)$ is a Pre-subspace of $\mathscr{Lip}^0(E,F)$. □

Next we want to apply the previous corollary in order to identify the topology of $\mathscr{Lip}^k(X,F)$ for a manifold X with the classically considered topology. For this we introduce manifolds modelled on convenient vector spaces; see also [Wegenkittl, 1987] where with the aid of jets various topologies on spaces of smooth maps between such manifolds are discussed.

4.4.42 Definition. A *chart* for a set X is an injective map $u: U \to X$, where U is M-open in some convenient vector space.

A $\mathcal{Lip}^k$*-atlas* on a set X is a family $\mathcal{A}$ of charts such that the images of all charts in $\mathcal{A}$ cover X and for any two charts $u_1: U_1 \to X$ and $u_2: U_2 \to X$ in $\mathcal{A}$ the domain $u_1^{-1}(u_2(U_2))$ is M-open and $u_2^{-1} \circ u_1: u_1^{-1}(u_2(U_2)) \to u_2^{-1}(u_1(U_1)) \subseteq U_2$ is a $\mathcal{Lip}^k$-map.

Two $\mathcal{Lip}^k$-atlases on a set X are called $\mathcal{Lip}^k$-equivalent if the union is a $\mathcal{Lip}^k$-atlas.

A $\mathcal{Lip}^k$*-manifold* (*modelled on convenient vector spaces*) is a set X together with an equivalence class of $\mathcal{Lip}^k$-atlases on X, or, equivalently, with a maximal $\mathcal{Lip}^k$-atlas.

A map $f: X_1 \to X_2$ between two $\mathcal{Lip}^k$-manifolds X_1 and X_2 is called of *class* $\mathcal{Lip}^k$ if for any chart u_1 of X_1 and u_2 of X_2 the domain $u_1^{-1}(f^{-1}(u_2(U_2)))$ is M-open and the composite $u_2^{-1} \circ f \circ u_1: u_1^{-1}(f^{-1}(u_2(U_2))) \to u_2^{-1}(f(u_1(U_1))) \subseteq U_2$ is $\mathcal{Lip}^k$.

On every $\mathcal{Lip}^k$-manifold we will consider the final $\mathcal{Lip}^k$-structure induced by the family of charts of the maximal $\mathcal{Lip}^k$-atlas (or, equivalently, of an equivalent $\mathcal{Lip}^k$-atlas).

On every $\mathcal{Lip}^k$-space X, and in particular on every $\mathcal{Lip}^k$-manifold, we will consider the final topology generated by the $\mathcal{Lip}^k$-curves of X.

Remark. If X is a $\mathcal{Lip}^k$-manifold, the structure-functions of the considered $\mathcal{Lip}^k$-structure on X are precisely the maps $f: X \to \mathbb{R}$ which composed with charts of X are $\mathcal{Lip}^k$, i.e. are exactly the functions of class $\mathcal{Lip}^k$. However, there may be more $\mathcal{Lip}^k$-curves for the $\mathcal{Lip}^k$-structures on such manifolds than curves of class $\mathcal{Lip}^k$. The following proposition gives equivalent conditions when this does not occur.

4.4.43 Proposition. *Let X be a $\mathcal{Lip}^k$-manifold modelled on convenient vector spaces with (maximal) $\mathcal{Lip}^k$-atlas $\mathcal{A}$. Then the following statements are equivalent:*

(1) *The structure curves of the natural $\mathcal{Lip}^k$ structure of X are precisely the curves in X of class $\mathcal{Lip}^k$;*
(2) *The $\mathcal{Lip}^k$-maps from any $\mathcal{Lip}^k$-manifold into X are precisely the maps of class $\mathcal{Lip}^k$;*
(3) *The image of each chart is open;*
(4) *The topology of X is the classical one, i.e. the final one induced by the charts of $\mathcal{A}$.*

Proof. (1⇒4) Since by assumption (1) the $\mathcal{Lip}^k$-structure curves factor locally over the charts, the final topology induced by the $\mathcal{Lip}^k$-curves and the one induced by the charts coincide.

(4⇒3) For any chart u: $U \to X$ we have to show that the image $u(U)$ is open, i.e. the inverse image of $u(U)$ under any other chart v: $V \to X$ is M-open. This is obvious, since by definition of an atlas $v^{-1}(u(U))$ is M-open in V.

(3⇒2) Let Y be another $\mathscr{Lip}^k$-manifold and let g: $Y \to X$ be a $\mathscr{Lip}^k$-map. We have to show that g is of class $\mathscr{Lip}^k$. So let u: $U \to X$ and v: $V \to Y$ be charts. We first show that $v^{-1}(g^{-1}(u(U)))$ is M-open in V. Obviously g is continuous with respect to the final topologies generated by the $\mathscr{Lip}^k$-structure-curves and by assumption $u(U)$ is open in X. Thus $g^{-1}(u(U))$ is open in Y, so $c^{-1}(v^{-1}(g^{-1}(u(U))))$ is open in $\mathbb{R}$ for every $\mathscr{Lip}^k$-curve c: $\mathbb{R} \to V$, since $v \circ c$ is a $\mathscr{Lip}^k$-curve in Y. This shows at the same time that $u^{-1} \circ g \circ v \circ c$ is a $\mathscr{Lip}^k$-curve, i.e. $u^{-1} \circ g \circ v$ is $\mathscr{Lip}^k$ and thus g is of class $\mathscr{Lip}^k$.

(2⇒1) is obvious by taking as second manifold $\mathbb{R}$. □

Now let us show that in all interesting cases the equivalent conditions of the previous proposition are satisfied.

4.4.44 Proposition. *Let X be a $\mathscr{Lip}^k$-manifold modelled on convenient vector spaces such that the topology of X is regular and the* M*-closure topology of each of the modelling vector spaces is the initial one induced by its $\mathscr{Lip}^k$-functions. Then X satisfies the equivalent conditions of* (4.4.43).

Remarks. (i) For a Banach space the topology is the initial one induced by its smooth functions if and only if there exists a smooth non-zero function with bounded support [Bonic, Frampton, 1966].

(ii) Examples of Fréchet spaces which carry the initial topology induced by the smooth functions are: all nuclear ones [Michor, 1983], all function spaces $C^\infty(Z, E)$ with Z a finite-dimensional separable manifold and E a Fréchet space satisfying the condition; and in particular $C^\infty(Z, \mathbb{R}^n)$ with Z as before.

(iii) If the M-closure topology of a convenient vector space coincides with the locally convex topology then it also coincides with the initial topology induced by the $\mathscr{Lip}^0$-functions, cf. (6.4.4).

Proof. We will verify condition (3) of (4.4.43). So let $u{:}U \to X$ be a chart and consider a $\mathscr{Lip}^k$-structure-curve $c{:}\mathbb{R} \to X$. We have to show that $c^{-1}(u(U))$ is open in $\mathbb{R}$. So let $t_0 \in c^{-1}(u(U))$. Since the topology of X was assumed to be regular we can choose an open neighborhood V of $x_0 := c(t_0)$ such that its closure is contained in $u(U)$. By assumption on the topology of the modelling vector space $E \supseteq U$ there exists a $\mathscr{Lip}^k$-function $h{:}E \to \mathbb{R}$ with $h(u^{-1}(x_0)) > 0$ and $h \le 0$ outside of $u^{-1}(V)$. By composing with an appropriately chosen smooth function $\mathbb{R} \to \mathbb{R}$ we obtain a $\mathscr{Lip}^k$-function $h_1{:}E \to \mathbb{R}$ satisfying $h_1 = 1$ on a neighborhood of $u^{-1}(x)$ and $h_1 = 0$ outside $u^{-1}(V)$. Now we define the global function $f{:}X \to \mathbb{R}$ by $f(x) := h_1(u^{-1}(x))$ for $x \in u(U)$ and $f_1(x) := 0$ for $x \notin \bar{V}$. It is well-defined and of class $\mathscr{Lip}^k$ since composed with the chart u it is just h_1 and composed with another chart $u_1{:}U_1 \to X$ it is locally either 0 or the $\mathscr{Lip}^k$-

composite $h_1 \circ u^{-1} \circ u_1$. Thus $f \circ c$ has to be $\mathcal{L}ip^k$ and hence $\{t \in \mathbb{R}; f(c(t)) > 0\}$ is an open neighborhood of t_0 contained in $c^{-1}(V) \subseteq c^{-1}(u(U))$. □

4.4.45 Proposition. *Let X be a $\mathcal{L}ip^k$-manifold with countable atlas modelled on duals of Fréchet–Schwartz spaces and let F be metrizable. Then the locally convex topology of $\mathcal{L}ip^k(X, F)$ is metrizable.*

Proof. This follows from (4.4.41) by applying (4.4.6). □

Remark. It also follows from (4.4.41) that for finite-dimensional smooth manifolds X the locally convex topology of $C^\infty(X, F)$ is the classical 'C^∞-compact-open' topology, which is roughly speaking the topology of uniform convergence on compact sets of all the derivatives separately, cf. [Hirsch, 1976, p. 34].

4.5 Partial differentiability

4.5.1 Proposition. *Let $k \in \mathbb{N}_{0,\infty}$; X be a $\mathcal{L}ip^k$-space; Y a ℓ^∞-space; and E a convenient vector space. The map $f \mapsto \tilde{f}$, where $\tilde{f}(y)(x) := f(x)(y)$ constitutes an isomorphism of convenient vector spaces*

$$\mathcal{L}ip^k(X, \ell^\infty(Y, E)) \cong \ell^\infty(Y, \mathcal{L}ip^k(X, E)).$$

Proof. Let $f \in \mathcal{L}ip^k(X, \ell^\infty(Y, E))$. Then for $y \in Y$ one has $\tilde{f}(y) = \mathrm{ev}_y \circ f$ showing that $\tilde{f}(y)$ is a $\mathcal{L}ip^k$-map and one thus obtains a map $\tilde{f}: Y \to \mathcal{L}ip^k(X; E)$.

Conversely, let $g \in \ell^\infty(Y, \mathcal{L}ip^k(X, E))$. Then for $x \in X$ one has $\tilde{g} = \mathrm{ev}_x \circ g$ showing that $\tilde{g}(x)$ is an ℓ^∞-map and one thus obtains a map $\tilde{g}: X \to \ell^\infty(Y, E)$.

Now we show that the correspondence is a bijection: Let f be a map from X to the space of mappings from Y to E and $g := \tilde{f}$ the corresponding map from Y to the space of mappings from X to E.

$g: Y \to \mathcal{L}ip^k(X, E)$ is an ℓ^∞-morphism;
⇔ $g \circ e: \mathbb{N} \to \mathcal{L}ip^k(X, E)$ is ℓ^∞ for every ℓ^∞-map $e: \mathbb{N} \to Y$;
⇔ $c^* \circ g \circ e: \mathbb{N} \to \mathcal{L}ip^k(\mathbb{R}, E)$ is ℓ^∞ for all e and all $\mathcal{L}ip^k$-curves $c: \mathbb{R} \to X$;
⇔ $[\delta^i(e^* \circ f \circ c)]^\sim = \delta^i \circ c^* \circ g \circ e: \mathbb{N} \to \ell^\infty(\mathbb{R}^{\langle i \rangle}, E)$ is ℓ^∞ for all $i < k+2$ and all c and e;
⇔ $\delta^i(e^* \circ f \circ c)$: $\mathbb{R}^{\langle i \rangle} \to \ell^\infty(\mathbb{N}, E)$ is ℓ^∞ for all $i < k+2$, c and e (according to (1.2.8));
⇔ $e^* \circ f \circ c: \mathbb{R} \to \ell^\infty(\mathbb{N}, E)$ is $\mathcal{L}ip^k$ for all c and e;
⇔ $f \circ c: \mathbb{R} \to \ell^\infty(Y, E)$ is $\mathcal{L}ip^k$ for all c;
⇔ $f: X \to \ell^\infty(Y, E)$ is $\mathcal{L}ip^k$.

Let $\varphi: \mathcal{L}ip^k(X, \ell^\infty(Y, E)) \to \ell^\infty(Y, \mathcal{L}ip^k(X, E))$ be the map defined by $f \mapsto \tilde{f}$ and $\psi: \ell^\infty(Y, \mathcal{L}ip^k(X, E)) \to \mathcal{L}ip^k(X, \ell^\infty(Y, E))$ the map defined by $g \mapsto \tilde{g}$. Let us show that φ is a morphism: Since $\ell^\infty(Y, E)$ is a convenient vector space by (ii) in (3.6.1) so is $\mathscr{Lip}^k(X, \ell^\infty(Y, Z))$ by (4.4.2); and since φ is linear it is by the bornological uniform boundedness principle (3.6.6) enough to show that $\mathrm{ev}_y \circ \varphi\colon \mathscr{Lip}^k(X, \ell^\infty(Y, E)) \to \mathscr{Lip}^k(X, E)$ is a morphism. This holds according to the differentiable uniform boundedness principle (4.4.7) provided that $\mathrm{ev}_x \circ \mathrm{ev}_y \circ \varphi\colon \mathscr{Lip}^k(X, \ell^\infty(Y, E)) \to E$ is a morphism. This is true since $\mathrm{ev}_x \circ \mathrm{ev}_y \circ \varphi = \mathrm{ev}_y \circ \mathrm{ev}_x$.

The proof that ψ is a morphism is analogous. □

4.5.2 Proposition. *Let $k, j \in \mathbb{N}_{0,\infty}$; X be a $\mathscr{Lip}^k$-space, Y a $\mathscr{Lip}^j$-space; and E a convenient vector space. Then the map $f \mapsto \tilde{f}$ constitutes an isomorphism between convenient vector spaces:*

$$\mathscr{Lip}^k(X, \mathscr{Lip}^j(Y, E)) \cong \mathscr{Lip}^j(Y, \mathscr{Lip}^k(X, E)).$$

Proof. Let $f \in \mathscr{Lip}^k(X, \mathscr{Lip}^j(Y, E))$. Since $\tilde{f}(y) = \mathrm{ev}_y \circ f$ one has $\tilde{f}(y) \in \mathscr{Lip}^k(X, E)$ and thus obtains a map $\tilde{f}\colon Y \to \mathscr{Lip}^k(X, E)$.

Let us show that $\tilde{f}$ is $\mathscr{Lip}^j$:

$f\colon X \to \mathscr{Lip}^j(Y, E)$ is $\mathscr{Lip}^k$;
$\Rightarrow$ $e^* \circ f\colon X \to \mathscr{Lip}^j(\mathbb{R}, E)$ is $\mathscr{Lip}^k$ for all $\mathscr{Lip}^j$-curves $e\colon \mathbb{R} \to Y$;
$\Rightarrow$ $\delta^i \circ e^* \circ f\colon X \to \ell^\infty(\mathbb{R}^{\langle i \rangle}, E)$ is $\mathscr{Lip}^k$ for all e and all $i < j+2$;
$\Rightarrow$ $\delta^i(\tilde{f} \circ e) = [\delta^i \circ e^* \circ f]^\sim\colon \mathbb{R}^{\langle i \rangle} \to \mathscr{Lip}^k(X, E)$ is ℓ^∞ for all $i < j+2$ and all e;
$\Rightarrow$ $\tilde{f} \circ e\colon \mathbb{R} \to \mathscr{Lip}^k(X, E)$ is $\mathscr{Lip}^j$ for all e;
$\Rightarrow$ $\tilde{f}\colon Y \to \mathscr{Lip}^k(X, E)$ is $\mathscr{Lip}^j$.

Next we show that $\varphi\colon \mathscr{Lip}^k(X, \mathscr{Lip}^j(Y, E)) \to \mathscr{Lip}^j(Y, \mathscr{Lip}^k(X, E))$ defined by $f \mapsto \tilde{f}$ is a morphism: using the differentiable uniform boundedness principle (4.4.7) twice it is enough to show that $\mathrm{ev}_y \circ \mathrm{ev}_x = \mathrm{ev}_x \circ \mathrm{ev}_y \circ \varphi$ is a morphism for all $y \in Y$ and $x \in X$. This is trivially the case.

The converse direction follows from symmetry. □

Remark. One may call $f\colon X \times Y \to E$ a $\mathscr{Lip}^{k,j}$-map if $f^\vee \in \mathscr{Lip}^k(X, \mathscr{Lip}^j(Y, E))$. This notion, however, depends on the factorization of the domain in a product as shown by the following

Example. Let $f\colon \mathbb{R}^2 \to \mathbb{R}$ be defined by $f(t,s) = |t| + |s|$. Then f is $\mathscr{Lip}^{0,0}$, since

$$|\delta_1 f(t, t'; s)| = \left| \frac{f(t,s) - f(t',s)}{t - t'} \right| = \left| \frac{|t| - |t'|}{t - t'} \right| \leq 1$$

and by symmetry $|\delta_2 f(t; s, s')| \leq 1$ and finally

$$\delta_1 \delta_2 f(t, t'; s, s') = \frac{f(t,s) - f(t,s') - f(t',s) + f(t',s')}{(t-t')(s-s')} = 0.$$

On the other hand, $f \circ A$ is not $\mathscr{Lip}^{0,0}$, if A is the linear isomorphism given by the matrix $\begin{pmatrix} 1 & 1 \\ -1 & 1 \end{pmatrix}$. This follows since for $0<h<1$ the difference quotient

$$\delta_1^1\delta_2^1 f(0,h;0,h)=\frac{(f\circ A)(0,0)-(f\circ A)(0,h)-(f\circ A)(h,0)+(f\circ A)(h,h)}{h^2}$$

$$=\frac{f(0,0)-f(h,h)-f(h,-h)+f(2h,0)}{h^2}=\frac{-2}{h}$$

is unbounded.

4.5.3 Definition. Let $\mathscr{S}\subseteq F'$ be point separating. A map $f:\Pi_{j\in J}E_j\supseteq U\to F$ is called *partially $\mathscr{S}$-differentiable* iff for all $j\in J$ and all $x\in\Pi_j E_j$ the map $f\circ h_x^j$ is $\mathscr{S}$-differentiable, where $h_x^j: E_j\to\Pi_{i\in J}E_i$ is defined by $(\mathrm{pr}_i\circ h_x^j)(z):=\mathrm{pr}_i(x)$ for $j\neq i$, $z\in E_j$ and $\mathrm{pr}_j\circ h_x^j=\mathrm{id}_{E_j}$. The *partial differentials* $\mathrm{d}_j f: U\sqcap E_j\to F$ are then defined as $\mathrm{d}_j f(x,v):=\mathrm{d}(f\circ h_x^j)(\mathrm{pr}_j(x),v)$.

f is recursively called $(k+1)$-*times partially $\mathscr{S}$-differentiable* iff f is partially $\mathscr{S}$-differentiable and all the partial differentials $\mathrm{d}_j f$ $(j\in J)$ are k-times partially $\mathscr{S}$-differentiable. For $j_1,\dots,j_{k+1}\in J$ the *partial differentials* $\mathrm{d}_{j_{k+1}}\dots\mathrm{d}_{j_1}f: U\sqcap E_{j_1}\sqcap\dots\sqcap E_{j_{k+1}}\to F$ *of order* $k+1$ are then defined recursively as

$$\mathrm{d}_{j_{k+1}}\dots\mathrm{d}_{j_1}f(x;v_{j_1},\dots,v_{j_{k+1}}):=\mathrm{d}_{j_{k+1}}[\mathrm{d}_{j_k}\dots\mathrm{d}_{j_1}f(\_;v_{j_1},\dots,v_{j_k})](x,v_{j_{k+1}}).$$

A map $f:\Pi_{j\in J}E_j\supseteq U\to F$ is called *partially strongly differentiable* iff for all $j\in J$ and all $x\in\Pi_j E_j$ the map $f\circ h_x^j$ is strongly differentiable, where h_x^j is defined as above. The *partial derivatives* $\partial_j: U\to L(E_j,F)$ are then defined as $\partial_j f(x)(v):=(f\circ h_x^j)'(\mathrm{pr}_j(x))(v)$.

Remark. If $f:\Pi_{j\in J}E_j\supseteq U\to F$ is a $\mathscr{S}$-differentiable map then f is obviously partially $\mathscr{S}$-differentiable and $\mathrm{d}_j f=\mathrm{d}f\circ(\mathrm{id}_U\sqcap\mathrm{in}_j)$, where $\mathrm{in}_j: E_j\to\coprod_{i\in J}E_i\to\Pi_{i\in J}E_i$ denotes the natural injection.

If f is strongly differentiable then f is obviously partially strongly differentiable and $\partial_j f=(\mathrm{in}_j)^*\circ f'$.

4.5.4 Theorem. *Let $f:E\sqcap\mathbb{R}\supseteq U\to F$ be a $\mathscr{Lip}^k$-map. Then the domain of definition $\{x\in E;(x,t)\in U$ for all $t\in[0,1]\}$ of the map $x\mapsto g(x):=\int_0^1 f(x,t)\mathrm{d}t$ is* M*-open, g is also $\mathscr{Lip}^k$ and for $k>0$ one has $\mathrm{d}g(x,v)=\int_0^1\mathrm{d}_1 f(x,t;v)\mathrm{d}t$, or equivalently $g'(x)=\int_0^1\partial_1 f(x,t)\mathrm{d}t$.*

Proof. In (4.3.11) it was shown that the domain of definition of g, $W:=\{x\in E;\{x\}\times[0,1]\subseteq U\}$ is M-open in E and that the proposition holds for $k=0$. So let $k>0$ and suppose the proposition holds for $k-1$. It is enough to show that g is $\mathscr{Lip}$-differentiable and $\mathrm{d}g(x,y)=\int_0^1\mathrm{d}_1 f(x,t;y)\mathrm{d}t$, since by induction hypothesis this function is $\mathscr{Lip}^{k-1}$. In order to see this we use that f is $\mathscr{Lip}^1$ and thus the directional difference quotient has a $\mathscr{Lip}^0$ extension $\bar{\vartheta}f: U_{\bar{\vartheta}}\to F$. Recall that $U_{\bar{\vartheta}}\subseteq(E\sqcap\mathbb{R})^2\sqcap\mathbb{R}^2$, cf. (3) in (4.3.12). For $(x,y,t,s)\in W_{\vartheta}$ one has, cf. (iii) in (4.3.6),

$$\vartheta g(x, y, t, s) = \frac{g(x+ty) - g(x+sy)}{t-s} = \int_0^1 \frac{f(x+ty, r) - f(x+sy, r)}{t-s} \, dr$$

$$= \int_0^1 \bar{\vartheta} f(x, r; y, 0; t, s) \, dr.$$

Using (4.3.11) we conclude that ϑg has $\mathscr{Lip}^0$ extension to $W_{\bar{\vartheta}}$ given by $\bar{\vartheta} g(x, y, t, t) := \int_0^1 \bar{\vartheta} f(x, r; y, 0; t, t) \mathrm{d}r = \int_0^1 \mathrm{d}f(x+ty, r+0; y, 0) \mathrm{d}r = \int_0^1 \mathrm{d}_1 f(x+ty, r; y) \mathrm{d}r$. Thus g is $\mathscr{Lip}$-differentiable by (4.3.12), and $\mathrm{d}g(x, y) = \bar{\vartheta} g(x, y, 0, 0) = \int_0^1 \mathrm{d}_1 f(x, r; y) \mathrm{d}r$.

As a consequence we can characterize those maps which are derivatives: for every $\mathscr{Lip}^1$-map $g: U \to L(E, \ldots, E; F)$ whose values are k-linear and symmetric and for which the values of the derivative $g': U \to L(E, L(E, \ldots, E; F)) \cong L(E, \ldots, E; F)$ are also symmetric there exists a $\mathscr{Lip}^{k+1}$-map $f: U \to F$ with derivative of order k equal to g. In fact this can be easily deduced by induction from the following special case:

4.5.5 Proposition. *Let $g: U \to L(E, F)$ be a $\mathscr{Lip}^1$-map, where $U \subseteq E$ is convex,* M-*open and contains* 0. *Then there exists a $\mathscr{Lip}^2$-map $f: U \to F$ with $g = f'$ if and only if $g'(x)(v)(w) = g'(x)(w)(v)$ for all $x \in U$, v, $w \in E$.*

Proof. If $g = f'$ for some f the conditions just express the symmetry of the second derivative of f.

Conversely, let this symmetry condition be satisfied and define f by $f(x) := \int_0^1 g(tx)(x) \mathrm{d}t$. By (4.5.4) f is $\mathscr{Lip}^1$ and one has

$$f'(x)(v) = \int_0^1 (g'(tx)(tv)(x) + g(tx)(v)) \mathrm{d}t = \int_0^1 (t \cdot g'(tx)(x)(v) + g(tx)(v)) \mathrm{d}t$$

$$= \left(\int_0^1 (t \cdot g'(tx)(x) + g(tx)) \mathrm{d}t \right)(v).$$

With $c(t) = t \cdot g(tx)$ one has therefore $f'(x) = \int_0^1 (t \cdot g'(tx)(x) + g(tx)) \mathrm{d}t = \int_0^1 c^{\cdot}(t) \mathrm{d}t = c(1) - c(0) = g(x)$ which shows at the same time that f is $\mathscr{Lip}^2$. □

Remark. For a Poincaré lemma on differential forms in this setting see [Kriegl, 1983].

4.5.6 Corollary. *Let $f: E \supseteq U \to F$ be a map. Then the following statements are equivalent:*

(1) *f is $\mathscr{Lip}^{k+1}$;*
(2) *the directional difference quotient $\vartheta f: U_\vartheta \to F$ has a $\mathscr{Lip}^k$-extention $\bar{\vartheta} f: U_{\bar{\vartheta}} \to F$.*

Proof. (2⇒1) By (4.3.12) f is $\mathscr{Lip}$-differentiable and $\mathrm{d}f(x, v) = \bar{\vartheta} f(x, v, 0, 0)$. Hence $\mathrm{d}f$ is $\mathscr{Lip}^k$ and using (4.3.24) one concludes that f is $\mathscr{Lip}^{k+1}$.

(1⇒2) By (4.3.12) a $\mathscr{Lip}^0$-extension $\bar{\vartheta}f:U_{\bar{\vartheta}}\to F$ of the $\mathscr{Lip}^{k+1}$-map $\vartheta f:U_\vartheta\to F$ exists. It is given by $\bar{\vartheta}f(x,v,t,s):=\int_0^1 df(x+tv+r(t-s)v,v)\,dr$ in some neighborhood of $U_{\bar{\vartheta}}\setminus U_\vartheta$, thus is $\mathscr{Lip}^k$ on this neighborhood too by (4.5.4). □

4.5.7 Proposition. *Let $\mathscr{S}\subseteq F'$ be point separating and $f:\Pi_{j=1}^m E_j\supseteq U\to F$ be a partially $\mathscr{S}$-differentiable map. If all partial derivatives $d_jf: U\sqcap E_j\to F\,(j\in J)$ are $\mathscr{Lip}^k$ then f is $\mathscr{Lip}^{k+1}$.*

Proof. By (4.3.24) it is enough to show that f is $\mathscr{S}$-differentiable and $df(x;v_1,\ldots,v_m)=\sum_{j=1}^m d_jf(x,v_j)$ for all $x\in U$ and $v=(v_1,\ldots,v_m)\in\Pi_jE_j$. For $t\neq 0$ one has

$$\frac{f(x+tv)-f(x)}{t}=$$

$$=\sum_{j=1}^m \frac{f(x+t(v_1,\ldots,v_j,0,0\ldots,0))-f(x+t(v_1,\ldots,v_{j-1},0\ldots,0))}{t}$$

$$=\sum_{j=1}^m\int_0^1 d_jf(x+t(v_1,\ldots,v_{j-1},sv_j,0\ldots,0);v_j)ds.$$

Using (4.5.4) one concludes that the right side defines a $\mathscr{Lip}^k$-map and thus $(f(x+tv)-f(x))/t$ is M-convergent to

$$\sum_{j=1}^m\int_0^1 d_jf(x+0(v_1,\ldots,v_{j-1},sv_j,0\ldots,0);v_j)ds=\sum_{j=1}^m d_jf(x,v_j)\quad\text{for}\quad t\to 0.$$

□

4.5.8 Theorem. *Let $\mathscr{S}\subseteq F'$ be point separating and $f:\Pi_{j=1}^m E_j\supseteq U\to F$ be a map. Then the following statements are equivalent:*

(1) *f is $\mathscr{Lip}^k$;*
(2) *f is k-times partially $\mathscr{S}$-differentiable and the partial derivatives $d_{j_k}\ldots d_{j_1}f: U\sqcap E_{j_1}\sqcap\ldots\sqcap E_{j_k}\to F$ of order k are $\mathscr{Lip}^0$ for all $j_1,\ldots,j_k\in\{1,\ldots,m\}$.*

Proof. We show this by induction on k. For $k=0$ it is trivial and for $k=1$ it is contained in (4.5.7). So let now $k>1$. Then by (4.5.7) f is $\mathscr{Lip}^k$ iff f is partially $\mathscr{S}$-differentiable and d_j is $\mathscr{Lip}^{k-1}$ for all $j\in\{1,\ldots,m\}$. This is by induction hypothesis (applied to all the d_jf) equivalent with f being k-times partially $\mathscr{S}$-differentiable and $d_{j_k}\ldots d_{j_1}f$ being $\mathscr{Lip}^0$. □

4.6 Spaces of sections of vector bundles

In this section we want to prove that for quite general vector bundles the space of sections has a natural convenient vector space structure. These vector bundles can be considered as families of convenient vector spaces parametrized in a $\mathcal{Lip}^k$-way by a $\mathcal{Lip}^k$-space, triviality being assumed along $\mathcal{Lip}^k$-curves of the base space.

4.6.1 Remark. We shall use induced bundles and the Whitney sum of a bundle with itself. Since these are obtained as pullbacks we first describe explicitly pullbacks of $\mathcal{Lip}^k$-maps, cf. (8.3.5).

The pullback of two given $\mathcal{Lip}^k$-maps $X \xrightarrow{f} Z \xleftarrow{g} Y$ is given by the subspace $P := \{(x, y) \in X \sqcap Y; f(x) = g(y)\}$ of $X \sqcap Y$ and the restrictions to P of the two projections $X \xleftarrow{\mathrm{pr}_1} X \sqcap Y \xrightarrow{\mathrm{pr}_2} Y$.

Given two $\mathcal{Lip}^k$-maps $X \xleftarrow{f_1} Z_1 \xrightarrow{g_1} Y$ satisfying $f \circ f_1 = g \circ g_1$, the associated $\mathcal{Lip}^k$-map $Z_1 \to P$ shall be denoted by (f_1, g_1), since $(f_1, g_1)(z) = (f_1(z), g_1(z)) \in P \subseteq X \sqcap Y$ for all $z \in Z_1$.

If it is clear what the maps f and g are, then we write $X \underset{Z}{\sqcap} Y$ for P; and if we consider for a fixed given $g: Y \to Z$ the pullback for various maps $f: X \to Z$ then we use the traditional notation $f^*(Y) := P$ and $f^*(g) := \mathrm{pr}_1|_P$.

We remark that g surjective implies $f^*(g)$ surjective, and the fibre of $f^*(g): f^*(Y) \to X$ over a point $x \in X$ (i.e. the inverse image) is equal to the product of $\{x\}$ with the fibre of $g: Y \to Z$ over $f(x)$.

For vector bundles $\pi: E \to X$ the scalar multiplication is given fibre-wise but can be considered as a map $\mathbb{R} \sqcap E \to E$. In contrast the fibre-wise defined addition can be considered as a global map only on the Whitney sum, i.e. the pullback $E \underset{X}{\sqcap} E$ of the projection π with itself. Using the following lemma addition can be avoided in the definition of a $\mathcal{Lip}^k$-vector-bundle provided $k > 0$.

4.6.2 Lemma. *A convenient vector space is completely determined by its underlying set, its scalar multiplication and its $\mathcal{Lip}^k$-structure for some $k \in \mathbb{N}_\infty$.*

Proof. Let E_1 and E_2 be two convenient vector spaces having the same underlying set E, the same $\mathcal{Lip}^k$-structure and the same scalar multiplication. Then E_1 and E_2 certainly have the same zero-vector 0 obtained by $0 \cdot x$ for an arbitrary x. By symmetry it is enough to show that $\mathrm{id}_E: E_1 \to E_2$ is linear. Since $\mathrm{id}: E_1 \to E_2$ is $\mathcal{Lip}^k$, the derivative $\mathrm{id}'(0): E_1 \to E_2$ is linear; so we only have to prove that $\mathrm{id}'(0) = \mathrm{id}$. This is easy: $\mathrm{id}'(0)(v) = \lim_{t \to 0} tv/t = v$. □

Remark. This lemma fails for $k = 0$ as the following example shows. Consider $f: \mathbb{R}^2 \to \mathbb{R}^2$ defined by $f(0) := 0$ and $f(x) := (\|x\|_1 / \|x\|_\infty) x$ for $x \neq 0$. As one verifies easily f is a $\mathcal{Lip}^0$-map and preserves the scalar multiplication. But $f(1, 1) = (2/1)(1, 1) = (2, 2) \neq (1, 0) + (0, 1) = f(1, 0) + f(0, 1)$.

Using (4.6.2) we can shortly say that a set with given scalar multiplication and given $\mathcal{Lip}^k$-structure (for some $k \in \mathbb{N}_\infty$) is a convenient vector space if there exists some addition yielding a convenient vector space as characterized in (3) of (2.4.4).

4.6.3 Definition. A *$\mathscr{Lip}^k$-vector-bundle* π (for $k \geq 1$) is formed by two $\mathscr{Lip}^k$-morphisms $\pi: E \to X$ and $\mu: \mathbb{R} \sqcap E \to E$, written $\mu: (t, v) \mapsto t \cdot v$, subject to the following conditions:

(i) The scalar multiplication μ preserves the fibres, i.e. $\pi(t \cdot v) = \pi(v)$;
(ii) For each $x \in X$ the fibre $E_x := \pi^{-1}(x)$ with the restriction of μ to $\mathbb{R} \sqcap E_x \to E_x$ and the $\mathscr{Lip}^k$-structure inherited from E is a convenient vector space;
(iii) Triviality holds along each $\mathscr{Lip}^k$-curve $c: \mathbb{R} \to X$.

In case $k = 0$ one has to add a given family of maps $\alpha_x: E_x \sqcap E_x \to E_x$ ($x \in X$) such that all fibres E_x are convenient vector spaces with α_x as addition.

One calls X the base and E the total space of the bundle.

The meaning of (iii) is the following: for every $\mathscr{Lip}^k$-curve $c: \mathbb{R} \to X$ one considers the pullback $c^*(\pi): c^*(E) \to \mathbb{R}$ of c and π whose fibre over t is $(c^*(E))_t = \{t\} \sqcap E_{c(t)} \cong E_{c(t)}$. Then there should exist a $\mathscr{Lip}^k$-diffeomorphism $\iota_c: c^*(E) \to \mathbb{R} \sqcap E_{c(0)}$ which preserves the fibres (i.e. $\mathrm{pr}_1 \circ \iota_c = c^*(\pi)$) and the scalar multiplication (i.e. for all $(t, v) \in c^*(\pi)$ one has: $\mathrm{pr}_2(\iota_c(t, s \cdot v)) = s \cdot (\mathrm{pr}_2(\iota_c(t, v)))$). By lemma (4.6.2) this implies that ι_c induces for each $t \in \mathbb{R}$ a Con-isomorphism $E_{c(t)}$ to $E_{c(0)}$. The maps ι_c are called *trivializations* of π along c.

We remark furthermore that (ii) implies that π is onto.

4.6.4 Proposition. (*Induced bundles.*) *Let $\pi: E \to X$ be a $\mathscr{Lip}^k$-vector-bundle and $f: Y \to X$ a $\mathscr{Lip}^k$ map. Then the pullback $f^*(\pi): f^*(E) \to Y$ yields a $\mathscr{Lip}^k$-vector-bundle, the so-called induced bundle.*

Proof. The fibre of f^*E at y is $\{y\} \sqcap E_{f(y)}$. Thus one has fibre-wise a natural scalar multiplication on $f^*(\pi)$. The universal property of the pullback shows that it is in fact $\mathscr{Lip}^k$. Then the fibres $\{y\} \sqcap E_{f(y)} \cong E_{f(y)}$ are convenient vector spaces. And triviality along any curve $c: \mathbb{R} \to Y$ holds since $c^*(f^*(\pi)) = (f \circ c)^*(\pi)$, which is isomorphic to a trivial bundle because $f \circ c$ is a $\mathscr{Lip}^k$-curve into the base of π. □

Remark. In particular one concludes that $c^*(\pi): c^*(E) \to \mathbb{R}$ is a $\mathscr{Lip}^k$-vector-bundle for every $\mathscr{Lip}^k$-curve $c: \mathbb{R} \to X$ into the base of a $\mathscr{Lip}^k$-vector-bundle $\pi: E \to X$.

We now define vector bundle morphisms and triviality for vector-bundles. In particular the maps $\iota_c: c^*(E) \to \mathbb{R} \sqcap E_{c(0)}$ used in the definition of $\mathscr{Lip}^k$-vector-bundle are such vector bundle isomorphisms and the $c^*(\pi): c^*(e) \to \mathbb{R}$ are trivial vector bundles.

4.6.5 Definition. (i) Let $\pi_i: E_i \to X_i$ $(i = 1, 2)$ be two $\mathscr{Lip}^k$-vector-bundles. A $\mathscr{Lip}^k$-map $f: E_1 \to E_2$ is called a *vector bundle morphism* iff the following diagram commutes:

$$\begin{array}{ccc} \mathbb{R}\sqcap E_1 & \xrightarrow[\mu_1]{} & E_1 \\ \Big\downarrow{\scriptstyle \mathrm{id}\sqcap f} & & \Big\downarrow{\scriptstyle f} \\ \mathbb{R}\sqcap E_2 & \xrightarrow[\mu_2]{} & E_2 \end{array}$$

(ii) A $\mathscr{Lip}^k$-vector-bundle $\pi: E \to X$ is called (globally) *trivial* iff there exists a vector-bundle isomorphism onto a bundle $\mathrm{pr}_1: X\sqcap E_0 \to X$ where E_0 is some convenient vector space, and the scalar multiplication is fibre-wise given by that of E_0 (it is an easy exercise that $\mathrm{pr}_1: X\sqcap E_0 \to X$ thus becomes a $\mathscr{Lip}^k$-vector-bundle).

Remarks. (i) Every vector-bundle morphism $f: E_1 \to E_2$ induces a map $f_0: X_1 \to X_2$ between the bases, determined by the condition $f_0 \circ \pi_1 = \pi_2 \circ f$. Using the zero-section 0_1 of π_1, cf. (4.6.9), one has $f_0 = \pi_2 \circ f \circ 0_1$ and thus f_0 is a $\mathscr{Lip}^k$-map.

(ii) Every classical smooth vector bundle is a smooth vector bundle in the sense of (4.6.3), since local triviality implies triviality along smooth curves.

The following characterization of $\mathscr{Lip}^k$-maps into the total space of a $\mathscr{Lip}^k$-vector-bundle will be very useful.

4.6.6 Proposition. *Let $\pi: E \to X$ be $\mathscr{Lip}^k$-vector-bundle and $f: Y \to E$ a map. Then the following statements are equivalent:*

(1) *f is $\mathscr{Lip}^k$;*
(2) *$\pi \circ f: Y \to X$ is $\mathscr{Lip}^k$ and for every $\mathscr{Lip}^k$-curve $c: \mathbb{R} \to Y$ and trivialization $\iota_{\pi f c}$ of π along $\pi f c$ the map $\mathrm{pr}_2 \circ \iota_{\pi f c} \circ (\mathrm{id}, f \circ c): \mathbb{R} \to (\pi f c)^* E \to \mathbb{R}\sqcap E_{(\pi f c)(0)} \to E_{(\pi f c)(0)}$ is $\mathscr{Lip}^k$.*

Proof. ($\Rightarrow$) If f is $\mathscr{Lip}^k$ then obviously $\pi \circ f$ is $\mathscr{Lip}^k$ and $\mathrm{pr}_2 \circ \iota_{\pi f c}(\mathrm{id}, f \circ c)$ as well.

($\Leftarrow$) Let $c: \mathbb{R} \to Y$ be a $\mathscr{Lip}^k$-curve. Then by assumption $e := \pi \circ f \circ c$ is a $\mathscr{Lip}^k$-curve in X and thus $f \circ c = \mathrm{pr}_2 \circ \iota_e^{-1} \circ (\mathrm{id}, \mathrm{pr}_2 \circ \iota_e \circ (\mathrm{id}, f \circ c))$ is $\mathscr{Lip}^k$. □

4.6.7 Lemma. *Let $\pi: E \to X$ be a $\mathscr{Lip}^k$-vector-bundle. Then the fibre-wise determined addition $E_x \sqcap E_x \to E_x$ gives a $\mathscr{Lip}^k$-morphism $\alpha: E \underset{X}{\sqcap} E \to E$.*

Proof. Let $(c_1, c_2): \mathbb{R} \to E \underset{X}{\sqcap} E$ be a $\mathscr{Lip}^k$-curve, i.e. $c_i: \mathbb{R} \to E$ $(i = 1, 2)$ is $\mathscr{Lip}^k$ and $\pi \circ c_1 = \pi \circ c_2 =: c$. Using the previous proposition (4.6.6) and the identity $\pi \circ \alpha \circ (c_1, c_2) = c$ one concludes that it is enough to show that $\mathrm{pr}_2 \circ \iota_c \circ (\mathrm{id}, \alpha \circ (c_1, c_2)): \mathbb{R} \to E_{c(0)}$ is $\mathscr{Lip}^k$; but this is obviously the sum of the two $\mathscr{Lip}^k$-curves $\mathrm{pr}_2 \circ \iota_c \circ (\mathrm{id}, c_j): \mathbb{R} \to E_{c(0)}$ $(j = 1, 2)$. □

4.6.8 Definition. Let $\pi: E \to X$ be a $\mathscr{Lip}^k$-vector-bundle. The space $\Gamma^k(\pi)$ of *$\mathscr{Lip}^k$-sections* of the vector-bundle π is defined as $\Gamma^k(\pi) := \{s \in \mathscr{Lip}^k(X, E);\ \pi \circ s = \mathrm{id}\}$.

Remark. Let $\pi: E \to X$ be a $\mathcal{L}ip^k$-vector-bundle and $f: Y \to X$ be a $\mathcal{L}ip^k$-map. Then the set $\{s \in \mathcal{L}ip^k(Y, E); \pi \circ s = f\}$ of sections along f is in natural bijection to $\Gamma^k(f^*(\pi))$ via the map $s \mapsto (\mathrm{id}, s)$.

4.6.9 Lemma. *Let $\pi: E \to X$ be a $\mathcal{L}ip^k$-vector-bundle. Then the space $\Gamma^k(\pi)$ of $\mathcal{L}ip^k$-sections of π is a vector space with the pointwise operations.*

Proof. That the vector operations are defined pointwise means that $\Gamma^k(\pi)$ is considered as subspace of the vector space $\Pi_{x \in X} E_x$. We have to show that it is a linear subspace.

Let s be a section of π and $t \in \mathbb{R}$; then the map $x \mapsto t \cdot s(x)$ *is* $\mathcal{L}ip^k$, since it is the composite $\mu(t, \_) \circ s$, and is trivially a section.

Let s_1, s_2 be two sections; then the map $x \mapsto s_1(x) + s_2(x)$ is $\mathcal{L}ip^k$, since it is the composite $\alpha \circ (s_1, s_2)$, and is trivially a section.

The zero-section $x \mapsto 0_x$ is in $\Gamma^k(\pi)$, since a section s of π is $\mathcal{L}ip^k$ by (4.6.6) iff for every $\mathcal{L}ip^k$-curve $c: \mathbb{R} \to X$ the map $\mathrm{pr}_2 \circ \iota_c \circ (\mathrm{id}, s \circ c): \mathbb{R} \to E_{c(0)}$ is $\mathcal{L}ip^k$. For the zero-section this composite is the zero-map. □

4.6.10 Lemma. *Let $\pi: E \to X$ be a $\mathcal{L}ip^k$-vector-bundle. Then any $\mathcal{L}ip^k$-map $f: Y \to X$ induces a linear map $f^*: \Gamma^k(\pi) \to \Gamma^k(f^*(\pi))$* by setting $f^*(s) := (\mathrm{id}, s \circ f)$, *i.e.* $f^*(s)(y) = (y, s(f(y)))$.

Proof. Since $f \circ \mathrm{id} = \pi \circ (s \circ f)$ one concludes by the universal property of a pullback that $f^*(s)$ is a $\mathcal{L}ip^k$-map: $Y \to f^*(E)$. It is obviously a section. Finally the map f^* is linear, since the vector operations of the spaces of sections are defined pointwise and the fibres of $f^*(E)$ are mapped isomorphically onto the fibres of E. □

4.6.11 Lemma. *Let $\pi_i: E_i \to X$ $(i = 1, 2)$ be two $\mathcal{L}ip^k$-vector-bundles over the same base. Then any vector-bundle morphism $m: E_1 \to E_2$ induces a linear map $m_*: \Gamma^k(\pi_1) \to \Gamma^k(\pi_2)$ by setting $m_*(s) := m \circ s$.*

Proof. This is trivially verified. □

4.6.12 Lemma. *Let $\pi: X \sqcap E \to X$ be a trivial $\mathcal{L}ip^k$-vector-bundle. Then the map $f \mapsto (\mathrm{id}, f)$, $\mathcal{L}ip^k(X, E) \to \Gamma^k(\pi)$ is an isomorphism of vector spaces.*

Proof. The inverse map is given by $s \mapsto \mathrm{pr}_2 \circ s$. That both are well defined and linear is trivial. □

4.6.13 Definition. Let $\pi: E \to X$ be a $\mathcal{L}ip^k$-vector-bundle. Then $\Gamma^k(\pi)$ shall from now on denote the space of $\mathcal{L}ip^k$-sections of π together with the initial $\underline{\text{Pre}}$-structure induced by the linear maps $(\mathrm{pr}_2)_* \circ (\iota_c)_* \circ c^*: \Gamma^k(\pi) \to \Gamma^k(c^*(\pi)) \to \Gamma^k(\mathbb{R} \sqcap E_{c(0)}) \to \mathcal{L}ip^k(\mathbb{R}, E_{c(0)})$ with $c \in \mathcal{L}ip^k(\mathbb{R}, X)$.

4.6.14 Lemma. *Let $\pi: E \to X$ be a $\mathcal{Lip}^k$-vector-bundle. Then the* <u>Pre</u>*-structure of $\Gamma^k(\pi)$ does not depend on the choice of the vector-bundle isomorphisms ι_c.*

Proof. Given two trivializations $\iota_c^i: c^*(E) \to \mathbb{R} \sqcap E_{c(0)}$ $(i=1, 2)$ one has $\iota_c^2 = m \circ \iota_c^1$, where $m: \mathbb{R} \sqcap E_{c(0)} \to \mathbb{R} \sqcap E_{c(0)}$ is the $\mathcal{Lip}^k$-diffeomorphism $(\iota_c^2) \circ (\iota_c^1)^{-1}$. From this the result follows. □

4.6.15 Theorem. *Let $\pi: E \to X$ be a $\mathcal{Lip}^k$-vector-bundle. Then $\Gamma^k(\pi)$ is a convenient vector space.*

Proof. Since the maps $c^*: \Gamma^k(\pi) \to \Gamma^k(c^*(\pi))$ separate points and for trivial bundles the spaces of sections are separated, the preconvenient vector space $\Gamma^k(\pi)$ is separated. In order to prove that it is complete it is enough to show that the image in $\Pi_c \mathcal{Lip}^k(\mathbb{R}, E_{c(0)})$ is M-closed. This follows since an element $(s_c)_c$ of this product belongs to the image iff it satisfies the equations $\mathrm{pr}_2(\iota_{c_1})^{-1}(t_1, s_{c_1}(t_1)) = \mathrm{pr}_2(\iota_{c_2})^{-1}(t_2, s_{c_2}(t_2))$ for all $\mathcal{Lip}^k$-curves $c_1, c_2: \mathbb{R} \to X$ and reals t_1, t_2 with $c_1(t_1) = c_2(t_2)$. □

4.6.16 Proposition. (*Uniform Boundedness Principle for Bundle Sections.*) *Let $\pi: E \to X$ be a $\mathcal{Lip}^k$-vector-bundle. Then the structure of $\Gamma^k(\pi)$ is the coarsest convenient vector space structure making all evaluations $\mathrm{ev}_x: \Gamma^k(\pi) \to E_x$ $(x \in X)$ morphisms. In categorical language this means that $\{\mathrm{ev}_x; x \in X\}$ is an initial source in* <u>Con</u>.

Proof. This follows directly from the corresponding result (4.4.7) on $\mathcal{Lip}^k(\mathbb{R}, E_{c(0)})$. □

4.6.17 Proposition. *Let $\pi: E \to X$ be a $\mathcal{Lip}^k$-vector-bundle, and $f: Y \to X$ be a $\mathcal{Lip}^k$-map. Then the induced map $f^*: \Gamma^k(\pi) \to \Gamma^k(f^*(\pi))$ is a* <u>Con</u>*-morphism.*

Proof. Using (4.6.16) this follows from the equation $\mathrm{ev}_y \circ f^* = \mathrm{ev}_{f(y)}$. □

4.6.18 Proposition. *Let $\pi_i: E_i \to X$ $(i=1, 2)$ be two $\mathcal{Lip}^k$-vector-bundles over the same base and let $m: E_1 \to E_2$ be a vector-bundle morphism. Then the induced map $m_*: \Gamma^k(\pi_1) \to \Gamma^k(\pi_2)$ is a* <u>Con</u>*-morphism.*

Proof. Using (4.6.16) this follows from the equation $\mathrm{ev}_x \circ m_* = m_x \circ \mathrm{ev}_x$. □

4.6.19 Proposition. *Let $\pi: X \sqcap E \to X$ be a trivial $\mathcal{Lip}^k$-vector-bundle. Then the bijection of* (4.6.12) $\Gamma^k(\pi) \to \mathcal{Lip}^k(X, E)$ *is a* <u>Con</u>*-isomorphism.*

Proof. This follows since $(\mathrm{pr}_2)_* \circ c^*: \Gamma^k(\pi) \to \Gamma^k(c^*(\pi)) \to \mathcal{Lip}^k(\mathbb{R}, E)$, for $c \in \mathcal{Lip}^k(\mathbb{R}, X)$, is by definition an initial source and corresponds to the maps $c^*: \mathcal{Lip}^k(X, E) \to \mathcal{Lip}^k(\mathbb{R}, E)$ which form an initial source too. Thus the bijection is an initial morphism, hence an isomorphism. □

4.6.20 Proposition. *Let $\pi:E\to X$ be a $\mathscr{Lip}^k$-vector-bundle. Then* $\mathrm{ev}:\Gamma^k(\pi)\sqcap X\to E$ *is a $\mathscr{Lip}^k$-morphism.*

Proof. Consider first the case where $\pi:\mathbb{R}\sqcap E\to\mathbb{R}$ is a trivial bundle. Using (4.6.19) one shows that $\mathrm{ev}:\Gamma^k(\pi)\sqcap\mathbb{R}\to E$ is up to an isomorphism the $\mathscr{Lip}^k$-map $\mathrm{ev}:\mathscr{Lip}^k(\mathbb{R}, E)\sqcap\mathbb{R}\to E$. Now the general case. Let $(e, c):\mathbb{R}\to\Gamma^k(\pi)\sqcap X$ be a $\mathscr{Lip}^k$-curve. Then $\mathrm{ev}\circ(e,c)=\mathrm{pr}_2\circ\mathrm{ev}\circ(c^*\circ e, \mathrm{id})$, where the evaluation map on the right side is $\mathrm{ev}:\Gamma^k(c^*(\pi))\sqcap\mathbb{R}\to c^*(E)$. Since $c^*(\pi)$ is trivial the assertion follows. □

In case $k=\infty$ we have also on $\mathscr{Lip}^\infty(X,E)$ a $\mathscr{Lip}^\infty$-structure and then get the following simple description of the structure of $\Gamma^\infty(\pi)$.

4.6.21 Proposition. *Let $\pi:E\to X$ be a $\mathscr{Lip}^\infty$-vector-bundle. Then the smooth structure of $\Gamma^\infty(\pi)$ is the initial one induced by the inclusion $\Gamma^\infty(\pi)\subseteq\mathscr{Lip}^\infty(X, E)$.*

Proof. The inclusion is a morphism by cartesian closedness, cf. (4.4.13), since it is the map associated to the smooth map $\mathrm{ev}:\Gamma^\infty(\pi)\sqcap X\to E$. We further have to show that a curve $c:\mathbb{R}\to\Gamma^\infty(\pi)$ is smooth provided $c^\wedge$: $\mathbb{R}\sqcap X\to E$ is smooth. By (4.6.13) we have to consider for any smooth curve e: $\mathbb{R}\to X$ the composite $(\mathrm{pr}_2)_*\circ(\iota_e)_*\circ e^*\circ c$: $\mathbb{R}\to\Gamma^\infty(\pi)\to\Gamma^\infty(e^*(\pi))\to\Gamma^\infty(\mathbb{R}\sqcap E_{e(0)}\to\mathbb{R})\to C^\infty(\mathbb{R}, E_{e(0)})$. This curve is smooth because it is via cartesian closedness associated to the smooth map $\mathbb{R}\sqcap\mathbb{R}\to E_{e(0)}$, $(t,s)\mapsto\mathrm{pr}_2(\iota_e(c^\wedge(t, e(s))))$. □

4.6.22 Proposition. *Let $\pi:E\to X$ and $\pi_j:E_j\to X (j\in J)$ be $\mathscr{Lip}^k$-vector-bundles. Suppose the $\mathscr{Lip}^k$-structure of E is the initial one induced by $\mathscr{Lip}^k$-vector-bundle morphisms $m_j:E\to E_j$ $(j\in J)$. Then the Pre-structure of $\Gamma^k(\pi)$ is the initial one induced by the family $(m_j)_*:\Gamma^k(\pi)\to\Gamma^k(\pi_j)$ $(j\in J)$.*

Proof. This is obvious, since the structure of $\Gamma^k(\pi)$ is the initial one induced by the maps $(\mathrm{pr}_2\circ\iota_c)_*\circ c^*$ and the structure of $E_{c(0)}$ is induced by the maps $m_{c(0)}:E_{c(0)}\to(E_j)_{c(0)}$. □

4.6.23 Proposition. *Let $\pi:E\to X$ be a $\mathscr{Lip}^k$-vector-bundle and let $f_j:X_j\to X$ $(j\in J)$ be a family of $\mathscr{Lip}^k$-maps, such that for every $\mathscr{Lip}^k$-curve c: $\mathbb{R}\to X$ there exists a neighborhood U of 0 in $\mathbb{R}$, a $j\in J$ and a $\mathscr{Lip}^k$-curve $c_j:U\to X_j$, such that $f_j\circ c_j=c|_U$. Then the Pre-structure of $\Gamma^k(\pi)$ is the initial one induced by the maps $f_j^*:\Gamma^k(\pi)\to\Gamma^k(f_j^*(\pi))$ $(j\in J)$.*

Proof. Clearly the structure of $\Gamma^k(\pi)$ is the initial one induced by the maps $(\mathrm{incl}_U)^*\circ c^*:\Gamma^k(\pi)\to\Gamma^k(c^*(\pi))\to\Gamma^k((c|_U)^*(\pi))$. But this map equals $c_j^*\circ f_j^*$, which is a morphism. □

4.6.24 Proposition. *Let X be a Lindelöf smooth manifold modelled on duals of Fréchet Schwartz spaces and let $\pi:E\to X$ be a $\mathscr{Lip}^\infty$-vector-bundle that is locally trivial and has Fréchet spaces as fibres. Then $\Gamma^\infty(\pi)$ is a Fréchet space.*

Proof. We recall that a topological space is called Lindelöf iff for every open covering there exists a countable subcovering. Since the structure of $\Gamma^\infty(\pi)$ is induced by the maps $\Gamma^\infty(\pi|_U)$, where U run through a countable cover of open sets on which π is trivial, cf. (4.6.23), and since $\Gamma^\infty(\pi|_U)$ is isomorphic to $C^\infty(U, E_U)$ for some Fréchet space E_U, cf. (4.6.19), the assertion follows from (4.4.45). □

4.7 Certain function spaces are manifolds

In this section we will prove as main result that for finite-dimensional smooth manifolds X and Y, where X is supposed to be compact, the space $\mathrm{Emb}(X, Y)$ of embeddings of X in Y is a smooth principal fibre bundle over the space $\mathrm{Submf}(X, Y)$ of submanifolds of Y that are diffeomorphic to X, with the group $\mathrm{Diff}(X)$ of diffeomorphisms of X as typical fibre; cf. [Binz, Fischer, 1981].

In more detail this means: $\mathrm{Emb}(X, Y)$, $\mathrm{Diff}(X)$ and $\mathrm{Submf}(X, Y)$ are smooth manifolds modelled on convenient vector spaces; the group $\mathrm{Diff}(X)$ acts smoothly on $\mathrm{Emb}(X, Y)$ by composition; and the map $\mathrm{Emb}(X, Y)\to \mathrm{Submf}(X, Y)$ defined by $g\mapsto g(X)$ is a final smooth map, whose fibres are the orbits of $\mathrm{Diff}(X)$.

We already know that $\mathrm{Diff}(X)$ is a smooth group whose structure is the initial one induced by the inclusion in $C^\infty(X,X)$, cf. (1.4.8) and (4.7.4). Similarly $\mathrm{Emb}(X, Y)$ will be considered with the smooth structure inherited by the inclusion in $C^\infty(X, Y)$. Thus we begin the investigation with $C^\infty(X, Y)$.

4.7.1 Lemma. *Let X and Y be finite-dimensional smooth manifolds, $K\subseteq X$ compact, $W\subseteq Y$* open, *and $g_0\in C^\infty(X, Y)$ with $g_0(K)\subseteq W$. Then there exists a smooth function $f: C^\infty(X, Y)\to\mathbb{R}$ with $f(g_0)=1$ and such that $f(g)\neq 0$ implies $g(K)\subseteq W$.*

Proof. We begin with the special case where $X=I:=]-2,2[$, $K:=[-1,1]$, $Y:=\mathbb{R}$, $W:=\{t\in\mathbb{R};\, t>0\}$ and $g_0=1$. Let $h_0:\mathbb{R}\to\mathbb{R}$ be a smooth function with $h_0(t)\geq|t|$ for all t and $h_0(0)=\frac{1}{6}$, and let $h_1:\mathbb{R}^2\to\mathbb{R}$ be a smooth function with $h_1(1,\frac{2}{3})=1$ and such that $h_1(t,s)\neq 0$ implies that $t>\frac{1}{2}$ and $s>\frac{1}{2}$. Then $|g(s)-g(0)| =|\int_0^s g^\cdot(t)dt|\leq|\int_0^s|g^\cdot(t)|dt|\leq|\int_0^s h_0(g^\cdot(t))dt|\leq\int_{-1}^1 h_0(g^\cdot(t))dt$ for s with $|s|\leq 1$. Let $f: C^\infty(X,\mathbb{R})\to\mathbb{R}$ be the smooth map defined by $f(g):= h_1(g(0), 1-\int_{-1}^1 h_0(g^\cdot(t))dt)$. Then $f(g_0)=h_1(1,1-2h_0(0))=1$; and $f(g)=h_1(g(0), 1-\int_{-1}^1 h_0\circ g^\cdot)\neq 0$ implies $g(0)>\frac{1}{2}$ and $\int_{-1}^1 h_0(g^\cdot(t))dt<\frac{1}{2}$, hence $|g(s)-g(0)|<\frac{1}{2}$ and thus $g(s)>0$ for $s\in K$; i.e. $g(K)\subseteq W$.

Now we show by induction that there are maps $f_m: C^\infty(I^m,\mathbb{R})\to\mathbb{R}$ having the desired property for $K:=[-1,1]^m$, W and g_0 as above. For $m=1$ we have described such a map, namely $f_1:=f$. Assume we have f_m. Then let f_{m+1} be the map obtained by composing $C^\infty(I^{m+1},\mathbb{R})\cong C^\infty(I^m, C^\infty(I,\mathbb{R}))$, $(f_1)_*: C^\infty(I^m, C^\infty(I,\mathbb{R}))\to C^\infty(I^m,\mathbb{R})$ and $f_m: C^\infty(I^m,\mathbb{R})\to\mathbb{R}$, i.e. $f_{m+1}(g):=f_m(f_1\circ g^\vee)$. A trivial calculation shows that f_{m+1} has all the desired properties.

Next we consider a general X and K, but Y, W, g_0 still as above. For every point $x \in K$ let $u_x: I^m \to U_x$ be a diffeomorphism from the cube I^m onto a neighborhood U_x of x with $u_x(0)=x$. Then the open subsets V_x of U_x that correspond to the cube $]-1, 1[^m$ form a covering of K. Let $\{V_i; i=1 \dots n\}$ be a finite subcovering and u_i be the associated diffeomorphisms. Then $(u_i)^*: C^\infty(X, \mathbb{R}) \to C^\infty(I^m, \mathbb{R})$ is smooth, $(u_i)^*(1)=1$ and $(u_i)^*(g)(z)>0$ for all i and all $z \in [-1, 1]^m$ only if $g(x) \in W$ for $x \in K$. Choose a smooth map $h: \mathbb{R}^n \to \mathbb{R}$ with $h(1, \dots, 1)=1$ and $h(x)>0$ only if all coordinates of x are larger than 0. Then $f := h \circ (f_m \circ (u_1)^*, \dots, f_m \circ (u_n)^*)$ is the desired map.

Next we consider general X, K, Y and W but g_0 such that the closure $\overline{g_0(X)}$ is contained in W. Choose a smooth function $h: Y \to \mathbb{R}$ with $h|_{g_0(X)}=1$ and $h(y) \neq 0$ only for $y \in W$. Then $h_*: C^\infty(X, Y) \to C^\infty(X, \mathbb{R})$ is smooth, $h_*(g_0)=1$ and $h_*(g)(x)>0$ for all $x \in K$ implies $g(K) \subseteq W$. Thus the composite of h_* with f as obtained in the previous case has all claimed properties.

Finally we come to the general case. Let $W_1 \subseteq Y$ be open with $W_1 \supseteq g_0(K)$ and the closure $\overline{W_1} \subseteq W$. The open submanifold $X_1 := g_0^{-1}(W)$ of X contains K, and the restriction map $\text{incl}^*: C^\infty(X, Y) \to C^\infty(X_1, Y)$ is smooth. By the previous case applied to X_1, K, Y, W and the map $\text{incl}^*(g_0)$ we obtain a function f. Then the composition $f \circ \text{incl}^*$ has all the required properties. □

We shall study the following subspaces of $C^\infty(X, Y)$.

4.7.2 Definition

Onto(X, Y) denotes the space of surjective smooth maps from X to Y. *Imm*(X, Y) denotes the space of smooth immersions from X to Y. *Emb*(X, Y) denotes the space of smooth embeddings from X to Y.

Each of these spaces is considered with its initial smooth structure induced by the inclusion in $C^\infty(X, Y)$. *Submf*(X, Y) denotes the space of those *submanifolds* of Y that are diffeomorphic to X together with the final smooth structure induced by the map $\text{Emb}(X, Y) \to \text{Submf}(X, Y)$, $g \mapsto g(X)$.

We recall that a smooth map $g: X \to Y$ is called an *immersion* iff for all $x \in X$ the tangent map $T_x g: T_x X \to T_{g(x)} Y$ is injective. A smooth *embedding* is an injective smooth immersion (since X is compact the usual additional condition that it is a homeomorphism onto its image is automatically satisfied). Warning: Although the embeddings as defined here in a classical way are C^∞-embeddings in the sense of (8.8.1), an example of [Joris, 1982] shows that the converse fails.

4.7.3 Proposition. *Let X be a compact and Y any finite-dimensional smooth manifold. Then one has:*

(*i*) $\text{Imm}(X, Y)$ *is open in* $C^\infty(X, Y)$.
(*ii*) $\text{Emb}(X, Y)$ *is open in* $C^\infty(X, Y)$.
(*iii*) $\text{Diff}(X)$ *is open in* $C^\infty(X, X)$.
(*iv*) $\text{Onto}(X, Y)$ *is closed in* $C^\infty(X, Y)$.
(*v*) $\text{Diff}(X)$ *is closed in* $\text{Imm}(X, X)$.

Proof. (i) Let $c:\mathbb{R}\to C^\infty(X, Y)$ be a smooth curve with $c(0)\in\mathrm{Imm}(X, Y)$. Consider the smooth map $g:\mathbb{R}\sqcap X\to\mathbb{R}\sqcap X$ defined by $g(t, x):=(t, c(t)(x))$. The matrix representation of the tangent map $T_{(t,x)}g$ has the form

$$\begin{pmatrix}\mathrm{id} & 0\\ * & T_x c(t)\end{pmatrix}$$

Thus $T_{(0,x)}g$ is injective for all x. Hence there exists a neighborhood U of $\{0\}\times X$ such that $T_{(t,x)}g$ is injective for all $(t,x)\in U$. Since X is compact we may assume that $U=]-\varepsilon,\varepsilon[\times X$. From the injectivity of $T_{(t,x)}g$ we conclude, using again the matrix representation, that $T_x c(t)$ is injective for all $|t|<\varepsilon$ and all $x\in X$, i.e. $c(t)\in\mathrm{Imm}(X, Y)$ for all $|t|<\varepsilon$.

(ii) Let $c:\mathbb{R}\to C^\infty(X, Y)$ be a smooth curve with $c(0)\in\mathrm{Emb}(X, Y)$. Using that $\mathrm{Imm}(X, Y)$ is open in $C^\infty(X, Y)$ we know that $c(t)\in\mathrm{Imm}(X, Y)$ for all sufficiently small t. Since $\mathrm{Emb}(X, Y)$ consists of the injective immersions, we only have to show that $c(t)$ is injective for all sufficiently small t. We prove this indirectly. Assume that there are $t_n\to 0$, and $x_n, y_n\in X$ with $x_n\neq y_n$ and $c(t_n)(x_n)=c(t_n)(y_n)$. Since X is compact we may assume that $x_n\to x$ and $y_n\to y$. Continuity of $c^\wedge$ implies $c(0)(x)=c(0)(y)$, hence by injectivity of $c(0)$ one has $x=y$. The associated map $g:\mathbb{R}\sqcap X\to\mathbb{R}\sqcap X$ defined in (i) is certainly injective on some neighborhood $]-\varepsilon,\varepsilon[\times U_x$ of $(0, x)$. For n large enough one has $x_n, y_n\in U_x$ and $|t_n|<\varepsilon$, hence $g(t_n, x_n)=(t_n, c(t_n)(x_n))=(t_n, c(t_n)(y_n))=g(t_n, y_n)$, in contradiction with the injectivity of g on the considered neighborhood.

(iii) Let $c:\mathbb{R}\to C^\infty(X,X)$ be a smooth curve with $c(0)\in\mathrm{Diff}(X)$. By multiplying with $c(0)^{-1}$ we may assume $c(0)=\mathrm{id}$ and by (ii) we may assume that $c(t)\in\mathrm{Emb}(X,X)$ for all t (sufficiently small). Since $\mathrm{Diff}(X)$ consists exactly of the surjective embeddings we only have to show that $c(t)$ is surjective. So let $x\in X$ be arbitrary. The curve $c(\_)(x)$ connects $x=c(0)(x)$ with $c(t)(x)\in c(t)(X)$. The set $c(t)(X)$ is compact and open in X (since $c(t)$ is an immersion); hence it contains with every point all points belonging to the same connectivity component, which shows that $x\in c(t)(X)$.

(iv) Suppose $g_0\in C^\infty(X, Y)$ is not surjective. Then there exists an $y\notin g_0(X)$. Since $\{g\in C^\infty(X, Y); g(X)\subseteq Y\backslash\{y\}\}$ is open by (4.7.1) and disjoint with $\mathrm{Onto}(X, Y)$, one concludes that $C^\infty(X, Y)\backslash\mathrm{Onto}(X, Y)$ is open.

(v) Let $c:\mathbb{R}\to\mathrm{Imm}(X,X)$ be a smooth curve with $h:=c(0)\notin\mathrm{Diff}(X)$. Admit there exists $t_n\to 0$ with $h_n:=c(t_n)\in\mathrm{Diff}(X)$. Since $\mathrm{Diff}(X)$ consists exactly of the bijective immersions and since $\mathrm{Onto}(X,X)$ is closed in $C^\infty(X,X)$ we conclude that h is not injective. So let $y\in X$ be such that $h^{-1}(y)$ is not a single point. Since h is an immersion we find for any $x\in h^{-1}(y)$ a neighborhood U_x on which h is an embedding. Thus $U_x\cap h^{-1}(y)=\{x\}$. Using compactness of $h^{-1}(y)$ we conclude that $h^{-1}(y)$ is finite and the sets U_x can be chosen pairwise disjoint and diffeomorphic via h to some neighborhood W of y. Let W_1 be a connected neighborhood of y, such that the closure $\bar{W}_1$ of W_1 is contained in W and $\bar{W}_1\cap h(X\backslash\bigcup U_x)=\emptyset$. Then $h\in\{g;\ g(h^{-1}(y))\subseteq W_1$ and $g(X\backslash\bigcup U_x)\subseteq X\backslash\bar{W}_1\}$. Since this set is open by (4.7.1), we conclude that the diffeomorphisms h_n are contained in it for n sufficiently large. Thus $h_n(h^{-1}(y)) \subseteq W_1$ and $h_n(\bigcup U_x) \supseteq \overline{W_1}$, i.e. $h^{-1}(y) \subseteq h_n^{-1}(W_1) \subseteq \bigcup U_x$. This is a contradiction to the connectedness of $h_n^{-1}(W_1)$. □

4.7.4 Theorem. *If X is a regular smooth manifold modelled on Banach spaces which satisfy the equivalent conditions of* (4.4.43), *then the smooth structure of* Diff(X) *according to* (1.4.8) *is the initial one induced by the inclusion map* i:Diff(X)$\to C^\infty(X,X)$; *i.e. the inversion on* Diff(X) *is a C^∞-map if* Diff(X) *is considered as smooth subspace of $C^\infty(X,X)$.*

Proof. We only have to show that if $c:\mathbb{R}\to$Diff(X) is such that $i\circ c$ is a structure curve of $C^\infty(X,X)$, then the curve $j\circ c$ is also a structure curve, with j being the inversion map $f\mapsto f^{-1}$. By cartesian closedness of $\underline{C^\infty}$, (1.4.3) the map $f:=(i\circ c)^\wedge: \mathbb{R}\sqcap X\to X$ is C^∞ and hence by the equivalent conditions of (4.4.43) is of class C^∞ in the Fréchet sense. Furthermore, $g:=(j\circ c)^\wedge: \mathbb{R}\sqcap X\to X$ is the unique solution of the implicit equation $f(t,g(t,x))=x$. Since $f(t,\_)$ is a diffeomorphism, the second partial derivative of f is an isomorphism, and we may apply the implicit function theorem for the modelling Banach spaces to conclude that g is smooth. By cartesian closedness $g^\vee = j\circ c$ is smooth as well. □

4.7.5 Theorem. *The group* Diff(X) *of diffeomorphisms of a compact smooth manifold X is a smooth manifold modelled on convenient vector spaces; cf.* (4.7.9).

Proof. The smooth structure of Diff(X) was discussed in (4.7.4) and it was shown that Diff(X) is a smooth group, cf. (1.4.8). Since the multiplication is smooth it is enough to find a smooth chart at id_X. Consider an exponential map exp: $TX\to X$, cf. [Bröcker, Jänich, 1973, pp. 121]. Then (π, exp): $TX\to X\sqcap X$ is a diffeomorphism of a neighborhood U of the zero-section in TX onto a neighborhood V of the diagonal in $X\sqcap X$. The map $\exp_*:\Gamma(\pi)\to C^\infty(X,X)$ is obviously smooth and bijective from $U^\sim:=\{s\in\Gamma(\pi);\ s(X)\subseteq U\}$ onto $V^\sim:=\{h\in$Diff(X); graph(h)$\subseteq V\}$ with inverse map $h\mapsto(x\mapsto(\pi,\exp)^{-1}(x, hx))$, which is smooth as well. So it remains to show that $U^\sim$ and $V^\sim$ are open for the final topologies induced by the smooth curves.

Let $c:\mathbb{R}\to\Gamma\subseteq C^\infty(X, TX)$ be a smooth curve with $c(0)\in U^\sim$. Admit there exist $t_n\to 0$ with $c(t_n)\notin U^\sim$, i.e. there are $x_n\in X$ with $c^\wedge(t_n, x_n)\notin U$. X being compact we may assume $x_n\to x$. Then $c^\wedge(t_n, x_n)\to c^\wedge(0, x)\in U$, in contradiction with $c(0)\in U^\sim$.

Similarly, let $c:\mathbb{R}\to$Diff(X)$\subseteq C^\infty(X,X)$ be a smooth curve with $c(0)\in V^\sim$ but $c(t_n)\notin V^\sim$, i.e. there are $x_n\in X$ with $(x_n, c^\wedge(t_n, x_n))\notin V$. X being compact we may assume $x_n\to x$. Then $(x_n, c^\wedge(t_n, x_n))\to(x, c^\wedge(0, x))\in V$ in contradiction with $c(0)\in V^\sim$. □

4.7.6 Proposition. *Let X be a compact and Y any finite-dimensional smooth manifold. The space* Emb(X, Y) *of embeddings of X in Y is a smooth manifold modelled on convenient vector spaces; cf.* (4.7.9). *The smooth group* Diff(X) *acts by composition smoothly on* Emb(X, Y).

Proof. Let $e \in \mathrm{Emb}(X, Y)$. Choose a tubular neighborhood of e, i.e. a vector bundle $p: E \to X$ with a diffeomorphism u of E onto a neighborhood of $e(X)$ such that $u \circ 0_p = e$ (0_p denotes the zero section of p), cf. [Hirsch, 1976, p. 110]. Now set $U := \{g \in \mathrm{Emb}(X, Y);\ g(X) \subseteq u(E) \text{ and } p \circ u^{-1} \circ g \in \mathrm{Diff}(X)\}$.

In order to show that U is open in $\mathrm{Emb}(X, Y)$ we first prove that $\{g \in \mathrm{Emb}(X, Y); g(X) \subseteq u(E)\}$ is open in $\mathrm{Emb}(X, Y)$. Given a $g_0 \in C^\infty(X, Y)$ with $g_0(X) \subseteq u(E)$ we can choose an $f \in C^\infty(X, \mathbb{R})$ with $f|_{g_0(E)} = 1$ and $\mathrm{supp}(f) \subseteq u(E)$. Then $\{g; (f \circ g)(x) > 0 \text{ for all } x\}$ is an open neighborhood of g_0 in $\{g; g(X) \subseteq u(E)\}$. From this and the fact that $\mathrm{Diff}(X)$ is open in $C^\infty(X, X)$ it follows that U is open.

U is diffeomorphic to $\Gamma(p) \sqcap \mathrm{Diff}(X)$ by means of the map $g \mapsto (u^{-1} \circ g \circ (p \circ u^{-1} \circ g)^{-1},\ p \circ u^{-1} \circ g) \in \Gamma(p) \sqcap \mathrm{Diff}(X)$ whose inverse is $(\mathfrak{s}, h) \mapsto u \circ \mathfrak{s} \circ h \in U \subseteq \mathrm{Emb}(X, Y)$.

We have thus shown via the previous theorem that $\mathrm{Emb}(X, Y)$ admits a smooth atlas with charts having values in convenient vector spaces.

$\mathrm{Diff}(X)$ acts smoothly on $\mathrm{Emb}(X, Y)$ since the action is obtained by restricting the composition map $C^\infty(X, Y) \sqcap C^\infty(X, X) \to C^\infty(X, Y)$; cf. (1.4.6). □

4.7.7 Proposition. *Let X be a compact and Y any finite-dimensional smooth manifold. The space* Submf(X, Y) *of submanifolds of Y that are diffeomorphic to X is a smooth manifold modelled on convenient vector spaces.*

Proof. The structure of $\mathrm{Submf}(X, Y)$ is by definition the final one induced by the map $q: \mathrm{Emb}(X, Y) \to \mathrm{Submf}(X, Y)$, $g \mapsto g(X)$. We show first that the fibres of the map q are exactly the orbits under the action of $\mathrm{Diff}(X)$. In fact, if $g \in \mathrm{Emb}(X, Y)$ and $h \in \mathrm{Diff}(X)$, then $q(g \circ h) = q(g)$, i.e. g and $g \circ h$ are in the same fibre. Conversely, if $q(g_1) = q(g_2)$ for two embeddings g_1 and g_2, then g_1 and g_2 are diffeomorphisms onto their image, hence $h := g_2^{-1} \circ g_1 \in \mathrm{Diff}(X)$ and $g_1 - g_2 \circ h$ shows that g_1 and g_2 are in the same orbit.

We now show that $q(U)$ is open in $\mathrm{Submf}(X, Y)$, U being the set defined in the proof of (4.7.6). So let $X_0 \subseteq Y$ be a submanifold that is contained in $q(U)$. Using the diffeomorphism $U \cong \Gamma(p) \sqcap \mathrm{Diff}(X)$ of the proof of (4.7.6) one obtains an $\mathfrak{s}_0 \in \Gamma(p)$ and an $h_0 \in \mathrm{Diff}(X)$ with $X_0 = q(u \circ \mathfrak{s}_0 \circ h_0) = q(u \circ \mathfrak{s}_0)$. It is enough to find a smooth function $f: \mathrm{Emb}(X, Y) \to \mathbb{R}$ that is constant on orbits, satisfies $f(u \circ \mathfrak{s}_0) = 1$, and is such that $f(g) \neq 0$ only for $g \in U$. In fact, such an f factor smoothly over $\mathrm{Submf}(X, Y)$, and the corresponding map $\bar{f}$ on $\mathrm{Submf}(X, Y)$ has the property that $\bar{f}(X_0) = f(u \circ \mathfrak{s}_0) = 1$ and $\bar{f}(X_1) \neq 0$ only if $X_1 \in q(U)$. In order to obtain such a function we choose open relatively compact neighborhoods U_1 of $\mathfrak{s}_0(X)$ in E, U_2 of X in TX, U_3 of $T\mathfrak{s}_0(\overline{U_0})$ in TE and we define $U_0 \subseteq \Gamma(p)$ by $U_0 := \{\mathfrak{s} \in \Gamma(p); \mathfrak{s}(X) \subseteq U_1, T\mathfrak{s}(\overline{U_2}) \subseteq U_3\}$. By (4.7.1) there exists a smooth function $f_0: \Gamma(p) \to \mathbb{R}$ with $f_0(\mathfrak{s}_0) = 1$ and $f_0(\mathfrak{s}) \neq 0$ only if $\mathfrak{s} \in U_0$. Now define $f: \mathrm{Emb}(X, Y) \to \mathbb{R}$ by

$$f(g) := \begin{cases} f_0(u^{-1} \circ g \circ (p \circ u^{-1} \circ g)^{-1}) & \text{if } g \in U \\ 0 & \text{if } g \notin U \end{cases}$$

Obviously $f(u \circ \mathfrak{s}_0) = f_0(\mathfrak{s}_0) = 1$; $f(g) \neq 0$ implies $g \in U$; and f is constant on orbits.

It remains to prove that f is smooth. It is enough to show that for any smooth curve $c:\mathbb{R}\to\mathrm{Emb}(X, Y)$ the composite $f\circ c$ is smooth in a neighborhood of zero.

We first consider the case where $c(0)\in U$. Then $c(t)\in U$ for t in a 0-neighborhood V, hence $c(t)(X)\subseteq u(E)$ and thus $t\mapsto p\circ u^{-1}\circ c(t)$ is a smooth curve $V\to\mathrm{Diff}(X)$. Using that $\mathrm{Diff}(X)$ is a smooth group one concludes that also $t\mapsto(p\circ u^{-1}\circ c(t))^{-1}$ is smooth into $\mathrm{Diff}(X)$, and since $\mathrm{Diff}(X)$ acts smoothly on $\mathrm{Emb}(X, Y)$ the curve $t\mapsto s(t):=u^{-1}\circ c(t)\circ(p\circ u^{-1}\circ c(t))^{-1}$ is smooth from V into $\Gamma(p)$. Thus $f\circ c=f_0\circ s$ is smooth on V.

We now consider the case where $c(0)\notin U$. We want to prove that then $f(c(t))=0$ for sufficiently small t. If $c(0)(X)\nsubseteq u(\overline{U_1})$, i.e. there is an $x\in X$ with $c(0)(x)\notin u(\overline{U_1})$, then $c(t)\in\{g;g(x)\in Y\backslash u(\overline{U_1})\}$ and thus $c(t)\in U$ and $f(c(t))=0$ for sufficiently small t. If $c(0)(X)\subseteq u(\overline{U_1})$, then $c(t)\in\{g;g(X)\subseteq u(E)\}$ for t sufficiently small. Thus $h:=p\circ u^{-1}\circ c$ is a smooth (locally defined curve) into $C^\infty(X,X)$. Since $c(0)\notin U$ one has that $h_0:=h(0)\notin\mathrm{Diff}(X)$. We want to show that $h(t)\notin\mathrm{Diff}(X)$ for t sufficiently small. Suppose there are $t_n\to 0$ with $h_n:=h(t_n)\in\mathrm{Diff}(X)$. Let us show first that h_0 is an immersion: Admit $T_xh_0(\xi_x)=0$ for some $0\neq\xi_x\in T_xX$. Let $\vartheta_n:=u^{-1}\circ c(t_n)\circ h_n^{-1}\in\Gamma(p)$. Then $T\vartheta_n\cdot\mathrm{Th}_n\cdot\xi_x = T(\vartheta_n\circ h_n)(\xi_x)=T(u^{-1}\circ c(t_n))\cdot\xi_x\to T(u^{-1}\circ c(0))\cdot\xi_x=0$ for $n\to\infty$. Since $\mathrm{Th}_n\cdot\xi_x\to\mathrm{Th}\cdot\xi_x$, the fibre multiplication $\mathbb{R}\sqcap TX\to TX$ is continuous, and U_2 is an open neighborhood of X in TX, we may choose $r_n\to\infty$ with $r_n\cdot\mathrm{Th}_n\cdot\xi_x\in U_2$. Then $r_n\cdot T\vartheta_n\cdot\mathrm{Th}_n\cdot\xi_x=T\vartheta_n\cdot(r_n\cdot\mathrm{Th}_n\cdot\xi_x)\in U_3\subseteq\overline{U_3}$. Since $r_n\to\infty$ and $T\vartheta_n\cdot\mathrm{Th}_n\cdot\xi_x\to T(u^{-1}\circ c(0))\cdot\xi_x\neq 0$ this is a contradiction with the (sequential) compactness of $\overline{U_3}$.

Now we can use (*v*) of (4.7.3), namely that $\mathrm{Diff}(X)$ is closed in $\mathrm{Imm}(X,X)$. Thus h_0 is, as limit of the diffeomorphisms h_n, also a diffeomorphism. This is a contradiction, and hence $f(c(t))=0$ locally.

Since we just proved that $q(U)$ is open in $\mathrm{Submf}(X, Y)$ we may conclude that the map $U\to\Gamma(p)$ defined by $g\mapsto u^{-1}\circ g\circ(p\circ u^{-1}\circ g)^{-1}$ induces a diffeomorphism $q(U)\to\Gamma(p)$ with inverse $s\mapsto q(u\circ s)$. Thus $\mathrm{Submf}(X, Y)$ is a smooth manifold modelled on convenient vector spaces. □

4.7.8 Theorem. *Let X be a compact and Y any finite-dimensional smooth manifold. The map $q:\mathrm{Emb}(X, Y)\to\mathrm{Submf}(X, Y)$, $g\mapsto g(X)$ defines a smooth principal fibre bundle with typical fibre* $\mathrm{Diff}(X)$, *i.e. there is an open covering of* $\mathrm{Submf}(X, Y)$ *by sets W for which there exist diffeomorphisms $q^{-1}(W)\to W\sqcap\mathrm{Diff}(X)$ which composed with $\mathrm{pr}_1: W\sqcap\mathrm{Diff}(X)\to W$ gives $q|_{q^{-1}(W)}$.*

Proof. That the map q defines a fibre bundle follows immediately from the preceding proposition, where we proved that $q(U)$ is open in $\mathrm{Submf}(X, Y)$ and $q|_U:U\to q(U)$ corresponds via the diffeomorphism $U\to\Gamma(p)\sqcap\mathrm{Diff}(X)\to q(U)\sqcap\mathrm{Diff}(X)$ to the projection $\mathrm{pr}_1: q(U)\sqcap\mathrm{Diff}(X)\to q(U)$.

It is a principal fibre bundle since $\mathrm{Diff}(X)$ acts on $\mathrm{Emb}(X, Y)$ and the fibres are exactly the orbits of this action. □

4.7.9 Remark. All the manifolds of mappings that we considered in this chapter are in fact modelled on nuclear Fréchet spaces since the models are spaces of sections of finite-dimensional vector bundles. These spaces of sections are Fréchet spaces by (4.6.24). That they are nuclear follows from the fact that $C^\infty(U, \mathbb{R}^m)$ is nuclear if $U \subseteq \mathbb{R}^n$ is open, cf. [Jarchow, 1981, p. 498].

4.8 Theorems on inverse and implicit functions

In this section the differentiability properties of inverse and implicit functions are discussed. They depend on those of the inversion map $m \mapsto m^{-1}$, m invertible in $L(E, E)$, which we study first.

4.8.1 Definition. For a convenient vector space E we denote by $GL(E)$ the subset of the convenient vector space $L(E, E)$ formed by the Con-isomorphisms together with the bornology and $\mathscr{Lip}^k$-structures ($k \in \mathbb{N}_{0,\infty}$) induced by those of $L(E, E)$. With $\operatorname{inv}: GL(E) \to GL(E)$ we denote the map $m \mapsto m^{-1}$.

If $f: U \to V$ is a $\mathscr{Lip}^k$-diffeomorphism between M-open subsets U and V of E then differentiation of the equations $f \circ f^{-1} = \mathrm{id}$ and $f^{-1} \circ f = \mathrm{id}$ shows that $f'(x) \in GL(E)$ for all $x \in U$ and that $(f^{-1})': V \to L(E, E)$ is the composite of the maps $V \xrightarrow{f^{-1}} U \xrightarrow{f'} GL(E) \xrightarrow{\operatorname{inv}} GL(E) \subseteq L(E, E)$.

4.8.2 Lemma. *Let* $c: \mathbb{R} \to GL(E) \subseteq L(E, E)$ *be* $\mathscr{Lip}^k$ *and* $\operatorname{inv} \circ c: \mathbb{R} \to L(E, E)$ *be bornological, then* $\operatorname{inv} \circ c: \mathbb{R} \to L(E, E)$ *is* $\mathscr{Lip}^k$.

Proof. Let us prove the statement first for $k = 0$:

$$\frac{\operatorname{inv}(c(t)) - \operatorname{inv}(c(s))}{t - s} = -\operatorname{inv}(c(t)) \circ \left(\frac{c(t) - c(s)}{t - s}\right) \circ \operatorname{inv}(c(s))$$

$$= -\operatorname{comp}\left(\operatorname{inv}(c(t)), \frac{c(t) - c(s)}{t - s}, \operatorname{inv}(c(s))\right).$$

Using that comp is 3-linear and bornological and $\operatorname{inv} \circ c$ and $(t, s) \mapsto (c(t) - c(s))/(t - s)$ are bornological, we conclude that $\delta(\operatorname{inv} \circ c)$ is bornological, i.e. $\operatorname{inv} \circ c$ is $\mathscr{Lip}^0$.

Let now $k \geq 1$. We first show that $\operatorname{inv} \circ c$ is weakly differentiable: Consider for fixed s the map $t \mapsto -\operatorname{comp}(\operatorname{inv}(c(t+s)), \bar{\delta} c(s, t+s) \circ \operatorname{inv}(c(s)))$. It is $\mathscr{Lip}^0$ since both coordinates are so. Its value is $(\operatorname{inv}(c(t+s)) - \operatorname{inv}(c(s)))/t$ for $t \neq 0$ and $-\operatorname{inv}(c(s)) \circ c^{\cdot}(s) \circ \operatorname{inv}(c(s))$ for $t = 0$. Hence

$$\left\{\frac{1}{t}\left(\frac{\operatorname{inv}(c(t+s)) - \operatorname{inv}(c(s))}{t} + \operatorname{inv}(c(s)) \circ c^{\cdot}(s) \circ \operatorname{inv}(c(s))\right);\ 0 \neq |t| \leq 1\right\}$$

is bounded and therefore $(\operatorname{inv} \circ c)^{\cdot}(s)$ exists and equals -$\operatorname{inv}(c(s)) \circ c^{\cdot}(s) \circ \operatorname{inv}(c(s))$.

From $(\operatorname{inv} \circ c)^{\cdot} = -\operatorname{comp} \circ (\operatorname{inv} \circ c, c^{\cdot}, \operatorname{inv} \circ c)$ the statement follows now by induction: The map comp is 3-linear and bornological and by induction hypothesis the first and the last coordinate is $\mathscr{Lip}^{k-1}$, as is the middle one, hence the left side is $\mathscr{Lip}^{k-1}$ and consequently $\operatorname{inv} \circ c$ is $\mathscr{Lip}^k$. □

In the next theorem we will need the following

4.8.3 Lemma. *Let $f \in L(E, E)$ be such that there exists a unique $g \in L(E, E)$ with $g \circ f = \mathrm{id}$. Then $f \in GL(E)$.*

Proof. Obviously it is enough to show that also $f \circ g = \mathrm{id}$. Since $g \circ f = \mathrm{id}$ one has $f \circ g \circ f = f$. Thus $f \circ g = \mathrm{id}$ on $f(E)$. And since linear morphisms are continuous for the locally convex topology one even has $f \circ g = \mathrm{id}$ on the closure $\overline{f(E)}$ with respect to the locally convex topology. Thus it is enough to prove that $\overline{f(E)} = E$. Admit that there exists an $x_0 \in E \setminus \overline{f(E)}$. By the Hahn–Banach theorem there exists an $\ell \in E'$ with $\ell|_{f(E)} = 0$ and $\ell(x_0) \neq 0$. We put $g_1 := g + x_0 \cdot \ell$, where $(x_0 \cdot \ell)(x) := \ell(x) x_0$. Then $g_1 \in L(E, E)$ and $g_1 \circ f = g \circ f + x_0 \cdot (\ell \circ f) = \mathrm{id} + 0 = \mathrm{id}$, hence by the assumption that f has a unique left inverse we conclude that $x_0 \cdot \ell = g_1 - g = 0$. This is a contradiction to $\ell((x_0 \cdot \ell)(x_0)) = \ell(x_0)^2 \neq 0$. □

Now we can apply this to prove an inverse function theorem:

4.8.4 Theorem. *Let U and V be* M*-open in a convenient vector space E, $f: U \to V$ a bijective $\mathscr{Lip}^k$-map. If f^{-1} a $\mathscr{Lip}^0$-map and $f'(x)$ has a left inverse in $L(E, E)$ for all $x \in U$ then f^{-1} is a $\mathscr{Lip}^k$-map too, i.e. f is a $\mathscr{Lip}^k$-diffeomorphism.*

Proof. For $k = 0$ we have nothing to prove. So let $k \geq 1$ and $g := f^{-1}$.

First we want to show that g is weakly differentiable and dg is uniquely determined by $dg(y, \_) \circ f'(g(y)) = \mathrm{id}$. So let $y \in V$, $w \in E$, $x := g(y)$ and $m \in L(E, E)$ be an arbitrary left inverse of $f'(x)$, i.e. $m \circ f'(x) = \mathrm{id}$. We claim that g is weakly differentiable at y in direction w and $dg(y, w) = m(w)$. For this we have to consider the $\mathscr{Lip}^0$-curve $c: t \mapsto g(x + tw)$. Then

$$\frac{c(t) - c(0)}{t} - m(w) = (m \circ f'(x))\left(\frac{c(t) - c(0)}{t}\right) - m(w) = m\left(f'(x)\left(\frac{c(t) - c(0)}{t}\right) - w\right).$$

So it is enough to show that $r(t) := f'(x)\left(\frac{c(t) - c(0)}{t}\right) - w$ converges weakly to 0 for $t \to 0$. One has

$$r(t) = f'(x)\left(\frac{c(t) - c(0)}{t}\right) - \frac{x + tw - x}{t} = f'(x)\left(\frac{c(t) - c(0)}{t}\right) - \frac{f(c(t)) - f(c(0))}{t}$$

$$= \int_0^1 (f'(x) - f'(x + s(c(t)) - c(0)))\, ds \left(\frac{c(t) - c(0)}{t}\right).$$

Since the map $(t, s) \mapsto x + s(c(t) - c(0))$ is $\mathscr{Lip}^0$ into E and $(0, s) \mapsto x = c(0)$ we conclude that $g: (t, s) \mapsto f'(x) - f'(x + s(c(t) - c(0)))$ is defined in a neighborhood of $\{0\} \times [0, 1]$ and is $\mathscr{Lip}^0$ into $L(E, E)$. Thus $t \mapsto \int_0^1 g(t, s)\, ds$ is defined and $\mathscr{Lip}^0$

on a neighborhood of $\{0\}$. Hence $\int_0^1 g(t,s)ds$ is M-convergent to $\int_0^1 g(0,s)ds=0$ in $L(E,E)$ for $t\to 0$ and, in particular, it is uniformly M-convergent in E on the bounded set

$$\left\{\frac{c(r)-c(0)}{r}; 0\neq r\in I\right\},$$

where I is some bounded open interval containing 0. From this we conclude that

$$r(t)=\int_0^1 g(t,s)ds\left(\frac{c(t)-c(0)}{t}\right)$$

converges (Mackey, hence weakly) to 0 for $t\to 0$.

Since the differential $df(y,\_)$ is unique we conclude that the left inverse m of $f'(x)$ has to be unique and thus $f'(x)\in GL(E)$ by (4.8.3) and $dg(y,\_)=[f'(x)]^{-1}$.

Using that g is $\mathscr{Lip}^0$ and that the weak differential dg exists we conclude from (4.3.10) that $dg: V\sqcap E\to E$ is $\mathscr{Lip}^{-1}$. This implies that $g':=(dg)^\vee: V\to L(E,E)$ is $\mathscr{Lip}^{-1}$.

Finally we show by induction on k that $g': V\to L(E,E)$ is $\mathscr{Lip}^{k-1}$: Let us consider the factorization $g'=\text{inv}\circ f'\circ g$ and let $c:\mathbb{R}\to V$ be a $\mathscr{Lip}^{k-1}$-curve. Then by induction hypothesis $e:=f'\circ g\circ c:\mathbb{R}\to GL(E)$ is $\mathscr{Lip}^{k-1}$ and $\text{inv}\circ e=g'\circ c:\mathbb{R}\to L(E,E)$ is bornological. Hence by (4.8.2) one concludes that $g'\circ c=\text{inv}\circ e$ is $\mathscr{Lip}^{k-1}$, i.e. g' is $\mathscr{Lip}^{k-1}$, and thus g is $\mathscr{Lip}^k$. □

Let us give as corollary an implicit function theorem:

4.8.5 Theorem. *Let $f: I\sqcap U\to E$ be $\mathscr{Lip}^k$ $(k\geq 1)$, where U is* M*-open in E, and I an open interval in $\mathbb{R}$. If one has a $\mathscr{Lip}^0$-curve $c: I\to U$ solving the equation $f(t,c(t))=0$, and such that $\partial_2 f(t,c(t))\in GL(E)$ and $t\mapsto\partial_2 f(t,c(t))^{-1}$ is bornological, then c is $\mathscr{Lip}^k$.*

Proof. Consider the $\mathscr{Lip}^k$-map $\bar f:=(\text{pr}_1, f): I\sqcap U\to I\sqcap E$. Then $\bar f'(t,x)(s,v)=(s,\partial_1 f(t,x)s+\partial_2 f(t,x)v)$ and therefore $\bar f'(t,c(t))\in GL(\mathbb{R}\sqcap E)$, the inverse map being $(s,v)\mapsto(s,-\partial_2 f(t,c(t))^{-1}\cdot\partial_1 f(t,c(t))\cdot s+\partial_2 f(t,c(t))^{-1}\cdot v)$. Consequently $\text{inv}\circ\bar f'\circ(1,c)$ is bornological. Furthermore $\bar c:=(1_{\mathbb{R}},c): I\to I\sqcap U$ is $\mathscr{Lip}^0$ and $(\bar f\circ\bar c)(t)=\bar f(t,c(t))=(t,0)$ is smooth. By the proof of the inverse function theorem (4.8.4) we conclude that $\bar c$ is $\mathscr{Lip}^k$ and hence $c=\text{pr}_2\circ\bar c$ is also $\mathscr{Lip}^k$. □

5 DIFFERENTIABLE MAPS AND CATEGORICAL PROPERTIES

In section 5.1 it will be shown that there exist free convenient vector spaces over $\mathscr{Lip}^k$-spaces, i.e. that the respective forgetful functor has a left adjoint. This means that to every $\mathscr{Lip}^k$-space X one can associate a convenient vector space λX together with a $\mathscr{Lip}^k$-map $\iota_X: X \to \lambda X$ such that for any convenient vector space E the map $(\iota_X)^*: L(\lambda X, E) \to \mathscr{Lip}^k(X, E)$ is a bijection. The space λX can be obtained as the Mackey closure of the linear subspace spanned by the image of the canonical map $X \to \mathscr{Lip}^k(X, \mathbb{R})'$.

In the case where $k = \infty$ and X is a finite-dimensional smooth manifold we prove that the linear subspace generated by $\{\ell \circ \mathrm{ev}_x;\ x \in X,\ \ell \in E'\}$ is Mackey dense in $C^\infty(X, E)'$. From this we conclude that the free convenient vector space over a manifold X is the space of distributions with compact support on X.

The existence of free convenient vector spaces over ℓ^∞-spaces X is also proved, and in this case an explicit description of λX is given, namely as the space of those functions $f: X \to \mathbb{R}$ for which the support $\mathrm{supp}(f)$ is bounded and $\sum_x |f(x)| < \infty$, together with the bornology for which a set of functions f is bounded iff $\bigcup_f \mathrm{supp}(f)$ is bounded and $\sup_f \sum_x |f(x)| < \infty$. Since for any set X with its coarse ℓ^∞-structure this construction gives the usual Banach space $\ell^1 X$ it is natural to write $\ell^1 X$ instead of λX also for an arbitrary ℓ^∞-space X. On the product of $\ell^1 X$ with $\ell^\infty X := \ell^\infty(X, \mathbb{R})$ one can define a bilinear bornological function $(f, g) \mapsto \sum_x f(x) g(x)$ and show that it induces an isomorphism $(\ell^1 X)' \cong \ell^\infty X$. But already $\ell^1 X$ can be identified with the dual of some convenient vector space $c_0 X$ by a restriction of this bilinear function: one takes as $c_0 X$ the subspace of $\ell^\infty X$ formed by those functions $g: X \to \mathbb{R}$ for which $\mathrm{supp}(g)$ is countable and for which for any $\varepsilon > 0$ the set $\{x \in X; |g(x)| > \varepsilon\}$ has a finite intersection with every bounded subset of X. In the case that X is any set with its coarse ℓ^∞-structure one gets the usual Banach space $c_0 X$ associated to the set X.

In section 5.2 we consider convenient co-algebras. They are defined in a way dual to convenient algebras. We show that in the cases $\mathscr{M} = \ell^\infty$ and $\mathscr{M} = C^\infty$ the functor λ: $\underline{\mathcal{M}}\to\underline{\mathrm{Con}}$ constructed in section 5.1 lifts to a functor into the category $\underline{\mathrm{ConCoAlg}}$ of convenient co-algebras. The proof is based on the result that λ carries finite products to tensor products, and these are exactly the categorical products in the category $\underline{\mathrm{ConCoAlg}}$. Furthermore, we construct a right adjoint to λ: $\underline{\mathcal{M}}\to\underline{\mathrm{ConCoAlg}}$. It is obtained by taking for a given convenient co-algebra the subset of co-idempotent elements with its induced $\underline{\mathcal{M}}$-structure. The duality functor of $\underline{\mathrm{Con}}$ lifts to a functor from $\underline{\mathrm{ConCoAlg}}$ to the category $\underline{\mathrm{ConAlg}}$ of convenient algebras and the isomorphism between the convenient vector spaces $(\lambda X)'$ and $\mathcal{M}(X,\mathbb{R})$ becomes an isomorphism of algebras if $\mathcal{M}(X,\mathbb{R})$ is considered with the pointwise defined algebra operations. In case $\mathcal{M}=\ell^\infty$ we show that the functor ℓ^1 is even an embedding of $\underline{\ell^\infty}$ into $\underline{\mathrm{ConCoAlg}}$.

In section 5.3 we search for classes of convenient vector spaces that are still closed under the important constructions developed so far but whose spaces all have additional properties. For this we describe the smallest subclasses that are closed under certain constructions (like function spaces and finite products) and contain the finite-dimensional vector spaces. Then we give several examples showing that most of the properties one might look for (like existence of solutions of various types of equations) fail even for spaces in these smallest classes.

In section 5.4 we consider the concept of reflexivity associated to the internal duality functor described in section 3.9. We investigate its relationship to more classical reflexivity concepts. It is shown that for a finite-dimensional smooth manifold X the space of smooth functions $C^\infty(X,E)$ is reflexive if and only if E is reflexive. This is also generalized to certain infinite-dimensional manifolds X.

5.1 Free convenient vector spaces

5.1.1 Theorem. *For every $\mathcal{Lip}^k$-space X where $k\in\mathbb{N}_{0,\infty}$ (resp. ℓ^∞-space X) there exists a convenient vector space λX and a $\mathcal{Lip}^k$-map (resp. ℓ^∞-map) $\iota_X\colon X\to\lambda X$ with the universal property that every $\mathcal{Lip}^k$-map (resp. ℓ^∞-map) $g\colon X\to E$ from X into a convenient vector space E factors as $g=\bar g\circ\iota_X$ with a unique linear $\underline{\mathrm{Con}}$-morphism $\bar g\colon\lambda X\to E$. One calls λX therefore the free convenient vector space over X. It can be constructed as the* M*-closure of the linear subspace of $\mathcal{Lip}^k(X,\mathbb{R})'$ (resp. $\ell^\infty(X,\mathbb{R})'$) generated by the point evaluations* ev_x, *with* $\iota_X(x):=\mathrm{ev}_x$ *for $x\in X$. Categorically this means: one has a functor λ: $\underline{\mathcal{Lip}^k}\to\underline{\mathrm{Con}}$ (resp. λ: $\underline{\ell^\infty}\to\underline{\mathrm{Con}}$) which is left adjoint to the respective forgetful functor and ι is the unit of the adjunction.*

Proof. In order to treat both cases simultaneously we write $\mathcal{M}$ for either $\mathcal{Lip}^k$ or ℓ^∞. We first describe a functor λ_s: $\underline{\mathcal{M}}\to\underline{\mathrm{sPre}}$ which is left adjoint to the forgetful functor $\underline{\mathrm{sPre}}\to\underline{\mathcal{M}}$. For any $\mathcal{M}$-space X we consider the convenient vector space $\mathcal{M}(X,\mathbb{R})'$ (for $\mathcal{M}(X,\mathbb{R})$ see (3.6.1) if $\mathcal{M}=\ell^\infty$ and (4.4.3) if

$\mathcal{M}=\mathcal{Lip}^k$). The canonical map $\iota_X\colon X\to\mathcal{M}(X,\mathbb{R})'$, defined by $\iota_X(x)(g):=g(x)$ for $x\in X$ and $g\in\mathcal{M}(X,\mathbb{R})$, is an $\underline{\mathcal{M}}$-morphism as one easily verifies by using that the $\mathcal{M}$-structure of a dual of a convenient vector space is the initial one induced by the evaluation maps, cf. (3.9.2). Moreover, the $\mathcal{M}$-structure of X is the initial one induced by ι_X; in fact, if Z is any $\mathcal{M}$-space and $f\colon Z\to X$ a map such that $\iota_X\circ f\colon Z\to\mathcal{M}(X,\mathbb{R})'$ is an $\underline{\mathcal{M}}$-morphism, then so is $\mathrm{ev}_g\circ\iota_X\circ f=g\circ f$ for all $g\in\mathcal{M}(X,\mathbb{R})$, which implies that $f\colon Z\to X$ is an $\underline{\mathcal{M}}$-morphism.

One now defines $\lambda_s X$ to be the linear subspace of $\mathcal{M}(X,\mathbb{R})'$ generated by the image $\iota_X(X)$, together with the $\underline{\text{Pre}}$-structure induced from $\mathcal{M}(X,\mathbb{R})'$. Then $\lambda_s X$ is a separated preconvenient vector space and ι_X induces an initial $\underline{\mathcal{M}}$-morphism $\iota_X\colon X\to\lambda_s X$. We now verify that one has the universal property for $\lambda_s\colon \underline{\mathcal{M}}\to\underline{\text{sPre}}$ being a left adjoint to the forgetful functor, namely: any $\underline{\mathcal{M}}$-morphism $f\colon X\to E$ into a separated preconvenient vector space E factors in a unique way as $f=\bar f\circ\iota_X$ with some linear $\underline{\mathcal{M}}$-morphism (i.e. $\underline{\text{sPre}}$-morphism) $\bar f\colon\lambda_s X\to E$. Uniqueness is trivial since $\bar f(\iota_X(x))=f(x)$, $\{\iota_X(x);\,x\in X\}$ generates $\lambda_s X$ and $\bar f$ is linear. For the existence of $\bar f$ we use the $\underline{\text{sPre}}$-morphism $\iota_E\colon E\to\Pi_{E'}\mathbb{R}$, characterized by $\mathrm{pr}_\ell\circ\iota_E=\ell$ for all $\ell\in E'$, and which is initial by the special embedding lemma (2.5.5). Let $f_0\colon\mathcal{M}(X,\mathbb{R})'\to\Pi_{E'}\mathbb{R}$ be the $\underline{\text{sPre}}$-morphism characterized by $\mathrm{pr}_\ell\circ f_0=\mathrm{ev}_{\ell\circ f}$ for all $\ell\in E'$. Then $f_0\circ\iota_X=\iota_E\circ f$ as verified by composing with pr_ℓ. So $f_0(\iota_X(X))\subseteq\iota_E(E)$, and since f_0 is linear and $\lambda_s X$ is the linear subspace generated by $\iota_X(X)$ one has $f_0(\lambda_s X)\subseteq\iota_E(E)$. Therefore $f_0|_{\lambda_s X}\colon\lambda_s X\to\Pi_{E'}\mathbb{R}$ factors as $f_0|_{\lambda_s X}=\iota_E\circ\bar f$ where $\bar f\colon\lambda_s X\to E$ is the desired $\underline{\text{sPre}}$-morphism.

Since the completion functor $\tilde\omega\colon\underline{\text{sPre}}\to\underline{\text{Con}}$ is left adjoint to the inclusion $\underline{\text{Con}}\to\underline{\text{sPre}}$, cf. (2.6.5), the composite $\lambda:=\tilde\omega\circ\lambda_s\colon\underline{\mathcal{M}}\to\underline{\text{Con}}$ is left adjoint to the corresponding forgetful functor. In order to show that $\tilde\omega(\lambda_s E)$ can be obtained by taking the M-closure of $\lambda_s X$ in $\mathcal{M}(X,\mathbb{R})'$, we verify that condition (2) of (2.6.7) is satisfied. We show that every $\ell\in(\lambda_s X)'$ extends even to a morphism $\mathcal{M}(X,\mathbb{R})'\to\mathbb{R}$. So let $\ell\in(\lambda_s X)'$. Then $g:=\ell\circ\iota_X\in\underline{\mathcal{M}}(X,\mathbb{R})$, hence $\mathrm{ev}_g\in\mathcal{M}(X,\mathbb{R})''$. Since $\mathrm{ev}_g(\iota_X(x))=\iota_X(x)(g)=g(x)=\ell(\iota_X(x))$ one has $\mathrm{ev}_g|_{\iota_X(X)}=\ell|_{\iota_X(X)}$ and hence $\mathrm{ev}_g|_{\lambda X}=\ell|_{\lambda X}=\ell$. □

5.1.2 Corollary. *For any $\mathcal{Lip}^k$-space (respectively ℓ^∞-space) the map $\iota_X\colon X\to\lambda X$ into the free convenient vector space is an initial $\underline{\mathcal{Lip}^k}$-morphism (respectively $\underline{\ell^\infty}$-morphism).*

Proof. Again we treat both cases simultaneously. The structure of an $\mathcal{M}$-space X is the initial one induced by the family of morphisms $f\colon X\to\mathbb{R}$. Since all these morphisms factor by the universal property over ι_X it follows that ι_X is initial. □

5.1.3 Proposition. *The bijection $\mathcal{Lip}^k(X,E)\to L(\lambda X,E)$ (respectively $\ell^\infty(X,E)\to L(\lambda X,E)$) obtained according to the adjunction of the preceding theorem* (5.1.1) *is a $\underline{\text{Con}}$-isomorphism.*

Proof. Again we treat both cases simultaneously, with $\mathcal{M}=\mathcal{Lip}^k$ or $\mathcal{M}=\ell^\infty$. The map $h\colon \mathcal{M}(X, E)\to L(\lambda X, E)$, whose action $f\mapsto \tilde{f}$ was described in the proof of (5.1.1), is obviously linear. By (3.6.5) it is a $\underline{\mathrm{Con}}$-morphism provided $\mathrm{ev}_z\circ h\colon \mathcal{M}(X, E)\to E$ is so for all $z\in\lambda X$, or equivalently $\ell\circ\mathrm{ev}_z\circ h\colon \mathcal{M}(X, E)\to\mathbb{R}$ is so for all $z\in\lambda X$ and $\ell\in E'$. According to the construction of $\tilde{f}=h(f)$ one has $(\ell\circ\mathrm{ev}_z\circ h)(f)=z(\ell\circ f)=(z\circ\ell_*)(f)$ and the result follows since $z\in\lambda X\subseteq\mathcal{M}(X,\mathbb{R})'$ and $\ell_*\colon \mathcal{M}(X, E)\to\mathcal{M}(X,\mathbb{R})$ is a $\underline{\mathrm{Con}}$-morphism.

The verification that the inverse map h^{-1} is a $\underline{\mathrm{Con}}$-morphism is much simpler. Again it is enough to show (cf. (3.6.6) and (4.4.7)) that $\mathrm{ev}_x\circ h^{-1}\colon L(\lambda X, E)\to\mathcal{M}(X, E)\to E$ is a morphism for all $x\in X$. This is obvious, because $h^{-1}(g)=g\circ\iota_X=\iota_X^*(g)$ and hence $\mathrm{ev}_x\circ h^{-1}=\mathrm{ev}_{\iota_X(x)}$, which is a morphism. □

5.1.4 Corollary. *For any $\mathcal{Lip}^k$-space (respectively ℓ^∞-space) X the function space $\mathcal{Lip}^k(X,\mathbb{R})$ (resp. $\ell^\infty(X,\mathbb{R})$) is up to an isomorphism the dual of the convenient vector space λX.*

We have constructed the free convenient vector space λX over a $\mathcal{Lip}^k$-space X as the M-closure of the linear subspace generated by the point evaluations in $\mathcal{Lip}^k(X,\mathbb{R})'$. This is not very constructive, in particular since adding M-limits of sequences (or even nets) of a subspace does not always give its M-closure. In the case that X is a finite-dimensional smooth manifold we show, however, that not only $\lambda X=C^\infty(X,\mathbb{R})'$, but even that every element of λX is the M-limit of a sequence of linear combinations of point evaluations, and that $C^\infty(X,\mathbb{R})'$ is the space of distributions of compact support.

5.1.5 Proposition. *Let E be a convenient vector space and X a finite-dimensional smooth separable manifold. Then for every $\ell\in C^\infty(X, E)'$ there exists a compact set $K\subseteq X$ such that $\ell(f)=0$ for all $f\in C^\infty(X, E)$ with $f|_K=0$.*

Proof. Since X is separable its compact bornology has a countable basis of compact sets $\{K_n; n\in\mathbb{N}\}$. Assume now that no compact set has the claimed property. Then for every $n\in\mathbb{N}$ there has to exist a function $f_n\in C^\infty(X, E)$ with $f_n|_{K_n}=0$ but $\ell(f_n)\neq 0$. By multiplying f_n with $n/\ell(f_n)$ we may assume that $\ell(f_n)=n$. Since every compact subset of X is contained in some K_n one has that $\{f_n; n\in\mathbb{N}\}$ is bounded in $C^\infty(X, E)$, but $\ell(\{f_n; n\in\mathbb{N}\})$ is not; this contradicts the assumption that ℓ is a morphism. □

5.1.6 Remark. The proposition above remains true if X is a finite-dimensional smooth paracompact manifold with non-measurably many components. In order to show this generalization one uses that $C^\infty(X, E)$ is the product $\Pi_{j\in J}C^\infty(X_j, E)$, where $\{X_j; j\in J\}$ is the partition in the non-measurably many components and the fact that an ℓ belongs to the dual of such a product if it is a finite sum of elements of the duals of the factors, cf. (3.9.5). Now the result follows from (5.1.5) since the components of a paracompact manifold are paracompact and hence separable.

For such manifolds X the dual $C^\infty(X, \mathbb{R})'$ is the space of distributions with compact support. In fact, in case X is connected, $C^\infty(X, \mathbb{R})'$ is the space of all linear functionals which are continuous for the classically considered topology on $C^\infty(X, \mathbb{R})$ by corollary (4.4.41); and in case of an arbitrary X this result follows using (3.9.5) and the isomorphism $C^\infty(X, \mathbb{R}) \cong \prod_j C^\infty(X_j, \mathbb{R})$, where the X_j denote the connected components of X.

5.1.7 Theorem. *Let E be a convenient vector space and X a finite-dimensional separable smooth manifold. Then the linear subspace generated by $\{\ell \circ \mathrm{ev}_x; x \in X, \ell \in E'\}$ is* M*-dense in $C^\infty(X, E)'$.*

Proof. The proof is in several steps.

(Step 1) There exist $g_n \in C^\infty(\mathbb{R}, \mathbb{R})$ with

$$\mathrm{supp}(g_n) \subseteq \left[-\frac{2}{n}, \frac{2}{n}\right]$$

such that for every $f \in C^\infty(\mathbb{R}, E)$ the set $\{n \cdot (f - \sum_{k \in \mathbb{Z}} f(r_{n,k}) g_{n,k}); n \in \mathbb{N}\}$ is bounded in $C^\infty(\mathbb{R}, E)$, where $r_{n,k} := k/2^n$ and $g_{n,k}(t) := g_n(t - r_{n,k})$.

We choose a smooth $h: \mathbb{R} \to [0, 1]$ with $\mathrm{supp}(h) \subseteq [-1, 1]$ and $\sum_{k \in \mathbb{Z}} h(t-k) = 1$ for all $t \in \mathbb{R}$ and we define $Q^n: C^\infty(\mathbb{R}, E) \to C^\infty(\mathbb{R}, E)$ by setting $Q^n(f)(t) := \sum_k f(k/n) h(tn - k)$. Let $K \subseteq \mathbb{R}$ be compact. Then

$$n(Q^n(f) - f)(t) = \sum_k (f(k/n) - f(t)) \cdot n \cdot h(tn-k) \in B_1\left(f, K + \frac{1}{n}\mathrm{supp}(h)\right)$$

for $t \in K$, where $B_n(f, K_1)$ denotes the absolutely convex hull of the bounded set $\delta^n f(K_1^{\langle n \rangle})$.

To get similar estimates for the derivatives we use convolution. Let $h_1: \mathbb{R} \to \mathbb{R}$ be a smooth function with support in $[-1, 1]$ and $\int_{\mathbb{R}} h_1(s) ds = 1$. Then for $t \in K$ one has $(f * h_1)(t) := \int_{\mathbb{R}} f(t-s) h_1(s) ds \in B_0(f, K + \mathrm{supp}(h_1)) \cdot \| h_1 \|_1$, where $\| h \|_1 := \int_{\mathbb{R}} |h(s)| ds$. For smooth functions $f: \mathbb{R} \to \mathbb{R}$ one has $(f * h)^{(k)} = f * h^{(k)}$; one immediately deduces that the same holds for smooth functions $f: \mathbb{R} \to E$ and one obtains $(f * h_1)(t) - f(t) = \int_{\mathbb{R}} (f(t-s) - f(t)) h_1(s) ds \in \mathrm{diam}(\mathrm{supp}(h_1)) \cdot \| h_1 \|_1 \cdot B_1(f, K + \mathrm{supp}(h_1))$ for $t \in K$, where $\mathrm{diam}(S) := \sup\{|s|; s \in S\}$. Using now $h_n(t) := n \cdot h_1(nt)$ we obtain for $t \in K$:

$$\begin{aligned}
&(Q^m(f) * h_n - f)^{(k)}(t) = (Q^m(f) * h_n^{(k)} - f * h_n^{(k)})(t) + (f^{(k)} * h_n - f^{(k)})(t) = \\
&(Q^m(f) - f) * h_n^{(k)}(t) + (f^{(k)} * h_n - f^{(k)})(t) \in B_0(Q^m(f) - f, K + \mathrm{supp}(h_n)) \cdot \| h_n^{(k)} \|_1 + \\
&B_1(f^{(k)}, K + \mathrm{supp}(h_n)) \cdot \mathrm{diam}(\mathrm{supp}(h_n)) \cdot \| h_n \|_1 \subseteq \\
&\tfrac{1}{m} n^k \cdot B_1(f, K + \mathrm{supp}(h_n) + \tfrac{1}{m}\mathrm{supp}(h)) \cdot \| h_1^{(k)} \|_1 + n \cdot B_1(f^{(k)}, K + \mathrm{supp}(h_n)) \cdot \| h_n \|_1.
\end{aligned}$$

Let now $m := 2^n$ and $P^n(f) := Q^m(f) * h_n$. Then

$$n \cdot (P^n(f) - f)^{(k)}(t) \in n^{k+1} 2^{-n} \cdot B_1\left(f, K + \left(\frac{1}{n} + \frac{1}{2^n}\right)[-1, 1]\right) \| h_1^{(k)} \|_1$$

$$+ B_1(f^{(k)}, K + \frac{1}{n}[-1, 1]) \| h_1 \|_1$$

for $t\in K$ and this is uniformly bounded for $n\in\mathbb{N}$. With $g_n(t):=\int_{\mathbb{R}}h(s2^n-k)h_n(t+k2^{-n}-s)ds=\int_{\mathbb{R}}h(s2^n)h_n(t-s)ds$ we obtain $P^n(f)(t)=(Q^{2^n}(f)*h_n)(t)=\sum_k f(k2^{-n})\int_{\mathbb{R}}h(s2^n-k)h_n(t-s)ds=\sum_k f(k2^{-n})g_n(t-k2^{-n})$. Thus $r_{n,k}:=k2^{-n}$ and the g_n have all the claimed properties.

(Step 2) For every $m\in\mathbb{N}$ and every $f\in C^\infty(\mathbb{R}^m,E)$ the set $\{n\cdot(f-\sum_{k_1\in\mathbb{Z},\dots,k_m\in\mathbb{Z}}f(r_{n;k_1,\dots,k_m})g_{n;k_1,\dots,k_m});\, n\in\mathbb{N}\}$ is bounded in $C^\infty(\mathbb{R}^m,E)$, where $r_{n;k_1,\dots,k_m}:=(r_{n,k_1},\dots,r_{n,k_m})$ and $g_{n;k_1,\dots,k_m}(x_1,\dots,x_m):=g_{n,k_1}(x_1)\cdot\ldots\cdot g_{n,k_m}(x_m)$.

We prove this statement by induction on m. For $m=1$ it was shown in step 1. Now assume that it holds for m and $C^\infty(\mathbb{R},E)$ instead of E. Then by induction hypothesis applied to $f^\vee\in C^\infty(\mathbb{R}^m,C^\infty(\mathbb{R},E))$ we conclude that

$$\{n\cdot(f-\textstyle\sum_{k_1\in\mathbb{Z},\dots,k_m\in\mathbb{Z}}f(r_{n;k_1,\dots,k_m},\_)g_{n;k_1,\dots,k_m});\, n\in\mathbb{N}\}$$

is bounded in $C^\infty(\mathbb{R}^{m+1},E)$. Thus it remains to show that

$$\{n\cdot\textstyle\sum_{k_1,\dots,k_m}g_{n;k_1,\dots,k_m}(f(r_{n;k_1,\dots,k_m},\_)$$
$$-\textstyle\sum_{k_{m+1}}f(r_{n;k_1,\dots,k_m},r_{k_{m+1}})g_{n,k_{m+1}});\, n\in\mathbb{N}\}$$

is bounded in $C^\infty(\mathbb{R}^{m+1},E)$. Since the support of the $g_{n;k_1,\dots,k_m}$ is locally finite only finitely many summands of the outer sum are non-zero on a given compact set. Thus it is enough to consider each summand separately. By step 1 we know that the linear operators $h\mapsto n(h-\Sigma_k h(r_{n,k})g_{n,k})$ $(n\in\mathbb{N})$ are pointwise bounded. So they are bounded on bounded sets, by the linear uniform boundedness principle (3.6.4). Hence

$$\{n\cdot(f(r_{n;k_1,\dots,k_m},\_)-\textstyle\sum_{k_{m+1}}f(r_{n;k_1,\dots,k_m},r_{k_{m+1}})g_{n,k_{m+1}});\, n\in\mathbb{N}\}$$

is bounded in $C^\infty(\mathbb{R}^{m+1},E)$. Using that the multiplication $\mathbb{R}\sqcap E\to E$ is a morphism one concludes immediately that also the multiplication with a map $g\in C^\infty(X,\mathbb{R})$ is a morphism $C^\infty(X,E)\to C^\infty(X,E)$ for any smooth space X. Thus the proof of step (2) is complete.

(Step 3) For every $\ell\in C^\infty(X,E)'$ there exist $x_{n,k}\in X$ and $\ell_{n,k}\in E'$ such that $\{n(\ell-\sum_k\ell_{n,k}\circ\mathrm{ev}_{x_{n,k}});\, n\in\mathbb{N}\}$ is bounded in $C^\infty(X,E)'$, where in the sum only finitely many terms are non-zero. In particular the subspace generated by $\ell_E\circ\mathrm{ev}_x$ $(\ell_E\in E',\, x\in X)$ is M-dense.

By (5.1.5) there exists a compact set K with $f|_K=0$ implying $\ell(f)=0$. One can cover K by finitely many relatively compact $U_j\cong\mathbb{R}^m$ $(j=1\dots N)$. Let $\{h_j; j=0\dots N\}$ be a partition of unity subordinated to $\{X\setminus K, U_1,\dots,U_N\}$. Then $\ell(f)=\sum_{j=1}^N\ell(h_j\cdot f)$ for every f. By step 2 the set $\{n(h_jf-\sum h_jf(r_{n,k_1,\dots,k_m})g_{n,k_1,\dots,k_m});\, n\in\mathbb{N}\}$ is bounded in $C^\infty(U_j,E)$. Since supp(h_j) is compact in U_j this is even bounded in $C^\infty(X,E)$ and for fixed n only finitely many $r_{n,k_1,\dots,k_m}$ belong to supp(h_j). Thus the above sum is actually finite and the supports of all functions in the bounded subset of $C^\infty(U_j,E)$ are included in a common compact subset. Applying ℓ to this subset yields that $\{n(\ell(h_jf)-\sum\ell_{n,k_1,\dots,k_m}\circ\mathrm{ev}(r_{n,k_1,\dots,k_m}));\, n\in\mathbb{N}\}$ is bounded in $\mathbb{R}$, where $\ell_{n,k_1,\dots,k_m}(x):=\ell(h_j(r_{n,k_1,\dots,k_m})g_{n;k_1,\dots,k_m}\cdot x)$. To complete the proof one only has to take as $x_{n,k}$ all the $r_{n,k_1,\ldots,k_m}$ for the finitely many charts $U_j \cong \mathbb{R}^m$ and as $\ell_{n,k}$ the corresponding functionals $\ell_{n,k_1,\ldots,k_m} \in E'$. □

5.1.8 Corollary. *Let X be a finite-dimensional separable smooth manifold. Then the free convenient vector space λX over X is equal to $C^\infty(X, \mathbb{R})'$.*

Now we will give an explicit description of the free convenient vector space λX over an arbitrary ℓ^∞-space X. We shall show that every element of λX is M-limit of a sequence of linear combinations of point evaluations and we identify λX with a space $\ell^1 X$ of functions $X \to \mathbb{R}$. In general λX is strictly included in $\ell^\infty(X, \mathbb{R})'$.

We use infinite sums of reals and start by recalling their definition and basic properties.

5.1.9 Definition. For any set J we denote will $\mathscr{P}_f(J)$ the set of all finite subsets of J, directed by inclusion. And for a family $t_j (j \in J)$ of reals, $\sum_{j\in J} t_j$ denotes the limit of the net $\mathscr{P}_f(J) \to \mathbb{R}$ defined by $J_0 \mapsto \sum_{j \in J_0} t_j$ for $J_0 \in \mathscr{P}_f(J)$ (by definition $\sum_{j\in\varnothing} t_j = 0$).

Obviously any sum equals the sum of its non-zero terms.

5.1.10 Lemma.

(i) *If $\sum_{j\in J} t_j$ exists, then $\{j \in J; t_j \neq 0\}$ is countable.*
(ii) *If $\sum_{j\in J} |t_j|$ exists, then $\sum_{j\in J} t_j$ exists and $|\sum t_j| \leq \sum |t_j|$.*

Proof. (i) follows since for any $n \in \mathbb{N}$ the set $\{j \in J; |t_j| \geq 1/n\}$ has to be finite.
(ii) follows using the Cauchy condition and the triangle inequality. □

5.1.11 Definition. For an ℓ^∞-space X we define:

(i) $\ell^1 X$ denotes the space $\{f: X \to \mathbb{R};\ \operatorname{supp} f \subseteq X$ bounded and $\|f\|_1 < \infty\}$ with the bornology given by: $B \subseteq \ell^1 X$ is bounded iff $\bigcup_{f\in B} \operatorname{supp} f \subseteq X$ is bounded and $\{\|f\|_1; f \in B\} \subseteq \mathbb{R}$ is bounded; where $\|f\|_1 := \sum_{x\in X} |f(x)|$.

One easily verifies that $\ell^1 X$ is a convex bornological space.

(ii) $\ell^1_c X$ denotes the subspace $\{f: X \to \mathbb{R};\ \operatorname{supp} f$ is finite$\}$ of $\ell^1 X$ with the induced bornology.

Obviously $\ell^1_c X$ is a linear subspace and hence also a convex bornological space.

5.1.12 Proposition. *For any ℓ^∞-space X the map $\langle \_, \_ \rangle: \ell^1 X \sqcap \ell^\infty(X, \mathbb{R}) \to \mathbb{R}$ defined by $(f, g) \mapsto \langle f, g \rangle := \sum_{x\in X} f(x) \cdot g(x)$ is a bornological bilinear function.*

Proof. Let A be bounded in X and $f: X \to \mathbb{R}$ be such that $\operatorname{supp} f \subseteq A$. Let $g: X \to \mathbb{R}$ be bounded on A and $\|g|_A\|_\infty := \operatorname{supp}\{|g(x)|;\ x \in A\}$. Then $\sum |f(x)g(x)| \leq \|f\|_1 \cdot \|g|_A\|_\infty$. From this inequality it follows that $\langle \_, \_ \rangle$ is well-defined, cf. (ii) of (5.1.10). It is obviously bilinear and it is bornological by the same inequality. □

We will show that $\ell^1 X$ is a convenient vector space and that the bilinear function $\langle\_,\_\rangle$ of (5.1.12) induces an isomorphism $\ell^\infty(X, \mathbb{R}) \cong (\ell^1 X)'$. Furthermore, $\ell^1 X$ will turn out to be itself the dual of some other convenient vector space.

5.1.13 Definition. For any ℓ^∞-space X we define:

(i) $c_0 X$ denotes the $\underline{\text{Pre}}$-subspace $\{g\colon X \to \mathbb{R};\ \operatorname{supp} g$ is countable and for every $\varepsilon > 0$ and $A \subseteq X$ bounded the set $\{x \in A; |g(x)| > \varepsilon\}$ is finite$\}$ of $\ell^\infty(X, \mathbb{R})$.

(ii) $\ell_c^\infty X$ denotes the $\underline{\text{Pre}}$-subspace $\{g\colon X \to \mathbb{R};\ \operatorname{supp} g$ is finite$\}$ of $\ell^\infty(X, \mathbb{R})$.

5.1.14 Proposition. *For every ℓ^∞-space X the space $c_0 X$ is a convenient vector space.*

Proof. Using the closed embedding lemma (2.6.4) it is enough to show that $c_0 X$ is M-closed in $\ell^\infty(X, \mathbb{R})$. So let g be the M-limit in $\ell^\infty(X, \mathbb{R})$ of a sequence $(g_n) \subseteq c_0 X$. Since $\operatorname{supp} g \subseteq \bigcup_{n \in \mathbb{N}} \operatorname{supp} g_n$ and each g_n has countable support the same is true for g. Assume that for some bounded $A \subseteq X$ and some $\varepsilon > 0$ there exist infinitely many $x \in A$ with $|g(x)| \geq \varepsilon$. Since the functions g_n converge to g uniformly on A there exists an n such that $|g(x) - g_n(x)| < \varepsilon/2$ for all $x \in A$. Hence $|g_n(x)| > \varepsilon/2$ for infinitely many $x \in A$, in contradiction to $g_n \in c_0 X$. □

5.1.15 Corollary. *For every ℓ^∞-space X the $\underline{\text{Con}}$-structure of $c_0 X$ is the initial one induced by the family $\mathrm{ev}_x\colon c_0 X \to \mathbb{R}$ $(x \in X)$. Furthermore, the bornology of $c_0 X$ has a basis of $\sigma(c_0 X, \{\mathrm{ev}_x; x \in X\})$-closed sets.*

Proof. Both results follow from the corresponding ones of $\ell^\infty(X, \mathbb{R})$, i.e. (3.6.6) and the example (iv) of (4.1.21), and the initiality of the inclusion of $c_0 X$ in $\ell^\infty(X, \mathbb{R})$. □

5.1.16 Proposition. *Let $\langle\_,\_\rangle\colon \ell^1 X \sqcap \ell_c^\infty X \to \mathbb{R}$ be the restriction of the bilinear map in* (5.1.12). *The induced map $f \mapsto \langle f, \_\rangle$ is an isomorphism $\ell^1 X \cong (\ell_c^\infty X)'$.*

Proof. Proposition (5.1.12) implies that $\varphi\colon \ell^1 X \to (\ell_c^\infty X)'$ defined by $f \mapsto \langle f, \_\rangle$ is a linear bornological map.

Now we construct its inverse ψ. Let $\chi_x \in \ell_c^\infty X$ denote the characteristic function of $\{x\}$. Let $\ell \in (\ell_c^\infty X)'$; we define $\psi(\ell)\colon X \to \mathbb{R}$ by $\psi\ell(x) := \ell(\chi_x)$. Suppose B is a bounded subset of $(\ell_c^\infty X)'$. We claim that then

(i) $\bigcup_{\ell \in B} \operatorname{supp} \psi\ell \subseteq X$ is bounded;
(ii) $\{\sum_{x \in X} |\psi\ell(x)|; \ell \in B\} \subseteq \mathbb{R}$ is bounded.

Both statements are proved indirectly.

Assume first that (i) fails. Then by (1.2.5) there exist points $x_k \in \bigcup \operatorname{supp} \psi\ell$ $(k \in \mathbb{N})$, all different, and such that only the finite subsets of $\{x_k; k \in \mathbb{N}\}$ are bounded. One chooses $\ell_k \in B$ such that $t_k := \psi\ell_k(x_k) \neq 0$ and puts

$f_k := (k/t_k)\chi_{x_k}$. Then $\{f_k; k \in \mathbb{N}\} \subseteq \ell_c^\infty X$ is bounded since on any bounded subset of X this set of functions takes only finitely many values. Therefore $B\{f_k; k \in \mathbb{N}\}$ is bounded which is a contradiction because it contains for any $k \in \mathbb{N}$ the value

$$\ell_k(f_k) = \frac{k}{t_k}\ell_k(\chi_{x_k}) = \frac{k}{t_k}(\psi \ell_k)(x_k) = k.$$

Suppose now that (ii) fails. Then one can choose for $k \in \mathbb{N}$ elements $\ell_k \in B$ with $\sum_{x \in X} |\psi \ell_k(x)| > k$; and $A_k \subseteq X$ finite with $\sum_{x \in A_k} |\psi \ell_k(x)| > k$. Define $f_k(x) := \text{sign}(\psi \ell_k(x))$ for $x \in A_k$ and $f_k(x) := 0$ for $x \notin A_k$. Then $\{f_k; k \in \mathbb{N}\} \subseteq \ell_c^\infty X$ is bounded. Therefore $B\{f_k; k \in \mathbb{N}\}$ is bounded which is a contradiction because it contains for any $k \in \mathbb{N}$ the value

$$\ell_k(f_k) = \sum_{x \in A_k} \ell_k(f_k(x)\chi_x) = \sum_{x \in A_k} f_k(x)\ell_k(\chi_x) =$$
$$\sum_{x \in A_k} \text{sign}(\psi \ell_k(x))\psi \ell_k(x) = \sum_{x \in X} |\psi \ell_k(x)| > k.$$

(i) and (ii) together have two consequences: for any $\ell \in (\ell_c^\infty X)'$ one has $\psi \ell \in \ell^1 X$, i.e. ψ is a map $(\ell_c^\infty X)' \to \ell^1 X$; and this map is bornological.

It remains to verify that φ and ψ are inverse to each other. That $\psi \circ \varphi = \text{id}$ is easily verified: $\psi(\varphi(f))(x) = (\varphi f)(\chi_x) = \langle f, \chi_x \rangle = f(x)$. Now the converse composition: let $\ell \in (\ell_c^\infty X)'$ and $g \in \ell_c^\infty X$. Since g is a finite sum $\sum g(x)\chi_x$ one obtains $\varphi(\psi(\ell))(g) = \sum g(x)\varphi(\psi(\ell))(\chi_x) = \sum g(x)\psi\ell(x) = \sum g(x)\ell(\chi_x) = \ell(\sum g(x)\chi_x) = \ell(g)$. □

5.1.17 Corollary. *For any ℓ^∞-space X the space $\ell^1 X$ is convenient.*

5.1.18 Proposition. *For any ℓ^∞-space X the space $c_0 X$ is the completion $\bar{\omega}(\ell_c^\infty X)$ of $\ell_c^\infty X$.*

Proof. We first show that $\ell_c^\infty X$ is M-dense in $c_0 X$.

Start with a $g \in c_0 X$ for which there exists an $\varepsilon > 0$ such that $|g(x)| > \varepsilon$ for all $x \in \text{supp } g$. Let $a_1, a_2, \ldots$ be an enumeration of the countable support of g. Define $g_n \in \ell_c^\infty X$ by $g_n(x) := g(x)$ for $x \in \{a_1, \ldots, a_n\}$ and $g_n(x) := 0$ otherwise. Then $B := \{n(g - g_n); n \in \mathbb{N}\}$ is bounded in $c_0 X$. In fact, for any bounded $A \subseteq X$, the set $\{x \in A; |g(x)| > \varepsilon\}$ is finite and hence $B(A)$ has only finitely many values. Hence the g_n are M-convergent to g.

Let now $g \in c_0 X$ be arbitrary. Let $A_n := \{a \in X; |g(a)| \geq 1/n\}$. We define $g_n(x) := g(x)$ for $x \in A_n$ and $g_n(x) := 0$ otherwise. Then the functions g_n are of the special type discussed above, and thus are M-limits of elements in $\ell_c^\infty X$. Furthermore, by construction $B := \{n(g - g_n); n \in \mathbb{N}\}$ is bounded in $c_0 X$ because even $B(X)$ is bounded (by 1).

Using (2.6.7) it is now enough to show that every $\ell \in (\ell_c^\infty X)'$ extends to a morphism $c_0 X \to \mathbb{R}$. By (5.1.16) ℓ is of the form $\langle f, \_ \rangle$ for some $f \in \ell^1 X$, so by (5.1.12) ℓ even extends to $\ell^\infty(X, \mathbb{R})$. □

Remark. According to the proof every $g \in c_0 X$ can be written as $g = \text{M-lim}_{n\to\infty}(\text{M-lim}_{k\to\infty} g_{n,k})$, where $g_{n,k} \in \ell_c^\infty X$. However, this does not imply that g is an M-limit of a sequence in $\ell_c^\infty X$ as the example (6.3.1) will show.

5.1.19 Corollary. *For every ℓ^∞-space X the map $\langle \_,\_ \rangle : \ell^1 X \sqcap c_0 X \to \mathbb{R}$ induces an isomorphism $\ell^1 X \cong (c_0 X)'$.*

5.1.20 Proposition. *Let X be an ℓ^∞-space. Then the Con-structure of $\ell^1 X$ is the initial one induced by the family* $\text{ev}_x : \ell^1 X \to \mathbb{R}$ $(x \in X)$. *Furthermore the bornology of $\ell^1 X$ has a basis of $\sigma(\ell^1 X, \{\text{ev}_x; x \in X\})$-closed sets.*

Proof. For the initiality consider a linear map $m: E \to \ell^1 X = (c_0 X)'$ that is defined on a convenient vector space E, and such that $\text{ev}_x \circ m \in E'$ for all $x \in X$. Let $\tilde{m} : \ell_c^\infty X \to E'$ be the linear map characterized by $\tilde{m}(\chi_x) = \text{ev}_x \circ m$. For every $v \in E$ the composite $\text{ev}_v \circ \tilde{m}$ extends to the morphism $m(v) \in (c_0 X)'$. Thus we obtain a map $\bar{m}: c_0 X \to \Pi_E \mathbb{R}$ defined by $\text{pr}_v \circ \bar{m} = m(v)$ that is an extension of the map $\iota_{E'} \circ \tilde{m}$, where $\iota_{E'}$ denotes the initial Pre-morphism $E' \to \Pi_E \mathbb{R}$. Using the M-denseness of $\ell_c^\infty X$ in $c_0 X$ and the initiality of $\iota_{E'}$ we obtain a morphism again denoted $\tilde{m}: c_0 X \to E'$. The original map m corresponds to $\tilde{m}$ via the bijection $L(E, (c_0 X)') \cong L(c_0 X, E')$, thus is a morphism.

That the bornology of $\ell^1 X$ has a basis of $\sigma(\ell^1 X, \{\text{ev}_x; x \in X\})$-closed sets follows from (4.1.22), since the set $\{\chi_x; x \in X\} \subseteq c_0 X$ separates the points of $\ell^1 X = (c_0 X)'$. □

5.1.21 Proposition. *For any ℓ^∞-space X the space $\ell^1 X$ is the completion $\bar{\omega}(\ell_c^1 X)$ of $\ell_c^1 X$.*

Proof. We first show that the M-adherence of $\ell_c^1 X$ in $\ell^1 X$ is $\ell^1 X$. So let $f \in \ell^1 X \setminus \ell_c^1 X$, then the support of f is infinite countable. Let $\{a_n; n \in \mathbb{N}\}$ be an enumeration of supp f. We define $f_n \in \ell_c^1 X$ by $f_n(x) := f(x)$ for $x \in \{a_1, \ldots, a_n\}$ and $f_n(x) = 0$ otherwise. Let $t_n := (\sum_{k=n+1}^\infty |f(a_k)|)^{-1}$. Since $\|f\|_1 < \infty$ one has $\lim_{n\to\infty} t_n = \infty$. An easy calculation shows that $\{t_n(f - f_n); n \in \mathbb{N}\}$ is bounded in $\ell^1 X$. Thus the f_n are M-convergent to f.

Let us show next that every $\ell \in (\ell_c^1 X)'$ extends to an element of $(\ell^1 X)'$. Define a map $f(x) := \ell(\chi_x)$. Then $f \in \ell^\infty(X, \mathbb{R})$, since for every bounded $B \subseteq X$ the set $\{\chi_x; x \in B\}$ is bounded in $\ell_c^1 X$. Using again the map $\langle \_,\_ \rangle : \ell^1 X \sqcap \ell^\infty(X, \mathbb{R}) \to \mathbb{R}$ of (5.1.12) we obtain an $\bar{\ell} := \langle \_, f \rangle \in (\ell^1 X)'$ which extends ℓ. Thus it follows from (2.6.7) that $\ell^1 X$ is the completion of $\ell_c^1 X$. □

Remark. We have shown that every $f \in \ell^1 X$ is the M-limit of a sequence in $\ell_c^1 X$ which is in contrast to the analogous remark after (5.1.18).

5.1.22 Theorem. *Let X be an ℓ^∞-space and $\chi: X \to \ell^1 X$ be the map associating to $x \in X$ the characteristic function χ_x of $\{x\}$. Then χ is an ℓ^∞-map having the* *universal property that every ℓ^∞-map $g\colon X\to E$ into a convenient vector space E factors in a unique way as $g=\bar{g}\circ\chi$ with a $\underline{\text{Con}}$-morphism $\bar{g}\colon \ell^1X\to E$.*

Proof. We will show that the map $\chi\colon X\to\ell^1X$ is up to an isomorphism the map $\iota_X\colon X\to\lambda X$, which has the claimed universal property by (5.1.1). Let $\varphi\colon\ell^1X\to\ell^\infty(X,\mathbb{R})'$ be the morphism defined by $g\mapsto\langle g,\_\rangle$ and let the morphism $\psi\colon\ell^\infty(X,\mathbb{R})'\to\ell^1X$ be the composite $\ell^\infty(X,\mathbb{R})'\xrightarrow{\text{incl}^*}(c_0X)'\cong\ell^1X$, where incl: $c_0X\to\ell^\infty(X,\mathbb{R})$ denotes the inclusion and the isomorphism was described in (5.1.19). One easily checks the identities $\varphi\circ\chi=\iota_X$, $\psi\circ\iota_X=\chi$, $\psi\circ\varphi=\text{id}$ and, using the universal property of ι_X, one obtains that $\varphi\circ\psi|_{\lambda X}$ is the inclusion of λX in $\ell^\infty(X,\mathbb{R})'$. From this it follows that $\psi|_{\lambda X}$ is the desired isomorphism.

A more direct proof is along the following lines: let us define $\bar{g}$ by $\bar{g}(f):=\sum_{x\in X}f(x)g(x)$. One easily shows that the series is M-convergent in E and thus defines an element of E. It is also not difficult to show that $\bar{g}$ is linear and bornological, and satisfies $\bar{g}\circ\chi=g$.

The uniqueness is a direct consequence of proposition (5.1.21) □

5.1.23 Corollary. *Let $g\colon X\to Y$ be an ℓ^∞-map. Then there exists a unique linear morphism $\ell^1(g)\colon\ell^1X\to\ell^1Y$ with the property that for any $x\in X$ one has $\ell^1(g)(\chi_x)=\chi_{g(x)}$. The functor $\ell^1\colon\underline{\ell^\infty}\to\underline{\text{Con}}$ so obtained is isomorphic to the functor λ of (5.1.1). An explicit formula for $\ell^1(g)$ is $\ell^1(g)(f)(y)=\sum_{x\in g^{-1}(y)}f(x)$.*

5.1.24 Corollary. *For any ℓ^∞-space X and convenient vector space E one has a natural $\underline{\text{Con}}$-isomorphism $\ell^\infty(X,E)\cong L(\ell^1X,E)$.*

Proof. One either uses (5.1.3) and the isomorphism in (5.1.23); or one verifies directly that the bijection described in (5.1.22) is a linear bornological isomorphism. □

5.1.25 Corollary. *The bilinear function $\ell^1X\sqcap\ell^\infty(X,\mathbb{R})\to\mathbb{R}$ of (5.1.12) induces a $\underline{\text{Con}}$-isomorphism $\ell^\infty(X,\mathbb{R})\cong(\ell^1X)'$.*

5.1.26 Proposition. *Let X be a ℓ^∞-space. Together with the point-wise multiplication the space ℓ^1X is a non-unitary convenient algebra, i.e. a convenient vector space together with a commutative and associative multiplication which is a bilinear morphism, cf. (3.7.1). Every morphism $h\colon\ell^1X\to\mathbb{R}$, $h\neq0$, of non-unitary convenient algebras is of the form $h=\text{ev}_x$ for a unique $x\in X$.*

Remark. ℓ^1X has a unit iff X is finite.

Proof. That $\ell^1 X$ is a convenient algebra is trivial to verify. Let now $h: \ell^1 X \to \mathbb{R}$ be a morphism of non-unitary convenient algebras. Then by (5.1.25) there exists a $g \in \ell^\infty(X, \mathbb{R})$ with $h(f) = \langle f, g \rangle$ for all $f \in \ell^1 X$. Assume that the support of g is not single pointed. So let $x, y \in \mathrm{supp}(g)$ be different. Then $0 = h(0) = h(\chi_x \chi_y) = h(\chi_x) h(\chi_y) = g(x) g(y) \neq 0$, contradiction. Hence $g = t \cdot \chi_x$ and thus $h = t \cdot \mathrm{ev}_x$ for some $x \in X$ and $t \in \mathbb{R}$. Since $t = h(\chi_x) = h((\chi_x)^2) = (h(\chi_x))^2 = t^2$ we conclude that $t \in \{0, 1\}$. Thus $h \neq 0$ implies $h = \mathrm{ev}_x$ for some x. □

5.1.27 Remarks

(i) Since the functor $\underline{\ell^\infty} \to \underline{\mathrm{Born}}$ has a left adjoint (cf. (1.2.4)) one also obtains that the forgetful functor $\underline{\mathrm{Con}} \to \underline{\ell^\infty} \to \underline{\mathrm{Born}}$ has a left adjoint.

(ii) The inclusion functor $\underline{\mathscr{Lip}^\infty} \to \underline{\mathrm{Diff}}$, where $\underline{\mathrm{Diff}}$ denotes the category of diffeological spaces of [Nel, 1986] has also a left adjoint (one takes the $\mathscr{Lip}^\infty$-structure generated by the curves of the given diffeological structure). One similarly obtains that the forgetful functor $\underline{\mathrm{Con}} \to \underline{\mathrm{Diff}}$ has a left adjoint.

(iii) If X is an ℓ^∞-space whose structure is the coarse one (i.e. X itself is bounded), then the convenient vector spaces $c_0 X, \ell^1 X, \ell^\infty(X, \mathbb{R})$ are all Banach spaces and coincide with the spaces usually associated to the set X, cf. [Jarchow, 1981, p. 120] and [Jarchow, 1981, p. 26].

(iv) Every space of the form $\ell^1 X$ is up to an isomorphism the dual of a convenient vector space, namely $c_0 X$. For any convenient vector space E, the canonical embedding $\iota_E: E \to E''$ yields a retraction $\iota_E^*: E''' \to E'$ to $\iota_{E'}: E' \to E'''$. For $X = \mathbb{N}$ with the coarse bornology, $c_0(\mathbb{N}) = c_0$ is an example of a convenient vector space which is not isomorphic to a space of the form $\ell^1 X$, since $c_0'' = \ell^\infty$ and the inclusion $c_0 \to \ell^\infty$ admits no retraction, cf. [Jarchow, 1981, p. 207].

(v) For $k \geq 1$ let $\varphi: \underline{\mathrm{Con}} \to \underline{\mathscr{Lip}^k}$ be the forgetul functor and λ its left adjoint according to (5.1.1). For any convenient vector space E the identity map $E \to E$ has a unique factorization $1_E = \ell \circ \iota_E$, where ι_E is the (non-linear) $\mathscr{Lip}^k$-map $\iota_E: E \to \lambda \varphi E$ and ℓ is a $\underline{\mathrm{Con}}$-morphism $\ell: \lambda \varphi E \to E$. By differentiation we get, with $\ell_1 := (\iota_E)'(0)$, the equality $1_E = \ell \circ \ell_1$. We thus conclude that any convenient vector space E is a direct factor of one which is free over $\underline{\mathscr{Lip}^k}$.

5.2 Convenient co-algebras

5.2.1 Definition. (i) A *convenient algebra* E is a convenient vector space (also denoted E) together with a compatible algebra structure, or equivalently with two $\underline{\mathrm{Con}}$-morphisms

$\mu: E \tilde{\otimes} E \to E$ (called *multiplication*)
$\varepsilon: \mathbb{R} \to E$ (called *unit*)

such that with the isomorphisms mentioned in (3.8.4) one obtains the following commutative diagrams:

$$\begin{array}{ccc} E\tilde{\otimes}E & \xleftarrow{\mu\tilde{\otimes}\mathrm{id}} & (E\tilde{\otimes}E)\tilde{\otimes}E \cong E\tilde{\otimes}(E\tilde{\otimes}E) \\ \downarrow{\scriptstyle\mu} & & \downarrow{\scriptstyle\mathrm{id}\tilde{\otimes}\mu} \\ E & \xleftarrow{\mu} & E\tilde{\otimes}E \end{array}$$

$$\begin{array}{ccc} E\tilde{\otimes}E & \cong & E\tilde{\otimes}E \\ {\scriptstyle\mu}\downarrow & & {\scriptstyle\mu}\downarrow \\ E & = & E \end{array} \qquad \begin{array}{ccc} E & \xleftarrow{\cong} & \mathbb{R}\tilde{\otimes}E \\ {\scriptstyle\mu}\uparrow & & \downarrow{\scriptstyle\varepsilon\tilde{\otimes}\mathrm{id}} \\ E\tilde{\otimes}E & = & E\tilde{\otimes}E \end{array}$$

In words, the multiplication has to be associative and commutative, and ε has to be a unit with respect to μ.

(ii) The category <u>*ConAlg*</u> has as objects the convenient algebras and as morphisms $g: E \to F$ between convenient algebras E and F the <u>Con</u>-morphisms for which the following diagrams commute:

$$\begin{array}{ccc} E\tilde{\otimes}E & \xrightarrow{g\tilde{\otimes}g} & F\tilde{\otimes}F \\ {\scriptstyle\mu_E}\downarrow & & {\scriptstyle\mu_F}\downarrow \\ E & \xrightarrow{g} & F \end{array} \qquad \begin{array}{ccc} \mathbb{R} & \xrightarrow{\mathrm{id}} & \mathbb{R} \\ \downarrow{\scriptstyle\varepsilon_E} & & \downarrow{\scriptstyle\varepsilon_F} \\ E & \xrightarrow{g} & F \end{array}$$

(iii) A *convenient co-algebra* E is a convenient vector space (also denoted E) together with a compatible co-algebra structure, i.e. two <u>Con</u>-morphisms

$\mu: E \to E\tilde{\otimes}E$ (called *co-multiplication*)
$\varepsilon: E \to \mathbb{R}$ (called *co-unit*)

such that with the isomorphisms mentioned in (3.8.4) one obtains the following commutative diagrams:

$$\begin{array}{ccc} E\tilde{\otimes}E & \xrightarrow{\mu\tilde{\otimes}\mathrm{id}} & (E\tilde{\otimes}E)\tilde{\otimes}E \cong E\tilde{\otimes}(E\tilde{\otimes}E) \\ \uparrow{\scriptstyle\mu} & & \uparrow{\scriptstyle\mathrm{id}\tilde{\otimes}\mu} \\ E & \xrightarrow{\mu} & E\tilde{\otimes}E \end{array}$$

$$\begin{array}{ccc} E\tilde{\otimes}E & \cong & E\tilde{\otimes}E \\ {\scriptstyle\mu}\uparrow & & {\scriptstyle\mu}\uparrow \\ E & = & E \end{array} \qquad \begin{array}{ccc} E & \xrightarrow{\cong} & \mathbb{R}\tilde{\otimes}E \\ {\scriptstyle\mu}\downarrow & & \uparrow{\scriptstyle\varepsilon\tilde{\otimes}\mathrm{id}} \\ E\tilde{\otimes}E & = & E\tilde{\otimes}E \end{array}$$

In words, the co-multiplication has to be co-associative and co-commutative, and ε has to be a co-unit with respect to μ.

(iv) The category *ConCoAlg* has as objects the convenient co-algebras and as morphisms $g: E \to F$ between convenient co-algebras E and F the Con-morphisms for which the following diagrams commute:

$$
\begin{array}{ccc}
E \tilde{\otimes} E & \xrightarrow{g \tilde{\otimes} g} & F \tilde{\otimes} F \\
\uparrow{\scriptstyle \mu_E} & & \uparrow{\scriptstyle \mu_F} \\
E & \xrightarrow{g} & F
\end{array}
\qquad
\begin{array}{ccc}
\mathbb{R} & \xrightarrow{\mathrm{id}} & \mathbb{R} \\
\uparrow{\scriptstyle \varepsilon_E} & & \uparrow{\scriptstyle \varepsilon_F} \\
E & \xrightarrow{g} & F
\end{array}
$$

In the following $\mathcal{M}$ shall always denote either C^∞ or ℓ^∞. We want to show that the functor $\lambda: \mathcal{M} \to \mathrm{Con}$ of (5.1.1) lifts to a functor also denoted by $\lambda: \mathcal{M} \to \mathrm{ConCoAlg}$.

5.2.2 Lemma. *Let X and Y be $\mathcal{M}$-spaces, F a convenient vector space and $f: \lambda X \sqcap \lambda Y \to F$ a bilinear morphism, cf.* (3.7.1). *If $f \circ (\iota_X \sqcap \iota_Y) = 0$ then $f = 0$.*

Proof. For any fixed $y \in Y$ the map $f(\_, \iota_Y(y)): \lambda X \to F$ is a linear morphism and $f(\_, \iota_Y(y)) \circ \iota_X = 0$. Hence $f(\_, \iota_Y(y)) = 0$ by the universal property of ι_X: $X \to \lambda X$. Let now $\bar{x} \in \lambda X$ be arbitrary. Then $f(\bar{x}, \_): \lambda Y \to F$ is a linear morphism and $f(\bar{x}, \_) \circ \iota_Y = 0$ by the first part of the proof. Hence $f(\bar{x}, \_) = 0$ by the universal property of ι_Y: $Y \to \lambda Y$. Since $\bar{x} \in \lambda X$ was arbitrary we conclude that $f = 0$. □

5.2.3 Lemma. *Let X and Y be $\mathcal{M}$-spaces. There exists a unique bilinear morphism $m_{X,Y}: \lambda X \sqcap \lambda Y \to \lambda(X \sqcap Y)$ that satisfies $m_{X,Y} \circ (\iota_X \sqcap \iota_Y) = \iota_{X \sqcap Y}$.*

Proof. By the previous lemma (5.2.2) uniqueness is clear. To show the existence consider $\iota_{X \sqcap Y}: X \sqcap Y \to \lambda(X \sqcap Y)$. By the universal property of $X \to \lambda X$ the associated $\mathcal{M}$-map $X \to \mathcal{M}(Y, \lambda(X \sqcap Y))$ extends to a linear morphism $\lambda X \to \mathcal{M}(Y, \lambda(X \sqcap Y))$. By the universal property of $Y \to \lambda Y$ we have an isomorphism $\mathcal{M}(Y, \lambda(X \sqcap Y)) \cong L(\lambda Y, \lambda(X \sqcap Y))$. Thus we obtain a linear morphism $\lambda X \to L(\lambda Y, \lambda(X \sqcap Y))$ which corresponds by (3.7.3) to a bilinear bornological map $m_{X,Y}: \lambda X \sqcap \lambda Y \to \lambda(X \sqcap Y)$. □

5.2.4 Proposition. *Let X and Y be $\mathcal{M}$-spaces. Then $m_{X,Y}: \lambda X \sqcap \lambda Y \to \lambda(X \sqcap Y)$ as defined in* (5.2.3) *has the universal property of the tensor product of λX with λY, i.e. $\lambda(X \sqcap Y) \cong \lambda X \tilde{\otimes} \lambda Y$.*

Proof. Let $f: \lambda X \sqcap \lambda Y \to F$ be a bilinear bornological map into a convenient vector space. Then $f \circ (\iota_X \sqcap \iota_Y): X \sqcap Y \to F$ is an $\mathcal{M}$-map and by the universal property of $X \sqcap Y \to \lambda(X \sqcap Y)$ there exists a unique linear morphism $f^\sim: \lambda(X \sqcap Y) \to F$ satisfying $f^\sim \circ \iota_{X \sqcap Y} = f \circ (\iota_X \sqcap \iota_Y)$. This map $f^\sim$ satisfies also

$\tilde{f} \circ m_{X,Y} = f$, since the composites of both sides with $\iota_X \sqcap \iota_Y$ coincide and thus $\tilde{f} \circ m_{X,Y} = f$ by lemma (5.2.2). Uniqueness is obvious, since any solution $\tilde{f}$ of $\tilde{f} \circ m_{X,Y} = f$ satisfies also $\tilde{f} \circ \iota_{X \sqcap Y} = \tilde{f} \circ m_{X,Y} \circ (\iota_X \sqcap \iota_Y) = f \circ (\iota_X \sqcap \iota_Y)$ and hence has to be unique by the universal property of $\iota_{X \sqcap Y}: X \sqcap Y \to \lambda(X \sqcap Y)$. □

Remark. This proposition is rather surprising, since the corresponding statement: $C^\infty(X \sqcap Y, \mathbb{R}) \cong C^\infty(X, \mathbb{R}) \tilde{\otimes} C^\infty(Y, \mathbb{R})$ fails to be true in general, cf. (vi) in (7.4.5).

Now we are able to show that λX is always a convenient co-algebra.

5.2.5 Proposition. *The functor $\lambda: \underline{\mathcal{M}} \to \underline{Con}$ lifts to a functor $\lambda: \underline{\mathcal{M}} \to \underline{ConCoAlg}$. For any $\mathcal{M}$-space X the convenient co-algebra structure on λX is as follows: The co-multiplication is $\lambda X \xrightarrow{\lambda(\Delta)} \lambda(X \sqcap X) \cong \lambda(X) \tilde{\otimes} \lambda(X)$, where Δ denotes the diagonal map $\Delta: X \to X \sqcap X$, $x \mapsto (x, x)$. The co-unit is $\lambda X \xrightarrow{\lambda(*)} \lambda(\{*\}) \cong \mathbb{R}$, where $\{*\}$ denotes a singleton with its unique $\mathcal{M}$-structure ((i) of (1.1.6)) and where $*$ denotes the constant map $*: X \to \{*\}$, $x \mapsto *$.*

Proof. We show first that λX is a convenient co-algebra. Co-associativity follows from the commuting diagram:

$$\begin{array}{ccc} X \sqcap X & \xrightarrow{\Delta \sqcap \mathrm{id}} & X \sqcap X \sqcap X \\ \uparrow \Delta & & \uparrow \mathrm{id} \sqcap \Delta \\ X & \xrightarrow{\Delta} & X \sqcap X. \end{array}$$

Co-commutativity follows from the commuting diagram:

$$\begin{array}{ccc} X \sqcap X & \cong & X \sqcap X \\ \uparrow \Delta & & \uparrow \Delta \\ X & = & X, \end{array}$$ where the isomorphism is $(x, y) \mapsto (y, x)$.

That $\lambda(*)$ is a co-unit follows from the commutative diagram:

$$\begin{array}{ccc} X & \xrightarrow{\cong} & \{*\} \sqcap X \\ \downarrow \Delta & & \uparrow * \sqcap \mathrm{id} \\ X \sqcap X & = & X \sqcap X. \end{array}$$

In order to show that λ lifts to a functor it is enough to verify that λf is a convenient co-algebra morphism. This follows immediately from $(f \sqcap f) \circ \Delta = \Delta \circ f$ and $* \circ f = *$. □

Now we are going to construct an adjoint to $\lambda: \underline{\mathcal{M}} \to \underline{\text{ConCoAlg}}$. For this we need the

5.2.6 Definition. Let E be a convenient co-algebra. An element $x \in E$ is called *co-idempotent* if $\varepsilon(x)=1$ and $\mu(x)=x \otimes x$. The set of all co-idempotent elements of E is denoted by $I(E)$.

5.2.7 Proposition. *For every convenient co-algebra E the map ev_1: $\underline{ConCoAlg}$ $(\mathbb{R}, E) \to E$ yields a bijective map onto the subset $I(E)$ of the co-idempotent elements of E.*

Proof. The inverse map is given by $x \mapsto (t \mapsto tx)$. These two maps define obviously a bijection between $L(\mathbb{R}, E)$ and E and it remains to show that x is co-idempotent iff $f_x: t \mapsto tx$ is a co-algebra morphism: $\mu(x)=x \otimes x$ gets translated into $(\mu \circ f_x)(t)=t\mu(x)=t(x \otimes x)=(tx \otimes x)=(f_x \otimes f_x)(t \otimes 1)=(f_x \otimes f_x)(\mu(t))$. And $\varepsilon(x)=1$ gets translated into $(\varepsilon \circ f_x)(t)=t\varepsilon(x)=t=\varepsilon(t)$. □

5.2.8 Proposition. *The map $E \mapsto I(E)$ extends to a functor I: $\underline{ConCoAlg} \to \underline{\mathcal{M}}$.*

Proof. For a convenient co-algebra we put on $I(E)$ the initial $\mathcal{M}$-structure induced by the inclusion $I(E) \to E$. This extends clearly to a functor since $\varepsilon(f(x))=f(\varepsilon(x))$ and $\mu(f(x))=(f \tilde{\otimes} f)(\mu(x))$ for any convenient co-algebra morphism. □

5.2.9 Theorem. *Let $\mathcal{M}$ be either C^∞ or ℓ^∞. The functor λ: $\underline{\mathcal{M}} \to \underline{ConCoAlg}$ is left-adjoint to the functor I: $\underline{ConCoAlg} \to \underline{\mathcal{M}}$.*

Proof. The unit of the adjunction is given by $\iota_X: X \to I(\lambda X)$. That $\iota_X: X \to \lambda X$ really factors over $I(\lambda X)$ is obvious, since $\varepsilon \circ \iota_X = \iota_{\{*\}} \circ * = 1: X \to \mathbb{R}$ and $\mu \circ \iota_X = \iota_{X \sqcap X} \circ \Delta$ and hence $\varepsilon(\iota_X(x))=1$ and $\mu(\iota_X(x))=\iota_{X \sqcap X}(x, x)=x \otimes x$.

The co-unit of the adjunction is given by the map $\eta: \lambda(I(E)) \to E$ associated to the inclusion $I(E) \to E$ using the universal property of $I(E) \to \lambda(I(E))$, i.e. η is determined by $\eta \circ \iota_{I(E)} = \mathrm{incl}: I(E) \to E$. We have to show that this map η is a convenient co-algebra morphism. Composed with ε it has to be the map associated to the composition of ε with the inclusion $I(E) \to E$. Since this composite has by definition of $I(E)$ constant value 1, the composite $\varepsilon \circ \eta$ is the co-unit of $\lambda(I(E))$. Next consider the composite $\mu \circ \eta$. Composed with $\iota_{I(E)}$ it yields $\mu|_{I(E)}$ which coincides by the definition of $I(E)$ with the map $x \mapsto x \otimes x$, i.e. with

$$\otimes \circ \Delta = \otimes \circ ((\eta \circ \iota_{I(E)}) \sqcap (\eta \circ \iota_{I(E)})) \circ \Delta = \otimes \circ (\eta \sqcap \eta) \circ (\iota_{I(E)} \sqcap \iota_{I(E)}) \circ \Delta =$$

$$(\eta \otimes \eta) \circ \otimes \circ (\iota_{I(E)} \sqcap \iota_{I(E)}) \circ \Delta = (\eta \otimes \eta) \circ \iota_{I(E) \sqcap I(E)} \circ \Delta = (\eta \tilde{\otimes} \eta) \circ \mu \circ \iota_{I(E)},$$

thus $\mu \circ \eta = (\eta \tilde{\otimes} \eta) \circ \mu$.

Now let $f: X \to I(F)$ be a smooth map, F being a convenient co-algebra. Then there exists a unique linear morphism $f^\sim: \lambda X \to F$ that satisfies $f^\sim \circ \iota_X = f$. It can be factorized as $f^{\sim}=\eta\circ\lambda f$ $(\lambda f\circ\iota_X=\iota_{I(F)}\circ f$, hence $\eta\circ\lambda f\circ\iota_X=\eta\circ\iota_{I(F)}\circ f=f$ and $\eta\circ\lambda f=f^{\sim})$ and thus it is a convenient co-algebra morphism as composite of convenient co-algebra morphisms. Its well defined restriction $f^{\sim}: I(\lambda X)\to I(F)$ composed with ι_X gives also f. Thus we verified the universal property of $X\to I(\lambda(X))$. □

5.2.10 Corollary. *The functor $\lambda: \underline{\mathcal{M}}\to \underline{ConCoAlg}$ preserves colimits.*

5.2.11 Proposition. *$\mathbb{R}\cong\lambda(\{*\})$ is a convenient co-algebra with co-unit $\mathrm{id}:\mathbb{R}\to\mathbb{R}$ and co-multiplication $\mathbb{R}\cong\mathbb{R}\tilde{\otimes}\mathbb{R}$, $t\mapsto t\cdot(1\otimes 1)$. Furthermore, the co-unit $\varepsilon: E\to\mathbb{R}$ of any convenient co-algebra is a convenient co-algebra morphism.*

Proof. That the co-unit and the co-multiplication of $\lambda(\{*\})$ correspond via the isomorphism $\mathbb{R}\cong\lambda(\{*\})$ to the mentioned maps is easily checked. The map $\varepsilon: E\to\mathbb{R}$ preserves the co-unit since $\mathrm{id}\circ\varepsilon=\varepsilon$; and that it preserves the co-multiplication μ follows by composing the diagram which expresses that ε is a co-unit with respect to μ with the natural isomorphism $(\_)\cong\mathbb{R}\tilde{\otimes}(\_)$. □

5.2.12 Proposition. *For any two convenient co-algebras E_1 and E_2 the tensor product $E_1\tilde{\otimes}E_2$ is a convenient co-algebra with co-unit $E_1\tilde{\otimes}E_2\xrightarrow{\varepsilon_1\tilde{\otimes}\varepsilon_2}\mathbb{R}\tilde{\otimes}\mathbb{R}\cong\mathbb{R}$ and co-multiplication $E_1\tilde{\otimes}E_2\xrightarrow{\mu_1\tilde{\otimes}\mu_2}(E_1\tilde{\otimes}E_1)\tilde{\otimes}(E_2\tilde{\otimes}E_2)\cong(E_1\tilde{\otimes}E_2)\tilde{\otimes}(E_1\tilde{\otimes}E_2)$. Furthermore, the co-multiplication $\mu: E\to E\tilde{\otimes}E$ of any convenient co-algebra is a convenient co-algebra morphism.*

Proof. That $E_1\tilde{\otimes}E_2$ is a convenient co-algebra follows by forming the tensor product of the corresponding diagrams which express that E_1 and E_2 are convenient co-algebras.

That $\mu: E\to E\tilde{\otimes}E$ preserves the co-unit follows from the diagram which expresses that ε is a co-unit with respect to μ. That μ preserves the co-multiplication can be obtained using the co-associativity twice. □

5.2.13 Proposition. *The functor $(\_)\tilde{\otimes}(\_): \underline{Con}\times\underline{Con}\to\underline{Con}$ lifts to a functor $(\_)\tilde{\otimes}(\_): \underline{ConCoAlg}\times\underline{ConCoAlg}\to\underline{ConCoAlg}$.*

Proof. That this functor is well defined on objects is (5.2.12). So let $f: E_1\to E_2$ and $g: F_1\to F_2$ be two convenient co-algebra morphisms. Since one has $f\tilde{\otimes}g=(f\tilde{\otimes}F_2)\circ(E_1\tilde{\otimes}g)$ it is enough to show that $f\tilde{\otimes}F$ is a convenient co-algebra morphism. The $\underline{\text{Con}}$-morphism $f\tilde{\otimes}F:=f\tilde{\otimes}\mathrm{id}_F$ preserves the co-unit and the co-multiplication since f does it. □

5.2.14 Theorem. *A product in $\underline{ConCoAlg}$ of two arbitrary convenient co-algebras E_1 and E_2 is given by $E_1\tilde{\otimes}E_2$ with the projections*

$E_1 \tilde{\otimes} E_2 \xrightarrow{E_1 \tilde{\otimes} \varepsilon_2} E_1 \tilde{\otimes} \mathbb{R} \cong E_1$ *and* $E_1 \tilde{\otimes} E_2 \xrightarrow{\varepsilon_1 \tilde{\otimes} E_2} \mathbb{R} \tilde{\otimes} E_2 \cong E_2$. *The functor* $\lambda: \underline{\mathcal{M}} \to \underline{ConCoAlg}$ *preserves also finite products, cf.* (5.2.10).

Proof. By (5.2.13) we know that $E_1 \tilde{\otimes} E_2$ is a convenient co-algebra and the described projections are convenient co-algebra morphisms. So it remains to show the universal property. Let E be an arbitrary convenient co-algebra and let $f_j: E \to E_j$ ($j \in \{1, 2\}$) be two convenient co-algebra morphisms. Define $f: E \to E_1 \tilde{\otimes} E_2$ by $f = (f_1 \tilde{\otimes} f_2) \circ \mu: E \to E \tilde{\otimes} E \to E_1 \tilde{\otimes} E_2$. By (5.2.12) and (5.2.13) this yields a convenient co-algebra morphism. That its composites with the projections give f_1 and f_2 follows since the $(f_j)^*$ carries the co-unit ε_j to ε, the co-unit with respect to $\mu: E \to E \tilde{\otimes} E$.

It remains to show the uniqueness. So let $f: E \to E_1 \tilde{\otimes} E_2$ be any convenient co-algebra morphism with the property that its composites with the projections give f_1 and f_2. Up to natural isomorphisms one has $f \cong f \circ (E \tilde{\otimes} \varepsilon) \circ \mu \cong ((E_1 \tilde{\otimes} E_2) \tilde{\otimes} \varepsilon) \circ (f \tilde{\otimes} f) \circ \mu \cong ((E_1 \tilde{\otimes} \varepsilon_2) \tilde{\otimes} (\varepsilon_1 \tilde{\otimes} E_2)) \circ (f \tilde{\otimes} f) \circ \mu \cong (f_1 \tilde{\otimes} f_2) \circ \mu$.

That λ preserves finite products is just a reformulation of (5.2.4) since the co-algebra structure on the tensor product is the unique one making the projections onto the factors convenient co-algebra morphisms. □

5.2.15 Proposition. *The duality functor* $(\_)': \underline{Con}^{op} \to \underline{Con}$ *lifts to a faithful functor* $(\_)': \underline{ConCoAlg}^{op} \to \underline{ConAlg}$.

For a given convenient co-algebra E with co-unit ε and co-multiplication μ the dual E' is a convenient algebra with unit $\mathbb{R} \cong \mathbb{R}' \xrightarrow{\varepsilon^*} E'$ *and multiplication* $E' \tilde{\otimes} E' \to (E \tilde{\otimes} E)' \xrightarrow{\mu^*} E'$, *where the map* $E' \tilde{\otimes} E' \to (E \tilde{\otimes} E)'$ *is the one associated to* $(E' \sqcap E') \sqcap (E \sqcap E) \cong (E' \sqcap E) \sqcap (E' \sqcap E) \xrightarrow{\text{ev} \sqcap \text{ev}} \mathbb{R} \sqcap \mathbb{R} \xrightarrow{\text{mult}} \mathbb{R}$.

Proof. The axioms for a convenient algebra are expressed by commutative diagrams. For E' these can be obtained by applying the duality functor to those expressing that E is a convenient co-algebra and composing where necessary with the natural transformation $E' \tilde{\otimes} E' \to (E \tilde{\otimes} E)'$. The map $E \mapsto E'$ certainly extends to a functor that is faithful since any morphism $f: E \to F$ is just the restriction of $f^{**}: E'' \to F''$ via the initial morphisms $\iota_E: E \to E''$ and $\iota_F: F \to F''$. □

Remark. See (ii) in (7.4.5) for an example where $E' \tilde{\otimes} E' \to (E \tilde{\otimes} E)'$ is not an isomorphism.

5.2.16 Proposition. *The functor* $\mathcal{M}(\_, \mathbb{R}): \underline{\mathcal{M}}^{op} \to \underline{Con}$ *lifts to a functor* $\mathcal{M}(\_, \mathbb{R}): \underline{\mathcal{M}}^{op} \to \underline{ConAlg}$ *such that for any $\mathcal{M}$-space X the algebra operations on $\mathcal{M}(X, \mathbb{R})$ are induced pointwise by those of $\mathbb{R}$, i.e.* $\text{ev}_x: \mathcal{M}(X, \mathbb{R}) \to \mathbb{R}$ *is an algebra morphism for all* $x \in X$.

Proof. That the pointwise defined multiplication on $\mathcal{M}(X, \mathbb{R})$ is a Con-morphism follows immediately since $\mathrm{ev}_x: \mathcal{M}(X, \mathbb{R}) \to \mathbb{R}$ $(x \in X)$ is an initial family. Furthermore, $\mathcal{M}(f, \mathbb{R})$ is an algebra morphism since composed with ev_x it is the algebra morphism $\mathrm{ev}_{f(x)}$. □

5.2.17 Corollary. *The functors $\mathcal{M}(\_, \mathbb{R})$: $\underline{\mathcal{M}}^{\mathrm{op}} \to \underline{ConAlg}$ and $(\_)' \circ \lambda$: $\underline{\mathcal{M}}^{\mathrm{op}} \to \underline{ConCoAlg}^{\mathrm{op}} \to \underline{ConAlg}$ are naturally isomorphic.*

Proof. In (5.1.3) it was proved that these two functors considered as Con-valued functors are naturally isomorphic, the isomorphism $(\lambda X)' \to \mathcal{M}(X, \mathbb{R})$ being given by $\ell \mapsto \ell \circ \iota_X$. Thus it remains to show that this isomorphism is an algebra morphism, i.e. $\mathrm{ev}_{\iota(x)}: \ell \mapsto \mathrm{ev}_x(\ell \circ \iota_X)$ is an algebra morphism for all $x \in X$. This follows easily since $\iota(x)$ is co-idempotent in $\lambda(X)$, hence defines a co-algebra morphism $\mathbb{R} \to \lambda X$ whose dual $(\lambda X)' \to \mathbb{R}' \cong \mathbb{R}$ is just $\mathrm{ev}_{\iota(x)}$. □

5.2.18 Lemma. *Any co-idempotent element $e \in \lambda X$ is an algebra morphism $\mathcal{M}(X, \mathbb{R}) \to \mathbb{R}$.*

Proof. Being co-idempotent, e satisfies $\varepsilon(e) = 1$ and $\mu(e) = e \otimes e$. When e is considered as a functional on $\mathcal{M}(X, \mathbb{R})$ this means the following: $e(1) = 1$ and $e(x \mapsto h(x, x)) = e(x \mapsto e(h(x, \_)))$. Choosing $h(x, y) = f(x)g(y)$ one obtains $e(fg) = e(f)e(g)$, i.e. $e: \mathcal{M}(X, \mathbb{R}) \to \mathbb{R}$ has to be an algebra morphism. This can also be seen by applying the duality functor $(\_)'$: $\underline{\mathrm{ConCoAlg}}\,(\mathbb{R}, \lambda X) \to \underline{\mathrm{ConAlg}}\,(\mathcal{M}(X, \mathbb{R}), \mathbb{R})$ to the co-algebra morphism $\mathbb{R} \to \lambda X$, $t \mapsto te$. □

5.2.19 Remark. For which $\mathcal{M}$-objects X does one have $I(\lambda X) = X$? Equivalently: when is every co-idempotent element of $\lambda X \subseteq \mathcal{M}(X, \mathbb{R})'$ of the form ev_x for some $x \in X$? According to the previous lemma this is true for those X for which every convenient algebra morphism $\mathcal{M}(X, \mathbb{R}) \to \mathbb{R}$ is the evaluation at some point of X. We will show in (5.2.22) that this holds for all X in case $\mathcal{M} = \ell^\infty$. For the case $\mathcal{M} = C^\infty$ a large class of spaces for which even every algebra morphism $C^\infty(X, \mathbb{R}) \to \mathbb{R}$ is a point evaluation (often called Milnor's exercise) was given in [Kriegl, Michor, Schachermayer, 1985].

As in section 5.1 we are able to obtain more results in the case where $\mathcal{M} = \ell^\infty$. First we describe the map $m_{X,Y}$ more explicitly. We recall that $x \mapsto \chi_x$, $X \to \ell^1 X$ is a description of the free convenient vector space over an ℓ^∞-space X, cf. (5.1.22).

5.2.20 Proposition. *For any two ℓ^∞-spaces X and Y the map $m_{X,Y}$: $\ell^1 X \sqcap \ell^1 Y \to \ell^1(X \sqcap Y)$ is given by $(f, g) \mapsto ((x, y) \mapsto f(x)g(y))$.*

Proof. The map $m_{X,Y}$ was defined in (5.2.3) as the unique map m that satisfies $m \circ (\iota_X \sqcap \iota_Y) = \iota_{X \sqcap Y}$. Thus $m(\chi_x, \chi_y) = \chi_{(x,y)}$. Every $f \in \ell^1 X$ can be written as the M-convergent sum (i.e. as M-limit of the finite subsums) $f = \sum_x f(x)\chi_x$, see the proof of (5.1.21); similarly for $g \in \ell^1 Y$. So we obtain $m(f, g) = m\left(\sum_x f(x)\chi_x, \sum_y g(y)\chi_y\right)$

$=\sum_{x,y} f(x)g(y)m(\chi_x, \chi_y)=\sum_{x,y} f(x)g(y)\chi_{(x,y)}$, using that m is a bilinear morphism and hence has to commute with M-convergent sums. □

Next we describe explicitly the convenient co-algebra structure on $\ell^1 X$.

5.2.21 Proposition. *The co-unit $\varepsilon: \ell^1 X \to \mathbb{R}$ of the co-algebra $\ell^1 X$ is given by $\varepsilon(f):=\Sigma_{x\in X} f(x)$. The co-multiplication $\mu: \ell^1(X) \to \ell^1 X \tilde{\otimes} \ell^1 X \cong \ell^1(X \sqcap X)$ is given by $\mu(f)(x, x):=f(x)$ and $\mu(f)(x, y)=0$ if $x \neq y$.*

Proof. Recall that $\iota: X \to \lambda(X)$ has only co-idempotent values. Thus $\varepsilon(\chi_x)=1$ and $\mu(\chi_x)=\chi_x \otimes \chi_x$. Using again that $f=\sum_x f(x)\chi_x$ for all $f\in\ell^1 X$ one obtains $\varepsilon(f)=\varepsilon(\sum_x f(x)\chi_x)=\sum_x f(x)\varepsilon(\chi_x)=\sum_x f(x)$ and $\mu(f)=\mu(\sum_x f(x)\chi_x)=\sum_x f(x)\mu(\chi_x)=\sum_x f(x)\chi_x \otimes \chi_x=\sum_x f(x)\chi_{(x,x)}$. □

Finally we show that $\underline{\ell^\infty}$ can be identified with a full coreflective subcategory of <u>ConCoAlg</u>.

5.2.22 Theorem. *The functor $\ell^1: \underline{\ell^\infty} \to$ <u>ConCoAlg</u> is full and faithful and has a right adjoint.*

Proof. Since $X \mapsto I(\lambda(X))$ is initial we only have to show that it is bijective. It is injective, since for any $x\in X$ one has $\chi_x \in \ell^\infty(X, \mathbb{R})$ and $\iota(y)(\chi_x)=\chi_x(y)=1$ iff $y=x$. Now we prove surjectivity. Let $e\in\ell^1 X$ be an arbitrary co-idempotent element. Then $\varepsilon(e)=1$, i.e. $\sum_x e(x)=1$; and $\mu(e)=e\otimes e$, i.e. $e(x)=e(x)e(x)$ and $0=e(x)e(y)$ for $x\neq y$. Thus $e(x)\in\{0, 1\}$ and for exactly one x one has $e(x)=1$, i.e. $e=\chi_x$. □

Remark. The functor ℓ^1 is not an equivalence of $\underline{\ell^\infty}$ with <u>ConCoAlg</u>. In order to see this, one can use that the dual of the algebra $\mathbb{C}$ is a convenient co-algebra which is not isomorphic to some $\ell^1(X)$, since the only element of $\mathbb{C}'$ that is co-idempotent is the real-part functional Re(_).

5.3 Cartesian closed categories of convenient vector spaces

5.3.1 Definition. With <u>*Con*$^\infty$</u> we denote the category having as objects the convenient vector spaces and as morphisms the smooth maps, i.e. the $\mathscr{Lip}^\infty$-maps.

5.3.2 Proposition. *The category <u>Con$^\infty$</u> is cartesian closed. Products are formed as in <u>Con</u>.*

Proof. Recall that the smooth structure of a product of convenient vector spaces E_j has as structure curves those curves $c: \mathbb{R} \to \Pi_{j\in J} E_j$ for which all projections $\mathrm{pr}_j \circ c: \mathbb{R} \to E_j$ are smooth. Thus a map $f: E \to \Pi_{j\in J} E_j$ is smooth iff $\mathrm{pr}_j \circ f: E \to E_j$ is smooth for every $j\in J$. This shows that the product is formed as in Con. Since for convenient vector spaces E and F the space $C^\infty(E, F)$ is a convenient vector space, the functor $C^\infty(\_,\_)$ which makes $\underline{C^\infty}$ cartesian closed lifts to a functor making $\underline{\text{Con}}^\infty$ cartesian closed. □

The reason for being interested in cartesian closed subcategories of $\underline{\text{Con}}^\infty$ is the following. We have discussed so far a rather maximal setting for differential calculus in vector spaces. For more restricted questions it could be of interest to use certain subclasses formed by convenient vector spaces having additional properties like, for example, nuclearity (implying the approximation property), or some stronger form of completeness, or even reflexivity. In order to be able to work inside a subcategory one should, of course, be assured that the important constructions developed so far (like the internal hom-functor) restrict to this subcategory. We shall discuss now the problems that arise in this connection.

Let us begin with a categorical lemma characterizing reasonable cartesian closed subcategories:

5.3.3 Lemma. *Let $\mathscr{A}$ be a non-void, full, replete (cf. (8.1.1)) subcategory of $\underline{\text{Con}}^\infty$. Then $\mathscr{A}$ is cartesian closed iff it is closed under the functors $\sqcap$ and $C^\infty(\_,\_)$ and contains $\{0\}$.*

Proof. ($\Rightarrow$) The space $\{0\}$ must belong to $\mathscr{A}$, since cartesian closedness requires a terminal object. The case where all objects in $\mathscr{A}$ are isomorphic to $\{0\}$ is trivial. So now let A_0 be a non-terminal object of $\mathscr{A}$. We will twice use the following fact: if there exists a smooth map $\varphi: E_1 \to E_2$ such that $\varphi_*: C^\infty(A_0, E_1) \to C^\infty(A_0, E_2)$ is a bijection, then $\varphi: E_1 \to E_2$ is a diffeomorphism. In order to see this, let $x \in A_0$, $x \neq 0$. Then the subspace generated by x is complemented and isomorphic to $\mathbb{R}$ (the linear functionals separate points). Using that φ_* commutes with f^* for any morphism f, one concludes that $\varphi_*: C^\infty(\mathbb{R}, E_1) \to C^\infty(\mathbb{R}, E_2)$ is a bijection as well, and thus $\varphi: E_1 \to E_2$ is a diffeomorphism.

Let us denote by $E \sqcap F$ with $\text{pr}_1: E \sqcap F \to E$ and $\text{pr}_2: E \sqcap F \to F$ (resp. $E \sqcap_{\mathscr{A}} F$ with $p_1: E \sqcap_{\mathscr{A}} F \to E$ and $p_2: E \sqcap_{\mathscr{A}} F \to F$) the product of E and F in $\underline{\text{Con}}^\infty$ (resp. in $\mathscr{A}$).

First we want to show that $E_1 \sqcap E_2 \in |\mathscr{A}|$ for all $E_i \in |\mathscr{A}|$. Using repleteness it is enough to show that $E_1 \sqcap E_2 \cong E_1 \sqcap_{\mathscr{A}} E_2$. By the universal property of $E \sqcap F$ there exists a unique morphism $\varphi: E \sqcap_{\mathscr{A}} F \to E \sqcap F$ satisfying $\text{pr}_j \circ \varphi = p_j$ for $j \in \{1, 2\}$. And by the universal property of the product in $\mathscr{A}$ (resp. in $\underline{\text{Con}}^\infty$) $((p_1)_*, (p_2)_*)$: $C^\infty(A_0, E \sqcap_{\mathscr{A}} F) \to C^\infty(A_0, E) \times C^\infty(A_0, F)$ (resp. $((\text{pr}_1)_*, (\text{pr}_2)_*)$: $C^\infty(A_0, E \sqcap F) \to C^\infty(A_0, E) \times C^\infty(A_0, F)$) is a bijection. Since $(\text{pr}_j)_* \circ \varphi_* = ((\text{pr}_j) \circ \varphi)_* = (p_j)_*$ one concludes that $\varphi_*: C^\infty(A_0, E \sqcap_{\mathscr{A}} F) \to C^\infty(A_0, E \sqcap F)$ is a bijection; so the claim follows from the fact stated above.

Since $E \sqcap_{\mathscr{A}} F$ and $E \sqcap F$ have just been proved to be isomorphic and the product is determined only up to isomorphisms we may assume that $\varphi: E \sqcap F \to E \sqcap_{\mathscr{A}} F$ is the identity map. Let $[E, \_]$ denote the adjoint functor to $(\_) \sqcap E: \mathscr{A} \to \mathscr{A}$ and let $\varepsilon: [E, F] \sqcap E \to F$ denote the counit of adjunction. By the universal property of ev: $C^\infty(E, F)\sqcap_{\mathscr{A}} E \to F$ there exists a morphism π: $[E, F] \to C^\infty(E, F)$ such that $\text{ev} \circ (\pi \sqcap E) = \varepsilon$. By a similar argument as for the product we conclude that φ_*: $C^\infty(A_0, [E, F]) \to C^\infty(A_0, C^\infty(E, F))$ is a bijection and hence φ is a diffeomorphism. Using repleteness we conclude that $C^\infty(E, F) \in |\mathscr{A}|$.

($\Leftarrow$) is trivial. □

In order to determine what additional properties can possibly hold for all objects of some non-trivial, full, replete, cartesian closed subcategory of $\underline{\text{Con}}^\infty$ we show that there is a smallest one and give a description of it.

5.3.4 Corollary. *The smallest cartesian closed, full, replete subcategory of* $\underline{\text{Con}}^\infty$ *that contains* $\mathbb{R}$ *is formed by those objects which are isomorphic to spaces obtained from* $\{0\}$ *by finitely many successive applications of the functors* $(\_)\sqcap(\_)$ *and* $C^\infty(\_, \mathbb{R})$.

Proof. Since any cartesian closed, replete, full subcategory has to be closed under these functors and has to contain a terminal object, the objects so described have to be contained in any category of that type. Since the full subcategory formed by these objects is replete and closed under $\_\sqcap\_$, it is, by (5.3.3), enough to prove that it is closed under $C^\infty(\_, \_)$. So let E and F be objects obtained in the described way. Using the equations $C^\infty(E, F_1 \sqcap F_2) \cong C^\infty(E, F_1) \sqcap C^\infty(E, F_2)$ and $C^\infty(E, C^\infty(F_1, \mathbb{R})) \cong C^\infty(E \sqcap F_1, \mathbb{R})$ one obtains by induction on the length of the expression for F that $C^\infty(E, F)$ can be represented in the same way. □

Most of the interesting additional properties for convenient vector spaces are inherited by subspaces or by quotients, hence in particular by complemented subspaces. Thus it is natural to consider the smallest non-trivial cartesian closed, full, replete subcategory of $\underline{\text{Con}}^\infty$ that is also closed with respect to complemented subspaces:

5.3.5 Proposition. *The smallest cartesian closed, full, replete subcategory of* $\underline{\text{Con}}^\infty$ *that contains* $\mathbb{R}$ *and is also closed under complemented subspaces is formed by those spaces which are isomorphic to complemented subspaces of spaces obtained from* $\{0\}$ *by finitely many successive applications of* $C^\infty(\_, \mathbb{R})$. *This category is in addition closed under* $L(\_, \_)$ *and* $C^\infty(X, \_)$ *for every separated separable finite-dimensional smooth manifold* X.

Proof. Again these spaces have to be contained in every category of the described type. Let conversely $\mathscr{A}_n$ be defined inductively by $\mathscr{A}_0 := \{\{0\}, \mathbb{R}\}$ and $\mathscr{A}_{n+1} := \{E; E$ is isomorphic to a complemented subspace of a space $C^\infty(E_1, \mathbb{R})$ with $E_1 \in \mathscr{A}_n\}$. The class of objects described above is exactly the union of all $\mathscr{A}_n$. So it remains to show that this union gives a cartesian closed full subcategory. For this it is enough to verify that it is closed under the functors $(\_)\sqcap(\_)$ and $C^\infty(\_, \_)$. We use the symbol $\underset{\oplus}{\leq}$ for complemented subspace.

Since $\{0\} \in \mathscr{A}_n$ and $E \cong C^\infty(\{0\}, E)$ we obtain for all $n \in \mathbb{N}_0$ that $\mathscr{A}_n \subseteq \mathscr{A}_{n+1}$.

We prove now by induction on n that any finite product ΠE_j of N spaces $E_j \in \mathscr{A}_n$ belongs to $\mathscr{A}_{n+1}$.

($n=0$) Then $\Pi E_j \underset{\oplus}{\leq} \mathbb{R}^N \underset{\oplus}{\leq} \mathbb{R}^{\mathbb{N}}$ and $\mathbb{R}^{\mathbb{N}} \underset{\oplus}{\leq} C^\infty(\mathbb{R}, \mathbb{R})$. The last inclusion is given by $(x_n) \mapsto \sum_{n\in\mathbb{N}} x_n \cdot h((\_)-n)$, where $h: \mathbb{R} \to [0, 1]$ is a smooth function with support in $[-1, 1]$ and with $h(0)=1$. A left inverse to the inclusion is given by $f \mapsto (f(n))_{n\in\mathbb{N}}$.
($n>0$) Then $E_j \underset{\oplus}{\leq} C^\infty(F_j, \mathbb{R})$ with $F_j \in \mathscr{A}_{n-1}$. Hence $\Pi E_j \underset{\oplus}{\leq} \Pi C^\infty(F_j, \mathbb{R}) \underset{\oplus}{\leq} C^\infty(\Pi F_j, \mathbb{R}^N) \underset{\oplus}{\leq} C^\infty(\Pi F_j, C^\infty(\mathbb{R}, \mathbb{R})) \cong C^\infty(\mathbb{R} \sqcap (\Pi F_j), \mathbb{R})$, and $\mathbb{R} \sqcap (\Pi_j F_j) \in \mathscr{A}_n$ by induction hypothesis.

Let now $E \in \mathscr{A}_n$ and $F \in \mathscr{A}_{n+1}$, i.e. $F \underset{\oplus}{\leq} C^\infty(F_1, \mathbb{R})$ with some $F_1 \in \mathscr{A}_n$. Then $C^\infty(E, F) \underset{\oplus}{\leq} C^\infty(E, C^\infty(F_1, \mathbb{R})) \cong C^\infty(E \sqcap F_1, \mathbb{R})$, and $E_1 \sqcap F_1 \in \mathscr{A}_{n+1}$, i.e. $C^\infty(E, F) \in \mathscr{A}_{n+1}$.

In order to show closedness under $L(\_, \_)$ one uses that $L(E_1, \ldots, E_n; F)$ is isomorphic to $L(E_1; L(\ldots; L(E_n; F)\ldots))$ and $L(E, F)$ is a complemented subspace of $C^\infty(E, F)$, a retraction being given by $f \mapsto f'(0)$.

For the closedness under $C^\infty(X, \_)$ one uses that any separable manifold X embeds into some $\mathbb{R}^m$ and has a tubular neighborhood in this $\mathbb{R}^m$. One then obtains an embedding of $C^\infty(X, F)$ into $C^\infty(\mathbb{R}^m, F)$ by taking a smooth function $h: \mathbb{R}^m \to \mathbb{R}$ with support in the tubular neighborhood and equal to 1 on X and defining a linear morphism $C^\infty(X, F) \to C^\infty(\mathbb{R}^m, F)$ by $f \mapsto (f \circ p) \cdot h$, where p is the projection of the tubular neighborhood onto X. A left inverse is given by $f \mapsto f|_X$.

□

We will show now that several properties one might hope for cannot hold for all objects of a cartesian closed full replete subcategory of $\underline{\text{Con}^\infty}$ by verifying that they already fail for one of the spaces $C^\infty(\mathbb{R}, \mathbb{R})$, $C^\infty(C^\infty(\mathbb{R}, \mathbb{R}), \mathbb{R})$, $C^\infty(C^\infty(C^\infty(\mathbb{R}, \mathbb{R}), \mathbb{R}), \mathbb{R})$.

5.3.6 Examples concerning the convenient vector space $E := C^\infty(\mathbb{R}, \mathbb{R})$. E is isomorphic to $s^{\mathbb{N}}$ [Mitiagin, 1961], where s denotes the Fréchet space of fast falling sequences, cf. [Jarchow, 1981, p. 28].

E is a universal nuclear Fréchet space, cf. [Jarchow, 1981, p. 502], i.e. every nuclear Fréchet space can be realized as subspace of E. Since the complemented subspace s is not strongly nuclear [Jarchow, 1981, p. 506] the same is true for E.

Let us show that already E has bad behavior with respect to linear differential equations, i.e. equations of form $c^{\cdot} = m \circ c$, $c(0) = c_0$, where $m: E \to E$ is a linear morphism and $c_0 \in E$ and a solution is a weakly differentiable (hence smooth) curve $c: \mathbb{R} \to E$. It is enough to find some complemented subspace E_1 with bad behavior with respect to linear differential equations, since the solutions of the differential equation on $E \cong E_1 \sqcap E_2$ that is given by the operator $m \sqcap 0$ and initial condition $0 \mapsto (x_0, y_0)$ are the curves $t \mapsto (c(t), y_0)$, where c is a solution of the differential equation on E_1 given by m and initial condition $0 \mapsto x_0$.

(i) An example where no solution exists: we consider on the complemented subpace s of E the linear operator $m: (x_1, x_2, \ldots) \mapsto (0, 1^2 x_1, \ldots, n^2 x_n, \ldots)$.

Suppose a solution c of $c^{\cdot}=m\circ c$ with $c(0):=(1, 0, \ldots, 0, \ldots)$ would exist; then $c(t)=(1, t, 2t^2, \ldots, n!\, t^n, \ldots)$, but this curve does not have values in s.

(ii) An example where solutions exist but are not unique: we consider the complemented subspace $\mathbb{R}^{\mathbb{N}}$ of E and the linear differential equation given by $m(x_1, x_2, \ldots):=(x_2, x_3, \ldots)$ and $c_0:=(0, 0, \ldots)$. Then c is a solution iff $c_n^{\cdot}=c_{n+1}$, hence the solutions are $(c_1, c_1^{\cdot}, c_1^{\cdot\cdot}, \ldots)$, where c_1 is infinitely flat at 0.

(iii) An example where a unique but non-analytic solution exists: consider $m: E\to E$ defined by $m(f):=f^{\cdot}$. Then one verifies that the unique solution c is given by $c(t)(s):=c_0(t+s)$. Pointwise the Taylor series of c at 0 is

$$\left(\sum \frac{c^{(k)}(0)}{k!} t^k\right)(s)=\sum \frac{c_0^{(k)}(s)}{k!} t^k.$$

Choose a c_0 such that

$$\left\{\frac{c_0^{(k)}(0)}{k!} t^k ; k\in\mathbb{N}_0\right\}$$

is unbounded for any $t\neq 0$. Then the Taylor series of c at 0 does not converge in E for any $t\neq 0$, since composed with ev_0 it does not converge.

It is well known that the inverse mapping theorem is wrong for E. The standard example is the smooth map $\exp_*: E\to E$. The derivative is $(\exp_*)'(g)(h)=h\cdot(\exp\circ g)$, hence $(\exp_*)'(g)\in GL(E)$ for all $g\in E$. Let $h: \mathbb{R}\to\mathbb{R}$ be a smooth function with compact support and $h(t)=1$ for t close to 0. Define $c(t)(s):=h(ts)$. Then c is a smooth curve through the constant function $1=\exp_*(0)$. But $c(t)\notin\exp_*(E)$ for all $t\neq 0$, since $0\in c(t)(\mathbb{R})$. Hence $\exp_*$ is not even locally surjective with respect to the Mackey-closure topology.

The following example showing that local injectivity may also not be satisfied arose in a discussion with Peter Michor. Let $E:=(\mathbb{R}^2)^{\mathbb{N}}$ and $f: E\to E$ be defined by $f((x_n, y_n)_n):=(h(x_n, y_n))_n$, where $h: \mathbb{R}^2\to\mathbb{R}^2$ is defined by $h(x, y):=(e^x\cos y, e^x\sin y)$. Obviously the derivative of h is invertible for all $(x, y)\in\mathbb{R}^2$ and thus $f'(x)\in GL(E)$ for $x\in E$. But $f((x_n, y_n)_n)=f((x_n, y_n')_n)$ provided $y_n'-y_n\in 2\pi\mathbb{Z}$. Infinitely many points with this property are contained in any non-void open subset of E.

In order to show that the group $GL(E)$ of automorphisms is not open with respect to the M-closure topology in $L(E, E)$, we consider the smooth curve $c: \mathbb{R}\to L(E, E)$ defined by $c(t)(f):=f+tf^{\cdot}$. One has $c(0)=\mathrm{id}\in GL(E)$, but

$$c(t)\left(s\mapsto k\cdot\exp\left(-\frac{s}{t}\right)\right)=\left(s\mapsto k\cdot\exp\left(-\frac{s}{t}\right)+t\cdot k\cdot\left(-\frac{1}{t}\right)\cdot\exp\left(-\frac{s}{t}\right)\right)=0 \text{ for } t\neq 0,$$

hence $c(t)$ is not injective and thus not contained in $GL(E)$ for $t\neq 0$.

The theorem of Borel (cf. (7.1.2)) does not generalize to functions $f: E\to\mathbb{R}$; i.e. given symmetric maps $f_n\in L(E, \ldots, E; \mathbb{R})$ there need not be a smooth function $f: E\to\mathbb{R}$ with derivatives $f^{(n)}(0)=f_n$ for all $n\in\mathbb{N}$. This was shown in [Colombeau, 1979] for $\mathbb{R}^{\mathbb{N}}$ and extends immediately to E.

The inversion map $\mathrm{inv}: GL(E) \to GL(E)$ is not continuous (for the final topology induced by the smooth curves): we show this first for the complemented subspace s of E. Define $c: \mathbb{R} \to GL(s)$ by $c(t)(x) := (x_1 - h_1(t)x_1, \ldots, x_n - h_n(t)x_n, \ldots)$ where $h_n(t) := (1 - 2^{-n})h(tn)$ for some smooth function h with compact support and $h(t) = 1$ for $|t| \leq 1$. Suppose $\mathrm{ev}_x \circ \mathrm{inv} \circ c: \mathbb{R} \to s$ is continuous. Using $x := (1, \frac{1}{2}, \ldots, (\frac{1}{2})^n, \ldots)$ yields $(\mathrm{inv} \circ c)(1/n)(x) = (c(1/n))^{-1}(x) = (\ldots, 2^n x_n, \ldots)$ which does not converge to $x = (\mathrm{inv} \circ c)(0)(x)$. (Use: $1 - h_n(1/n) = 1 - (1 - (\frac{1}{2})^n)h(1) = (\frac{1}{2})^n$). Now let E_2 be a complement to s, i.e. $E = s \oplus E_2$, and consider the smooth curve $\tilde{c}: t \mapsto (c(t) \oplus \mathrm{id}) \in GL(s) \oplus GL(E_2) \subseteq GL(E)$. Then $(\tilde{c}(t))^{-1} = (c(t))^{-1}, \mathrm{id})$ and hence $t \mapsto (\tilde{c}(t))^{-1}$ is not continuous.

5.3.7 The convenient vector space $E := C^\infty(C^\infty(\mathbb{R}, \mathbb{R}), \mathbb{R})$. Since $\mathbb{R}^{\mathbb{N}}$ is a complemented subspace of $C^\infty(\mathbb{R}, \mathbb{R})$, the dual $(\mathbb{R}^{\mathbb{N}})' = \mathbb{R}^{(\mathbb{N})}$ is a complemented subspace of $C^\infty(\mathbb{R}, \mathbb{R})'$ which is complemented in E. Since the locally convex topology of $\mathbb{R}^{(\mathbb{N})}$ is not Baire ([Jarchow, 1981, p. 97]) and the theorem of Borel is wrong for curves $c: \mathbb{R} \to \mathbb{R}^{(\mathbb{N})}$ (no smooth curve c can have, for all n, $c^{(n)}(0)$ equal to the nth unit vector), both statements also hold for E. Furthermore, one has the following chain of complemented subspaces

$$\mathbb{R}^{(\mathbb{N})} \oplus \mathbb{R}^{\mathbb{N}} \underset{\oplus}{\leq} (\mathbb{R}^{(\mathbb{N})})^{\mathbb{N}} = L(\mathbb{R}^{\mathbb{N}}, \mathbb{R})^{\mathbb{N}} = L(\mathbb{R}^{\mathbb{N}}, \mathbb{R}^{\mathbb{N}}) \underset{\oplus}{\leq} C^\infty(\mathbb{R}^{\mathbb{N}}, \mathbb{R}^{\mathbb{N}}) \underset{\oplus}{\leq}$$
$$C^\infty(\mathbb{R}^{\mathbb{N}}, C^\infty(\mathbb{R}, \mathbb{R})) \cong C^\infty(\mathbb{R}^{\mathbb{N}} \oplus \mathbb{R}, \mathbb{R}) \underset{\oplus}{\leq} C^\infty(C^\infty(\mathbb{R}, \mathbb{R}), \mathbb{R}) = E.$$

The M-closure topology of the first space is no topological vector space topology, cf. (i) of (6.2.8), and hence the same is true for E.

5.3.8 The convenient vector space $E := C^\infty(C^\infty(C^\infty(\mathbb{R}, \mathbb{R}), \mathbb{R}), \mathbb{R})$. Since the non-nuclear Fréchet space $C^\infty(C^\infty(\mathbb{R}, \mathbb{R})', \mathbb{R})$ (cf. [Meise, 1980]) is a complemented subspace of E, also E has to be non-nuclear. Furthermore, one has the following chain of complemented subspaces:

$$\begin{aligned}(\mathbb{R}^{\mathbb{N}})^{(\mathbb{N})} \oplus (\mathbb{R}^{(\mathbb{N})})^{\mathbb{N}} &\underset{\oplus}{\leq} (\mathbb{R}^{\mathbb{N}})^{(\mathbb{N})} \oplus ((\mathbb{R}^{\mathbb{N}})^{(\mathbb{N})})^{\mathbb{N}} \cong ((\mathbb{R}^{\mathbb{N}})^{(\mathbb{N})})^{\mathbb{N}} = L(\mathbb{R}, (\mathbb{R}^{\mathbb{N}})^{(\mathbb{N})})^{\mathbb{N}} \\ &= L(\mathbb{R}^{(\mathbb{N})}, (\mathbb{R}^{\mathbb{N}})^{(\mathbb{N})}) = L(\mathbb{R}^{(\mathbb{N})}, ((\mathbb{R}^{(\mathbb{N})})^{\mathbb{N}})') \cong L((\mathbb{R}^{(\mathbb{N})})^{\mathbb{N}}, (\mathbb{R}^{(\mathbb{N})})') \\ &\underset{\oplus}{\leq} C^\infty((\mathbb{R}^{(\mathbb{N})})^{\mathbb{N}}, C^\infty(\mathbb{R}, \mathbb{R})) \\ &\cong C^\infty((\mathbb{R}^{(\mathbb{N})})^{\mathbb{N}} \sqcap \mathbb{R}, \mathbb{R}) \cong C^\infty((\mathbb{R}^{(\mathbb{N})})^{\mathbb{N}}, \mathbb{R}) \\ &\underset{\oplus}{\leq} C^\infty(C^\infty(C^\infty(\mathbb{R}, \mathbb{R}), \mathbb{R}), \mathbb{R}) = E;\end{aligned}$$

for the last $\underset{\oplus}{\leq}$ compare with the proof of (5.3.5). The first space is not B_r-complete [Jarchow, 1981, p. 333] and hence the same is true for E.

It would be of great interest to know whether the spaces $\mathbb{R}^J$ and $\mathbb{R}^{(J)}$ for J with cardinality $2^{\aleph_0}$ are complemented subspaces of some iteration of $C^\infty(\_, \mathbb{R})$, because they fail to have several important topological properties (e.g. both spaces are not webbed [Jarchow, 1981, p. 98] and the second is not Schwartz [Jarchow, 1981, p. 202]).

After having shown that objects which fail to have certain nice properties can be found in every non-trivial cartesian closed subcategory of Con$^\infty$, we shall now prove that stronger completeness properties hold for all objects of certain subcategories.

5.3.9 Proposition. *The full subcategory of Con$^\infty$ having as objects all convenient vector spaces that are complemented subspaces of their biduals is cartesian closed. It is also closed under complemented subspaces and under $C^\infty(X,\_)$ for any smooth space X.*

Proof. We use (5.3.3). Closedness under products follows from the isomorphism $(E \sqcap F)'' \cong E'' \sqcap F''$. It remains to show closedness under $C^\infty(X,\_)$. Suppose the embedding ι_E of E in E'' admits a retraction $\rho: E'' \to E$. We have to show that $\iota_{C^\infty(X,E)}: C^\infty(X, E) \to C^\infty(X, E)''$ admits a retraction. A direct calculation shows that $\rho_* \circ \varphi$ suffices, where $\varphi: C^\infty(X, E)'' \to C^\infty(X, E'')$ is the map associated to $C^\infty(X, E)'' \sqcap X \sqcap E' \to \mathbb{R}$ given by $(\ell, x, y') \mapsto \ell(g \mapsto y'(g(x)))$. □

Remark. An example of a convenient vector space that is not complemented in its bidual is the Banach space c_0, cf. (iv) in (5.1.27).

5.3.10 Proposition. *The full subcategory of Con$^\infty$ whose objects are isomorphic to duals of convenient vector spaces is cartesian closed.*

Proof. Follows immediately from the isomorphisms $E' \sqcap F' \cong (E \sqcap F)'$, $C^\infty(X, E') \cong L(\lambda X, E') \cong (\lambda X \tilde{\otimes} E)'$, where λX denotes the free convenient vector space over the smooth space X. □

If one tries to restrict the convenient vector spaces by a stronger completeness condition the difficulties that show up are usually due to the non-invariance of such conditions under bornologification.

5.4 Reflexivity

For any separated preconvenient vector space E the canonical map $\iota_E: E \to E''$ is a Pre-embedding, cf. (3.9.3). If ι_E is an isomorphism then E is convenient, but the converse fails in general, i.e. not all convenient vector spaces are reflexive. Since we want to compare this natural reflexivity notion with more classical ones of the theory of locally convex spaces we shall use throughout this section the characterization of preconvenient vector spaces by their locally convex topology, i.e. we identify them with the bornological locally convex spaces; cf. (2.4.3) or (2.4.4).

We shall make use of the following duality functors:

5.4.1 Definition.

(i) $(\_)' : \underline{\mathrm{bLCS}}^{\mathrm{op}} \to \underline{\mathrm{bLCS}}$.

This is up to the isomorphism $\underline{\mathrm{bLCS}} \cong \underline{\mathrm{Pre}}$ the duality functor considered in (3.9.1).

(ii) $(\_)^b : \underline{\mathrm{LCS}}^{\mathrm{op}} \to \underline{\mathrm{CBS}}$.

To a locally convex space E one associates the vector space of all continuous linear functionals together with the so-called equicontinuous bornology, for which a basis is obtained by the polars $U^0 := \{\ell \in E^b ; \ell(U) \subseteq [-1, 1]\}$ with U running through a basis of the 0-neighborhoods of E.

(iii) $(\_)^t : \underline{\mathrm{CBS}}^{\mathrm{op}} \to \underline{\mathrm{LCS}}$.

To a convex bornological space E one associates the vector space of all bornological linear functionals together with the topology of uniform convergence on bounded sets, for which a basis for the 0-neighborhoods is obtained by the polars $B^0 := \{\ell \in E^t ; \ell(B) \subseteq [-1, 1]\}$ with B running through a basis of the bornology of E.

(iv) $(\_)^s : \underline{\mathrm{LCS}}^{\mathrm{op}} \to \underline{\mathrm{LCS}}$.

To a locally convex space E one associates the so-called strong dual, i.e. the vector space of all continuous linear functionals with the strong topology, i.e. the topology of uniform convergence on bounded sets.

All four functors are defined as usual on morphisms: to a morphism $m : E_1 \to E_2$ one associates the map $m^* : \ell \mapsto \ell \circ m$ between the duals. The verifications that one thus gets contravariant functors as stated are trivial.

The next lemma will be useful to compare these duality functors.

5.4.2 Lemma. *The following diagram commutes:*

$$\begin{array}{ccc} \underline{\mathrm{CBS}}^{\mathrm{op}} & \xrightarrow{(\_)^t} & \underline{\mathrm{LCS}} \\ \downarrow{\scriptstyle \gamma} & & \downarrow{\scriptstyle \beta} \\ \underline{\mathrm{LCS}}^{\mathrm{op}} & \xrightarrow{(\_)^b} & \underline{\mathrm{CBS}} \end{array}$$

And a subset $B \subseteq \beta(E^t) = (\gamma E)^b$ is bounded iff $B(A) \subseteq \mathbb{R}$ is bounded for all bounded $A \subseteq E$.

Proof. For the commutativity of the diagram we have to show that $\beta(E^t) = (\gamma E)^b$ for every convex bornological space E. Since any linear function $\ell : E \to \mathbb{R}$ is bornological iff $\ell : \gamma E \to \mathbb{R}$ is continuous, cf. (2.1.10), the underlying vector spaces are the same. To finish the proof of the lemma we only have to show that the following three conditions for any subset $B \subseteq E^t$ are equivalent:

(i) $B \subseteq (\gamma E)^b$ is equicontinuous;
(ii) $B(A) \subseteq \mathbb{R}$ is bounded for every bounded $A \subseteq E$;
(iii) $B \subseteq \beta(E^t)$ is bounded, i.e. is absorbed by any 0-neighborhood of E^t.

(i⇒ii) By assumption there exists a 0-neighborhood U of γE with $B \subseteq U^0$. If $A \subseteq E$ is bounded there exists an $N > 0$ with $A \subseteq N \cdot U$, i.e. $(1/N)A \subseteq U$. Hence for $\ell \in B$ and $x \in A$ one has $|\ell(x)| = N \cdot |\ell(x/N)| \leq N$, and this shows that $B(A)$ is absolutely bounded by N.

(ii⇒iii) One has to show that B is absorbed by every 0-neighborhood of some basis of E^t. So let $A \subseteq E$ be bounded and choose an $N > 0$ with $B(A) \subseteq [-N, N]$. Then $B \subseteq N \cdot A^0$, i.e. B gets absorbed by A^0.

(iii⇒i) The set $U := \{x \in E;\, B(x) \subseteq [-1, 1]\}$ is obviously absolutely convex. It is bornivorous and thus a 0-neighborhood of γE, since for any bounded $A \subseteq E$, the 0-neighborhood A^0 of E^t absorbs B, i.e. $B \subseteq N \cdot A^0$ for some $N > 0$, and thus $A \subseteq N \cdot U$. Using $B \subseteq U^0$ property (i) follows. □

5.4.3 Lemma. *For any bornological locally convex space E one has:*

(*i*) $E^s = (\beta E)^t$;
(*ii*) $E^b = \beta(E^s) = \beta(E')$;
(*iii*) $E' = \gamma(E^b) = \gamma\beta(E^s)$;
(*iv*) $E' = E'_\eta$ *(see below for the definition of* E'_η*).*

Proof. We first remark that the underlying vector space is the same for all the duals considered above and that by definition (2.1.12) one has $\mathrm{E} = \gamma\beta E$.

(i) For any locally convex space E, the strong dual E^s is an LCS-subspace of $(\beta E)^t$ according to the definition.

(ii) One has, using lemma (5.4.2) and (i): $E^b = (\gamma\beta E)^b = \beta((\beta E)^t) = \beta E^s$. A subset B of the space $\beta((\beta E)^t)$ is by (5.4.2) bounded iff $B(A) \subseteq \mathbb{R}$ is bounded for all bounded $A \subseteq \beta E$. By (1.2.13) and since E' is by definition (3.6.2) a Pre-subspace of $\ell^\infty(E, \mathbb{R})$ this is equivalent to B bounded in E'. So the convex bornological vector spaces in (ii) are all identical.

(iii) Applying γ to the equation (ii) and using that E' is bornological, i.e. $\gamma\beta E' = E'$, one obtains (iii).

(iv) By definition the polars U^0 of the 0-neighborhoods U of E form a basis of the bornology of $E^b = \beta(E^s)$. Therefore by (2.1.19) $E' = \gamma\beta(E^s)$ is the inductive limit of the normed spaces $(E')_{U^0}$. This means that E' is the η-dual E'_η in the terminology of [Jarchow, 1981, p. 200, 280].

With the duality functors of (5.4.1) one can form several biduals for a given (bornological) locally convex space E such as E'', E^{ss} and E^{bt}. This leads to various reflexivity notions. The first one is the natural one for convenient vector spaces; the second is the usual one for locally convex spaces; for the third see [Hogbe-Nlend, 1977, p. 89]. In each case one has a natural linear map ι_E of E in the bidual defined by $\iota_E(x) := \mathrm{ev}_x$, i.e. $\iota_E(x)(\ell) = \ell(x)$.

5.4.4 Definition

(i) A preconvenient vector space E is called *reflexive* iff the canonical map $\iota_E\colon E \to E''$ is a Pre-isomorphism.

(ii) A locally convex space E is called *LCS-reflexive* iff the canonical map ι_E: $E \to E^{ss}$ is an LCS-isomorphism; cf. [Jarchow, 1981, p. 227].

(iii) A locally convex space E is called *completely reflexive* iff the canonical map ι_E: $E \to E^{bt}$ is an LCS-isomorphism; cf. [Hogbe-Nlend, 1977, p. 89].

(iv) A locally convex space E is called *η-reflexive* iff the canonical map ι_E: $E \to (E'_\eta)'_\eta$ is an LCS-isomorphism; cf. [Jarchow, 1981, p. 280].

The following proposition shows that quite often the surjectivity of the natural map ι_E is sufficient in order to obtain the corresponding reflexivity of E.

5.4.5 Proposition

(i) *A separated preconvenient vector space E is reflexive iff ι_E: $E \to E''$ is surjective.*

(ii) *A separated locally convex space E is completely reflexive iff ι_E: $E \to E^{bt}$ is surjective.*

(iii) *A separated bornological locally convex space is LCS-reflexive iff ι_E: $E \to E^{ss}$ is surjective.*

Proof. (i) holds, since ι_E: $E \to E''$ is always a Pre-embedding by (3.9.3).

(ii) It is enough to show that ι_E: $E \to E^{bt}$ is always an LCS-embedding, i.e. that the topology of E is the trace topology of E^{bt}.

Let first U be a closed absolutely convex 0-neighborhood of E. Then $B := U^0 \subseteq E^b$ is bounded and therefore $U^{00} = B^0$ is a 0-neighborhood in E^{bt}. By the bipolar theorem [Jarchow, 1981, p. 149] $U = \iota_E^{-1}(U^{00})$, hence U is a 0-neighborhood in the trace of the locally convex topology of E^{bt}.

Conversely let $U \subseteq E$ be a 0-neighborhood in the trace topology of E^{bt}, i.e. there exists a bounded $B \subseteq E^b$ with $U \supseteq \iota_E^{-1}(B^0)$. Furthermore, there has to exist a 0-neighborhood V in E with $B \subseteq V^0$. Now $V \subseteq \iota_E^{-1}(V^{00}) \subseteq \iota_E^{-1}(\mathrm{B}^0) \subseteq U$, hence U is a 0-neighborhood in E.

(iii) Again we have to show that ι_E: $E \to E^{ss}$ is an LCS-embedding provided E is bornological. This holds since E bornological implies E quasi-barrelled ([Jarchow, 1981, p. 222]) and E quasi-barrelled is equivalent with $E \to E^{ss}$ being an LCS-embedding [Jarchow, 1981, p. 273]. A direct verification is as follows. By (i) of (5.4.3) E^{ss} is an LCS-subspace of $(\beta(E^s))^t$ and by (ii) of (5.4.3) $(\beta(E^s))^t = E^{bt}$. So the topology on E^{ss} is the initial one induced by the inclusion into E^{bt} and $E \to E^{ss}$ is an embedding since $E \to E^{bt}$ is one, as just proved in (ii).

Now we are able to compare all the considered reflexivity concepts:

5.4.6 Theorem. *For any convenient vector space E the following statements are equivalent:*

(1) *E is reflexive;*
(2) *As locally convex space E is η-reflexive;*
(3) *As locally convex space E is completely reflexive;*

(4) *As locally convex space E is LCS-reflexive and the strong dual E^s is bornological;*
(5) *The Schwartzification (or the nuclearification) of the locally convex space E is a complete locally convex space.*

Proof. Using (5.4.5) we prove the following implications:

(1 ⇔ 2) according to (iv) of (5.4.3).

(1 ⇔ 3) since by (5.4.2) and (iii) of (5.4.3) one has $\gamma\beta(E^{bt})=\gamma((\gamma E^b)^b)=\gamma(E'^b)=E''$.

(4 ⇒ 1) Using twice $F'=\gamma\beta F^s$ of (iii) in (5.4.3), as well as the hypothesis $E^s=\gamma\beta E^s$ one obtains $E'=E^s$ and $E''=\gamma\beta E^{ss}$.

(1 ⇒ 4) We first claim that E^s is bornological. Let ℓ: $\beta E^s=E^b\to\mathbb{R}$ be a bornological linear functional, hence by the supposed reflexivity $\ell=\mathrm{ev}_x$ for some $x\in E$. Thus the topology of E^s is coarser than the Mackey topology of E^s determined by E. The explicit description of the Mackey topology [Jarchow, 1981, p. 155] shows that the converse inequality holds. Hence E^s has the Mackey topology and since the evaluations at points of E are continuous the claim follows from (2.1.22).

By the argument used in (4 ⇒ 1) one obtains $E''=\gamma\beta E^{ss}$ and this yields the surjectivity of $E\to E^{ss}$.

(2 ⇔ 5) For this we refer to [Jarchow, 1981, p. 280] and [Hogbe-Nlend, Moscatelli, 1981, p. 89]. □

5.4.7 Corollary

(i) *A Fréchet space is reflexive iff it is LCS-reflexive.*
(ii) *A convenient vector space E with a countable basis of its bornology is reflexive iff it is LCS-reflexive.*
(iii) *The locally convex topology of every reflexive convenient vector space is complete.*

Proof. For (i) and (ii) see [Jarchow, 1981, p. 280] and for (iii) use that the canonical mapping $E\to E^{bt}$ is an LCS-isomorphism and the fact that F^t is complete for any convex bornological space F. □

Let us next consider the hereditary properties.

5.4.8 Proposition. *A Con-subspace of a reflexive convenient vector space which is as locally convex space a closed LCS-subspace is also reflexive.*

Proof. Let $\iota: E\to F$ be the inclusion of such a Con-subspace of a reflexive convenient vector space. Let $\ell\in E''$. Then $\iota^{**}(\ell)\in F''$, hence by the reflexivity of F there exists a $y\in F$ with $\iota_F(y)=\iota^{**}(\ell)$, i.e. such that for all $y_1\in F'$ one has $y_1(y)=\iota_F(y)(y_1)=(\ell\circ\iota^*)(y_1)=\ell(\iota^*(y_1))=\ell(y_1\circ\iota)$. Let us show that $y\in E$. Assume $y\notin E$, then by the Hahn–Banach theorem (E is closed in F) there exists a $y_1\in F'$ with $y_1(y)=1$ and $y_1(E)=0$. Thus $1=y_1(y)=\ell(y_1\circ\iota)=\ell(0)=0$, which is a contradiction. It remains to show that $\iota_E(y)=\ell$. So let $x_1 \in E'$. By the Hahn–Banach theorem (E is a $\underline{\text{LCS}}$-subspace of F) there exists a $y_1 \in F'$ with $y_1 \circ \iota = x_1$ and thus $\iota_E(y)(x_1)=x_1(y)=(y_1\circ\iota)(y)=y_1(y)=\ell(y_1\circ\iota)=\ell(x_1)$. □

5.4.9 Corollary. *A convenient vector space E is reflexive iff E′ is reflexive and the locally convex topology of E is complete.*

Proof. ($\Rightarrow$) The completeness of E was shown in (iii) of (5.4.7) and the reflexivity of E' follows immediately by applying $(\_)'$ to $E \cong E''$.

($\Leftarrow$) Since E' is reflexive so is E''. Furthermore, $E \to E''$ is an $\underline{\text{LCS}}$-embedding (3.9.3) and E is as complete locally convex space closed. Hence E is reflexive by (5.4.8). □

5.4.10 Corollary. *A complemented subspace of a reflexive convenient vector space is reflexive.*

Proof. Since the embedding has a bornological and hence continuous inverse, it is a closed embedding with respect to the locally convex topologies. Now apply (5.4.8). □

Remark. Not every $\underline{\text{Con}}$-quotient of a reflexive convenient vector space is reflexive, as the example of a non-reflexive quotient of a Fréchet Montel space shows, see [Jarchow, 1981, p. 233].

Reflexivity is inherited by products and coproducts provided the index set is of non-measurable cardinality:

5.4.11 Proposition. *Let $E_j \neq \{0\}$ $(j \in J)$ be a family of convenient vector spaces with index set J of non-measurable cardinality.*

(i) *The coproduct $\coprod_{j\in J} E_j$ is reflexive iff all E_j are reflexive.*
(ii) *The product $\prod_{j\in J} E_j$ is reflexive iff all E_j are reflexive.*

Proof. One has only to use the isomorphisms of (3.9.4) and (3.9.5). □

We will now prove that the reflexivity of E implies the reflexivity of $C^\infty(M,E)$ for any finite-dimensional smooth manifold M. We first analyze the reflexivity of spaces of type $C^\infty(X,E)$.

5.4.12 Proposition. *Let $X \neq \varnothing$ be a smooth space and E a convenient vector space. Then $C^\infty(X,E)$ is reflexive iff E is reflexive and the linear subspace of $C^\infty(X,E)'$ generated by $\{\ell \circ \mathrm{ev}_x;\ x\in X,\ \ell \in E'\}$ is dense in the locally convex topology.*

Proof. Reflexivity of $C^\infty(X,E)$ implies reflexivity of E by (5.4.10) since E is isomorphic to the complemented subspace formed by the constant functions. So it remains to prove that for a reflexive E the reflexivity of $C^\infty(X,E)$ is equivalent to the stated density condition. We use the factorization of the isomorphism $(\iota_E)_*$ expressed by the commutative diagram

$$\begin{array}{ccc} C^\infty(X,E) & \xrightarrow{\iota} & C^\infty(X,E)'' \\ & {\searrow}^{(\iota_E)_*} \quad {\swarrow}^{\varphi} & \\ & C^\infty(X,E'') & \end{array}$$

Here φ denotes the linear morphism used in (5.3.9) and defined by $\varphi(\ell)(x)(y_1) := \ell(y_1 \circ \mathrm{ev}_x)$ for $\ell \in C^\infty(X,E)''$, $x \in X$ and $y_1 \in E'$. Since $(\iota_E)_*$ is bijective, $\iota = \iota_{C^\infty(X,E)}$: $C^\infty(X,E) \to C^\infty(X,E)''$ is surjective iff φ is injective. The surjectivity of $\iota_{C^\infty(X,E)}$ is equivalent to the reflexivity of $C^\infty(X,E)$ and injectivity of φ means $\{0\} = \ker(\varphi) = \{\ell; \ell(y_1 \circ \mathrm{ev}_x) = 0\}$, which is by the Hahn–Banach theorem equivalent to the given density condition. □

5.4.13 Theorem. *Let $X \neq \emptyset$ be a finite-dimensional separable smooth manifold. Then $C^\infty(X,E)$ is reflexive iff E is reflexive.*

Proof. This follows immediately from (5.4.12), since according to (5.1.7) the density condition is satisfied. □

Remark. Since (5.1.7) is true for finite-dimensional paracompact smooth manifolds having only non-measurably many components, the theorem (5.4.13) is also true for such manifolds.

Now we consider $C^\infty(X,\mathbb{R})$ for general smooth spaces X.

5.4.14 Proposition. *Let X be a smooth space. Then $C^\infty(X,\mathbb{R})$ is reflexive iff $X \to C^\infty(X,\mathbb{R})'$ is universal for smooth maps in convenient vector spaces whose locally convex topology is complete; i.e. for every such smooth map $f: X \to E$ there exists a unique linear morphism from $C^\infty(X,\mathbb{R})'$ into E which composed with the above map $X \to C^\infty(X,\mathbb{R})'$ is f.*

Proof. ($\Leftarrow$) The universal property for $E = \mathbb{R}$ gives a bijection $C^\infty(X,\mathbb{R}) \to C^\infty(X,\mathbb{R})''$ which has to be equal to $\iota_{C^\infty(X,\mathbb{R})}$.

($\Rightarrow$) Let a smooth $f: X \to E$ be given. Define $\tilde{f}$: $C^\infty(X,\mathbb{R})' \to E''$ by $\tilde{f}(\ell)(y_1) := \ell(y_1 \circ f)$. By (5.4.12) the subspace generated by $\{\mathrm{ev}_x; x \in X\}$ is dense in $C^\infty(X,\mathbb{R})$. Hence the image of $\tilde{f}$ is contained in $\overline{\iota_E(E)}$; and since E is complete and ι_E is an embedding with respect to the locally convex topologies, $\tilde{f}$ factors as $\tilde{f} = \iota_E \circ \bar{f}$ and $\bar{f}$ is the desired morphism into E. Its uniqueness is trivial. □

5.4.15 Theorem. *Let U be an* M*-open subset of the dual of a Fréchet Schwartz space E, cf.* (4.4.39), *and F a Fréchet Montel space. Then $C^\infty(U,F)$ is a Fréchet Montel space.*

Proof. Let $C^\infty_{co}(U,F)$ denote the space of smooth functions from U to F, with the topology of uniform convergence on b-compact subsets contained in $U \sqcap E^k$ of each derivative. For a compact space K let us denote by $C(K,F)$ the space of continuous functions with the topology of uniform convergence. The Arzela–Ascoli theorem [Engelking, 1968, p. 333] states that a subset $B \subseteq C(K,F)$ is relatively compact iff B is equicontinuous and pointwise relatively compact.

We claim that $C^\infty_{co}(U,F)$ is semi-Montel, i.e. every bounded subset is relatively compact. In order to see this, we use that the topology is the initial one induced by the mappings $(\iota_K)^* \circ d^k$: $C^\infty_{co}(U,F) \to C^\infty(U \sqcap E^k, F) \to C(K,F)$, where $k \in \mathbb{N}_0$ and $K \subseteq U \sqcap E^k$ is b-compact and ι_K denotes the inclusion of K into $U \sqcap E^k$. So we have to show that for a bounded $B \subseteq C^\infty_{co}(U,F)$ the image under these maps in $C(K,F)$ is relatively compact. Since d^k: $C^\infty_{co}(U,F) \to C^\infty_{co}(U \sqcap E^k, F)$ is bornological it is enough to show this for the map $(\iota_K)^*$ with $k=0$. Since ev_x: $C^\infty_{co}(U,F) \to F$ is bornological, the set $B|_K := (\iota_K)^*(B)$ is pointwise bounded in F and since F is Montel it is relatively compact. So by the Ascoli–Arzela theorem it remains to show that $B|_K$ is equicontinuous. Since $\{f'$: $U \to L(E,F); f \in B\}$ is bounded, the image by f' of the b-compact set $\{x+t(y-x); x, y \in K, t \in [0,1]\}$ is bounded and hence also $\{\int_0^1 f'(x+t(y-x))dt; x, y \in K, f \in B\}$ is bounded in $L(E,F)$ and thus equicontinuous in $L(E,F)$. Since $f(y)-f(x)=\int_0^1 f'(x+t(y-x))(y-x)dt$, *cf.* (4.1.14), this means that for every 0-neighborhood U of F there exists a 0-neighborhood V in E with $f(y)-f(x) \in V$ for $x, y \in K$ and $x-y \in U$. So we have shown that $C^\infty_{co}(U,F)$ is semi-Montel.

Using (4.4.41) and (4.4.39) the locally convex space associated to $C^\infty(U,F)$ is the Fréchet space $C^\infty_{co}(U,F)$. It is Montel since by definition this means bornological and semi-Montel. □

5.4.16 Corollary. *Let E be a Fréchet Schwartz space, U an* M*-open subset of E', F a Fréchet Montel space. Then $C^\infty(U,F)$ is a reflexive convenient vector space.*

Proof. We only have to use that every Montel space is a $\underline{\text{LCS}}$-reflexive, cf. [Jarchow, 1981, p. 230], and that for Fréchet spaces $\underline{\text{LCS}}$-reflexivity is equivalent to reflexivity by (i) of (5.4.7). □

Remark. It has been shown by [Colombeau, Meise, 1981] that $C^\infty_{co}(U,F)$ is Schwartz, provided $U \subseteq E_1$ is open, where E_1 is a Schwartz convex bornological space and F is a Schwartz locally convex space; hence, in particular, when E and F are Fréchet Schwartz spaces and $E_1 = E'$. Thus under these assumptions $C^\infty(U,F)$ is even a Fréchet Schwartz space.

In order to investigate further the general case $C^\infty(X,E)$ we will use the following

5.4.17 Lemma. *Let $X \neq \emptyset$ be a smooth space and $E \neq \{0\}$ a convenient vector space. Then $C^\infty(X,E)$ is reflexive iff E, $C^\infty(X,\mathbb{R})$ and $L(C^\infty(X,\mathbb{R})', E)$ are all reflexive.*

Proof. Since E and $C^\infty(X,\mathbb{R})$ are complemented subspaces of $C^\infty(X,E)$ they are reflexive. Then $C^\infty(X,E)\cong C^\infty(X,E'')\cong L(E', C^\infty(X,\mathbb{R}))\cong L(E', C^\infty(X,\mathbb{R})'')\cong L(C^\infty(X,\mathbb{R})', E'')\cong L(C^\infty(X,\mathbb{R})', E)$. □

We will combine this lemma with the following result concerning the reflexivity of a linear function space.

5.4.18 Proposition. *Let E and F be convenient vector spaces. Then $L(E,F)$ is reflexive iff F is reflexive and $E\otimes F'\subseteq L(E,F)'$ is dense in the locally convex topology.*

Proof. The proof is similar to that of (5.4.12). One uses that $(\iota_F)_*: L(E,F)\to L(E,F'')$ can be factorized as $\varphi\circ\iota_{L(E,F)}$, where φ is defined by the same formula as in (5.4.12). □

Now one easily obtains the following general reflexivity criterion for function spaces:

5.4.19 Proposition. *Let $X\neq\emptyset$ be a smooth space and $E\neq\{0\}$ a convenient vector space. Then $C^\infty(X,E)$ is reflexive iff E and $C^\infty(X,\mathbb{R})$ are reflexive and $C^\infty(X,\mathbb{R})\otimes E'\subseteq C^\infty(X,E)'$ is dense in the locally convex topology.*

Proof. By (5.4.17) we know that $C^\infty(X,E)$ is reflexive iff E, $C^\infty(X,\mathbb{R})$ and $L(C^\infty(X,\mathbb{R})', E)$ are reflexive. And $L(C^\infty(X,\mathbb{R})', E)$ is by (5.4.18) reflexive iff $C^\infty(X,\mathbb{R})''\otimes E'$ is dense in $C^\infty(X,E)'$. □

Remark. One difficulty in proving more general statements about $C^\infty(X,F)$ is due to the fact that the (bornological) locally convex topology on this space cannot be explicitly described in general.

Example. The locally convex topology of $C^\infty_{co}(\mathbb{R},\mathbb{R}^{(\mathbb{N})})$ is not bornological.

In order to see this we consider the linear functional $\ell: C^\infty_{co}(\mathbb{R},\mathbb{R}^{(\mathbb{N})})\to\mathbb{R}$ defined by $\ell(f):=\sum_{k\in\mathbb{N}}(\mathrm{pr}_k\circ f)^{(k)}(0)$. For any bounded subset $B\subseteq C^\infty(\mathbb{R},\mathbb{R}^{(\mathbb{N})})$ there exists an $N\in\mathbb{N}$ such that $B\subseteq C^\infty(\mathbb{R},\mathbb{R}^N)$. Hence on such a set B the functional ℓ is a finite sum of derivatives at 0 composed with projections pr_k, and thus ℓ is a morphism. But ℓ cannot be continuous with respect to the topology of $C^\infty_{co}(\mathbb{R},\mathbb{R}^{(\mathbb{N})})$, because otherwise there would exist an $N\in\mathbb{N}$ and a 0-neighborhood $U\subseteq\mathbb{R}^{(\mathbb{N})}$ such that $f^{(k)}(t)\in U$ for $k<N$ and $|t|\leq N$ would imply $|\ell(f)|\leq 1$. This is impossible, since among all functions f satisfying $f^{(k)}(t)\in U$ for $k<N$ and $|t|\leq N$ there are such with only the projection $f_N:=\mathrm{pr}_N\circ f$ unequal to 0 and the Nth derivative of f_N at 0 larger than 1.

6 THE MACKEY CLOSURE TOPOLOGY

In this chapter we investigate several questions related to the Mackey closure topology.

In section 6.1 we study relations with a few other natural topologies which lie between the locally convex and the Mackey closure topology of a convenient vector space, such as for example, the Kelleyfication of the locally convex topology. Conditions are given under which some of these topologies coincide. Particular interest is laid on the case where the locally convex topology is the final one induced by the smooth curves, i.e. the case where all these topologies coincide.

Section 6.2 mainly deals with the question whether the Mackey closure topology is a vector space topology, i.e. such that the vector space operations are continuous with respect to the product topology. It is shown that this in fact fails for large classes of convenient vector spaces, even for all non-trivial strict inductive limits of Banach spaces. We also show that for products of sufficiently many factors $\mathbb{R}$ the Mackey closure topology is not even completely regular.

A subset is closed for the Mackey closure topology iff it is closed under Mackey convergent sequences. However, as shown by examples given in section 6.3, taking the Mackey adherence, i.e. adding to a subset all limits of Mackey convergent sequences lying in the subset, does not always give the Mackey closure of the subset (i.e. the closure with respect to the Mackey closure topology).

For convex functions continuity with respect to the locally convex topology, continuity with respect to the Mackey closure topology, the Lipschitz property ($\mathcal{Lip}^0$) and boundedness on the M-convergent sequences ($\mathcal{Lip}^{-1}$) are all equivalent, as shown in section 6.4. Convex $\mathcal{Lip}^1$- (resp. $\mathcal{Lip}^2$-) functions are characterized by properties of their first (resp. second) derivative.

6.1 Comparison with other topologies

We first specify those refinements of the locally convex topology which will be compared with each other and with the Mackey closure topology.

6.1.1 Definition. Let E be a preconvenient vector space. If the topology under consideration on E is not explicitly mentioned, then all topological notions (such as continuous, convergent, closure, etc.) are meant with respect to the locally convex topology in this section.

(i) We denote with $\tau_k E$ the *Kelleyfication* of the locally convex topology of E, i.e. the vector space E together with the final topology induced by the inclusions of the subsets being compact for the locally convex topology.

(ii) We denote with $\tau_s E$ the vector space E with the final topology induced by the curves being continuous for the locally convex topology, or equivalently the sequences $\mathbb{N}_\infty \to E$ converging in the locally convex topology. The equivalence holds since the infinite polygon, through a converging sequence, can be continuously parametrized by a compact interval.

(iii) We recall that by $\tau_{\mathrm{M}} E$ we denote the vector space E with its M-closure topology, i.e. the final topology induced by the smooth curves.

Using that smooth curves are continuous and that converging sequences $\mathbb{N}_\infty \to E$ have compact images, the following identities are continuous: $\tau_{\mathrm{M}} E \to \tau_s E \to \tau_k E \to E$.

If the locally convex topology of E coincides with the topology of $\tau_{\mathrm{M}} E$, resp. $\tau_s E$, resp. $\tau_k E$ then we call E smoothly generated, resp. sequentially generated, resp. compactly generated.

6.1.2 Example. On $E = \mathbb{R}^J$ all these refinements are different, i.e. $\tau_{\mathrm{M}} E \neq \tau_s E \neq \tau_k E \neq E$, provided the cardinality of the index set J is at least that of the continuum.

Proof. It is enough to show this for J equipotent to the continuum, since $\mathbb{R}^{J_1}$ is a direct summand in $\mathbb{R}^{J_2}$ for $J_1 \subseteq J_2$.

($\tau_{\mathrm{M}} E \neq \tau_s E$) We may take as index set J the set c_0 of all real sequences converging to 0. Define a sequence (x^n) in E by $(x^n)_j := j_n$. Since every $j \in J$ is a 0-sequence we conclude that the x^n converge to 0 in the locally convex topology of the product, hence also in $\tau_s E$. Assume now that the x^n converge towards 0 in $\tau_{\mathrm{M}} E$. Then by (2.3.10) some subsequence converges Mackey to 0. Thus there exists an unbounded sequence of reals λ_n with $\{\lambda^n x_n; n \in \mathbb{N}\}$ bounded. Let j be a 0-sequence with $\{j_n \lambda_n; n \in \mathbb{N}\}$ unbounded (e.g. $(j_n)^{-2} := 1 + \max\{|\lambda_k|;\ k \leq n\}$). Then the jth coordinate $j_n \lambda_n$ of $\lambda_n x^n$ is not bounded with respect to n, contradiction.

($\tau_s E \neq \tau_k E$) Consider in E the subset $A := \{x \in \{0, 1\}^J;\ x_j = 1$ for at most countably many $j \in J\}$. It is clearly closed with respect to the converging sequences, hence closed in $\tau_s E$. But it is not closed in $\tau_k E$ since it is dense in the compact set $\{0, 1\}^J$.

($\tau_k E \neq E$) Consider in E the subsets $A_n := \{x \in E;\ |x_j| < n$ for at most n many $j \in J\}$. Each A_n is closed in E since its complement is the union of the open sets $\{x \in E;\ |x_j| < n$ for all $j \in J_0\}$, where J_0 runs through all subsets of J with $n+1$ elements. We show that the union $A := \bigcup_{n \in \mathbb{N}} A_n$ is closed in $\tau_k E$. So let K be a compact subset of E; then $K \subseteq \Pi\, \mathrm{pr}_j(K)$ and each $\mathrm{pr}_j(K)$ is compact, hence bounded in $\mathbb{R}$. Since the family $\{j \in J;\ \mathrm{pr}_j(K) \subseteq [-n, n]\}$ $(n \in \mathbb{N})$ covers J there has to exist an $N \in \mathbb{N}$ and infinitely many $j \in J$ with $\mathrm{pr}_j(K) \subseteq [-N, N]$. Thus $K \cap A_n = \emptyset$ for all $n > N$. And hence $A \cap K = \bigcup_{n \in \mathbb{N}} A_n \cap K$ is closed. Nevertheless A is not closed in E, since 0 is in $\bar{A}$ but not in A. □

Let us now describe several important situations where at least some of these topologies coincide. For the proof we will need the following

6.1.3 Lemma. [*Averbukh, Smolyanov*, 1968.] *For any separated locally convex space E the following statements are equivalent:*

(1) *The sequential closure of any subset is formed by all limits of sequences in the subset.*
(2) *For any given double sequence $(x_{n,k})$ in E with $x_{n,k}$ convergent to some x_k for $n \to \infty$ and k fixed and x_k convergent to some x, there are strictly increasing sequences $i \mapsto n(i)$ and $i \mapsto k(i)$ with $x_{n(i),k(i)} \to x$ for $i \to \infty$.*

Proof. (1⇒2) Take an $a_0 \in E$ different from $k \cdot (x_{n,k} - x)$ and from $k \cdot (x_k - x)$ for all k and n. Define $A := \{a_{n,k} := x_{n,k} - (1/k) \cdot a_0;\ n, k \in \mathbb{N},\ n \geq k\}$. Then x is in the sequential closure of A, since $x_{n,k} - (1/k) \cdot a_0$ converges to $x_k - (1/k) \cdot a_0$ as $n \to \infty$ and $x_k - (1/k) \cdot a_0$ converges to $x - 0 = x$ as $k \to \infty$. Hence by (1) there has to exist a sequence $i \mapsto (n(i), k(i))$ with $a_{n(i),k(i)}$ convergent to x. By passing to a subsequence we may suppose that $i \mapsto k(i)$ and $i \mapsto n(i) - k(i)$ is monotone increasing. Assume that $i \mapsto k(i)$ is bounded, hence finally constant. Then a subsequence of $x_{n(i),k} - (1/k(i)) \cdot a_0$ is converging to $x_k - (1/k) \cdot a_0 \neq x$ if $i \mapsto n(i)$ is unbounded and to $x_{n,k} - (1/k) \cdot a_0 \neq x$ if $i \mapsto n(i)$ is bounded, which both yield a contradiction.

Thus $i \mapsto k(i)$ can be chosen strictly increasing and thus also $i \mapsto n(i) = (n(i) - k(i)) + k(i)$ is strictly increasing.

(1⇐2) is obvious. □

6.1.4 Theorem. *For any convenient vector space E the following implications hold:*

(*i*) *$\tau_M E = E$ provided the closure of subsets in E is formed by all limits of sequences in the subset; hence in particular if E is metrizable.*
(*ii*) *$\tau_M E = E$ provided E is the dual of a Fréchet Schwartz space; cf.* (4.4.39).
(*iii*) *$\tau_M E = \tau_k E$ provided E is the strict inductive limit of a sequence of Fréchet spaces.*
(*iv*) *$\tau_M E = \tau_s E$ provided E satisfies the* M*-convergence condition, i.e. every sequence converging in the locally convex topology is* M*-convergent.*
(*v*) *$\tau_s E = E$ provided E is the dual of a Fréchet Montel space; cf.* (4.4.38).

Proof. (i) Using the above lemma one obtains that the closure and the sequential closure coincide, hence $\tau_s E = E$. It remains to show that $\tau_s E \to \tau_{\mathrm{M}} E$ is continuous. So suppose a sequence converging to x is given and let (x_n) be an arbitrary subsequence. Then $x_{n,k} := k(x_n - x) \to k \cdot 0 = 0$ for $n \to \infty$, and hence by lemma (6.1.3) there are subsequences $k(i)$, $n(i)$ with $k(i) \cdot (x_{n(i)} - x) \to 0$, i.e. $i \mapsto x_{n(i)}$ is M-convergent to x. Thus the original sequence converges in $\tau_{\mathrm{M}} E$ by (2.3.10). For metrizable spaces see also (2.4.5).

(iii) Let E be the strict inductive limit of the Fréchet spaces E_n. By [Jarchow, 1981, p. 84] every E_n carries the trace topology of E, hence is closed in E and every bounded subset of E is contained in some E_n. Thus every compact subset of E is contained as a compact subset in some E_n. Since E_n is a Fréchet space such a subset is even b-compact and hence compact in $\tau_{\mathrm{M}} E$. Thus the identity $\tau_k E \to \tau_{\mathrm{M}} E$ is continuous.

(iv) is valid, since the M-closure topology is the final one induced by the M-converging sequences.

(v) Let E be the dual of any Fréchet Montel space F. Recall that the locally convex topology on the dual of a reflexive space is the strong topology, cf. (5.4.6).

Fréchet Montel spaces have a reflexive dual by (3) in (4.4.38), hence are reflexive themselves by (5.4.9).

First we show that $\tau_k E = \tau_s E$. Let $K \subseteq E = F'$ be compact for the locally convex topology. Then K is bounded, hence equicontinuous by the linear uniform boundedness principle (3.6.4), and since F is separable by [Jarchow, 1981, p. 231] K is metrizable in the weak topology $\sigma(E, F)$ [Jarchow, 1981, p. 157]. Since K is compact the weak topology and the locally convex topology of E coincide on K, thus the topology on K is the initial one induced by the converging sequences. Hence the identity $\tau_k E \to \tau_s E$ is continuous and therefore $\tau_s E = \tau_k E$. It remains to show $\tau_k E = E$. Since F is reflexive E is the strong dual of F, cf. (5.4.6), and since F is Montel the locally convex topology of the strong dual coincides with the topology of uniform convergence on precompact subsets of F. Since F is metrizable this topology coincides with the so-called equicontinuous weak*-topology, cf. [Jarchow, 1981, p. 182], which is the final topology induced by the inclusions of the equicontinuous subsets. These subsets are by the Alaoglu-Bourbaki theorem [Jarchow, 1981, p. 157] relatively compact in the topology of uniform convergence on precompact subsets. Thus the locally convex topology of E is compactly generated.

(ii) By (v) and since Fréchet Schwartz spaces are Montel (cf. [Jarchow, 1981, p. 202] or [Horváth, 1966, p. 277]) we have $\tau_s E = E$ and it remains to show that $\tau_{\mathrm{M}} E = \tau_s E$. So let (x_n) be a sequence converging to 0 in E. Then $\{x_n; n \in \mathbb{N}\}$ is relatively compact and by (4.4.39) this set is relatively compact in some Banach space E_{B}. Hence at least a subsequence has to be convergent in E_{B}. Clearly its Mackey limit has to be 0. This shows that (x_n) is convergent to 0 in $\tau_{\mathrm{M}} E$ and hence $\tau_{\mathrm{M}} E = \tau_s E$. One can even show that E satisfies the Mackey convergence condition, cf. [Jarchow, 1981, p. 266]. □

We give now a non-metrizable example to which (i) applies.

6.1.5 Example. Let E denote the subspace of $\mathbb{R}^J$ of all sequences with countable support. Then the closure of subsets of E is given by all limits of sequences in the subset but for non-countable J the space E is not metrizable. This was proved in [Balanzat, 1960].

Remark. The conditions (i) and (ii) in (6.1.4) are rather disjoint since every locally convex space that has a countable basis of its von Neumann bornology and for which the sequential adherence of subsets is sequentially closed is normable as the following proposition shows:

6.1.6 Proposition. *Let E be a non-normable locally convex space that has a countable basis of its bornology. Then there exists a subset of E whose sequential adherence is not sequentially closed.*

Proof. Let $\{B_k; k\in\mathbb{N}_0\}$ be an increasing basis of the von Neumann bornology with $B_0=\{0\}$. Since E is non-normable we may assume that B_k does not absorb B_{k+1} for all k. Now choose $b_{n,k}\in(1/n)\,B_{k+1}$ with $b_{n,k}\notin B_k$. We consider the subset $A:=\{b_{k,0}-b_{n,k}; n, k\geq 1\}$. For fixed k the sequence $b_{n,k}$ converges by construction (in $E_{B_{k+1}}$) to 0 for $n\to\infty$. Thus $b_{k,0}-0$ is the limit of the sequence $b_{k,0}-b_{n,k}$ for $n\to\infty$ and $b_{k,0}$ converges to 0 for $k\to\infty$. Consequently 0 is contained in the sequential closure of A. It remains to show that 0 is not contained in the sequential adherence of A. Suppose $b_{k(i),0}-b_{n(i),k(i)}$, converges to 0. Thus it has to be bounded and so there must be an $N\in\mathbb{N}$ with $B_1-\{b_{k(i),0}-b_{n(i),k(i)};\ i\in\mathbb{N}\}\subseteq B_N$. Hence $b_{n(i),k(i)}=b_{k(i),0}-(b_{k(i),0}-b_{n(i),k(i)})$ $\in B_N$, i.e. $k(i)<N$. By passing to a subsequence we may assume that $k(i)=k$ for some k and all i and thus $b_{k,0}-b_{n(i),k}\to 0$, i.e. $b_{n(i),k}\to b_{k,0}$. Admit $n(i)$ is unbounded. Then a subsequence of $b_{n(i),k}$ converges to $0\neq b_{k,0}$, which is a contradiction. So we may assume that $n(i)=n$ for some n and all i. But then $b_{n,k}\notin B_1$ and hence cannot be equal to $b_{k,0}\in B_1$. □

6.2 Continuity of the addition and regularity

In this section we describe classes of spaces where $\tau_M E\neq E$ or where $\tau_M E$ is not even a topological vector space. Finally we give an example where the Mackey-closure topology is not completely regular.

We begin with the relationship between the Mackey closure topology and the locally convex topology on preconvenient vector spaces.

6.2.1 Lemma. [*Kriegl*, 1982] *Let E be a preconvenient vector space, $U\subseteq E$ an absolutely convex subset. Then U is a 0-neighborhood for the locally convex topology of E iff U is a 0-neighborhood for the Mackey closure topology.*

Proof. ($\Rightarrow$) Since the Mackey convergent sequences are convergent in the locally convex topology it follows that the Mackey closure topology is finer than the locally convex topology, cf. (6.1.1).

($\Leftarrow$) Let U be an absolutely convex 0-neighborhood with respect to the Mackey closure topology. By (2.1.12) it is enough to show that U is bornivorous, i.e. absorbs bounded subsets. Assume that some bounded B does not get absorbed by U. Then for every $n \in \mathbb{N}$ there exists a $b_n \in B$ with $b_n \notin nU$. Since $(1/n)b_n$ is Mackey convergent to 0, we conclude that $(1/n)b_n \in U$ for sufficiently large n. This yields a contradiction. □

6.2.2 Corollary. *For any preconvenient vector space the finest locally convex topology that is coarser than the Mackey closure topology is the (bornological) locally convex topology of that preconvenient vector space. And a convex subset is open in the locally convex topology iff it is open in the* M*-closure topology.*

6.2.3 Proposition. *Let E and F be convenient vector spaces. If there exists a bilinear morphism $m: E \sqcap F \to \mathbb{R}$ that is not continuous with respect to the locally convex topologies, then $\tau_M(E \sqcap F)$ is not a topological vector space.*

Proof. Suppose that addition $\tau_M(E \sqcap F) \times \tau_M(E \sqcap F) \to \tau_M(E \sqcap F)$ is continuous with respect to the product topology. Using the continuous inclusions $\tau_M E \to \tau_M(E \sqcap F)$ and $\tau_M F \to \tau_M(E \sqcap F)$ we can write m as composite of continuous maps as follows:

$$\tau_M E \times \tau_M F \to \tau_M(E \sqcap F) \times \tau_M(E \sqcap F) \xrightarrow{+} \tau_M(E \sqcap F) \xrightarrow{m} \mathbb{R}.$$

Thus for every $\varepsilon > 0$ there are 0-neighborhoods U and V with respect to the Mackey closure topology such that $m(U \times V) \subseteq]-\varepsilon, \varepsilon[$. Then also $m^{-1}(\langle U \rangle \times \langle V \rangle) \subseteq]-\varepsilon, \varepsilon[$, where $\langle \_ \rangle$ denotes the absolutely convex hull. By (6.2.1) one concludes that m is continuous with respect to the locally convex topology; contradiction. □

6.2.4 Corollary. *Let E be a convenient vector space whose locally convex topology is not normable. Then $\tau_M(E \sqcap E')$ is not a topological vector space.*

Proof. By (6.2.3) it is enough to show that $\mathrm{ev}: E \sqcap E' \to \mathbb{R}$ is not continuous; this was done in (3.3.4). □

In order to get a large variety of spaces where the Mackey closure topology is not a topological vector space topology the next three technical lemmas will be useful.

6.2.5 Lemma. *Let E be a preconvenient vector space. Suppose a double sequence $b_{n,k}$ in E exists which satisfies the following two conditions:*

- (b') *For every sequence $k \mapsto n(k)$ the sequence $k \mapsto b_{n(k),k}$ has no accumulation point in $\tau_M E$.*
- (b'') *For all k the sequence $n \mapsto b_{n,k}$ converges to 0 in $\tau_M E$.*

Suppose furthermore that a double sequence $c_{n,k}$ in E exists that satisfies the following two conditions:

(c′) *For every 0-neighborhood U in $\tau_M E$ there exists a k_0 such that $c_{n,k} \in U$ for all $k \geq k_0$ and all n.*

(c″) *For all k the sequence $n \mapsto c_{n,k}$ has no accumulation point in $\tau_M E$.*

Then $\tau_M E$ is not a topological vector space.

Proof. Here convergence is meant always with respect to $\tau_M E$. We may without loss of generality assume that $c_{n,k} \neq 0$ for all n, k, since by (c″) we may delete all those $c_{n,k}$ which are equal to 0. Then we consider $A := \{b_{n,k} + \varepsilon_{n,k} c_{n,k}; n, k \in \mathbb{N}\}$, where the $\varepsilon_{n,k} \in \{-1, 1\}$ are chosen in such a way that $0 \notin A$.

We first show that A is closed in $\tau_M E$: Let $b_{n(i),k(i)} \pm c_{n(i),k(i)} \to x$ and assume that $(k(i))$ is unbounded. By passing if necessary to a subsequence we may even assume that $i \mapsto k(i)$ is strictly increasing. Then $c_{n(i),k(i)} \to 0$ by (c′), hence by (2.3.12) $b_{n(i),k(i)} \to x$ which is a contradiction to (b′). Thus $(k(i))$ is bounded and we may assume constant. Now suppose that $(n(i))$ is unbounded. Then $b_{n(i),k} \to 0$ by (b″) and hence $\varepsilon_{n(i),k} c_{n(i),k} \to x$ and for a subsequence where ε is constant one has $c_{n(i),k} \to \pm x$, which is a contradiction to (c″). Thus $n(i)$ is bounded as well and we may assume constant. Hence $x = b_{n,k} + \varepsilon_{n,k} c_{n,k} \in A$.

Assume now that the addition $\tau_M E \times \tau_M E \to \tau_M E$ is continuous. Then there has to exist an open and symmetric 0-neighborhood U in $\tau_M E$ with $U + U \subseteq E \setminus A$. For K sufficiently large and n arbitrary one has $c_{n,K} \in U$ by (c′). For such a fixed K and N sufficiently large $b_{N,K} \in U$ by (b′). Thus $b_{N,K} \pm c_{N,K} \notin A$, which is a contradiction. □

Let us now show that many spaces have a double sequence $c_{n,k}$ as in the above lemma.

6.2.6 Lemma. *Let E be an infinite-dimensional convenient vector space whose locally convex topology is metrizable. Then a double sequence $c_{n,k}$ subject to the conditions* (c′) *and* (c″) *of* (6.2.5) *exists.*

Proof. If E is normable we choose a sequence c_n on the unit ball without accumulation point and define $c_{n,k} := (1/k) c_n$.

If E is not normable we take a countable increasing family of non-equivalent seminorms p_k generating the locally convex topology, and we chose $c_{n,k}$ with $p_k(c_{n,k}) = 1/k$ and $p_{k+1}(c_{n,k}) > n$. □

Next we show that many spaces have a double sequence $b_{n,k}$ as in lemma (6.2.5).

6.2.7 Lemma. *Let E be a non-normable convenient vector space having a countable basis of its bornology. Then a double sequence $b_{n,k}$ subject to the conditions* (b′) *and* (b″) *of* (6.2.5) *exists.*

Proof. Let B_n $(n \in \mathbb{N})$ be absolutely convex sets forming an increasing basis of the bornology. Since E is not normable the sets B_n can be chosen such that B_n does not absorb B_{n+1}. Now choose $b_{n,k} \in (1/n) B_{k+1}$ with $b_{n,k} \notin B_k$. □

Using these lemmas one obtains the

6.2.8 Proposition. *For the following convenient vector spaces the Mackey closure topology is not a vector space topology:*

(i) *Every convenient vector space that contains as* M*-closed subspaces an infinite-dimensional Fréchet space and a non-normable space with countable basis of its bornology.*
(ii) *Every strict inductive limit of a strictly increasing sequence of infinite-dimensional Fréchet spaces.*
(iii) *Every product for which at least $2^{\aleph_0}$ many factors are non-zero.*
(iv) *Every coproduct for which at least $2^{\aleph_0}$ many summands are non-zero.*

Proof. (i) follows directly from the last three lemmas.

(ii) Let E be the strict inductive limit of the spaces E_n $(n \in \mathbb{N})$. Then E contains the infinite-dimensional Fréchet space E_1 as subspace. The subspace generated by points $x_n \in E_{n+1} \setminus E_n$ $(n \in \mathbb{N})$ is isomorphic to $\mathbb{R}^{(\mathbb{N})}$, hence its bornology has a countable basis. Thus by (i) we are done.

(iii) Such a product E contains the Fréchet space $\mathbb{R}^{\mathbb{N}}$ as complemented subspace. We want to show that $\mathbb{R}^{(\mathbb{N})}$ is also a subspace of E. For this we may assume that the index set J is $\mathbb{R}^{\mathbb{N}}$ and all factors are equal to $\mathbb{R}$. Now consider the linear subspace E_1 of the product that is generated by the sequence $x^n \in E = \mathbb{R}^{\mathbb{N}}$, where $(x^n)_j := j(n)$ for every $j \in J = \mathbb{R}^{\mathbb{N}}$. The linear map $\mathbb{R}^{(\mathbb{N})} \to E_1 \subseteq E$ that maps the nth unit vector to x^n is injective, since for a given finite linear combination $\sum t_n x^n = 0$ the jth coordinate for $j(n) := \operatorname{sign}(t_n)$ equals $\sum |t_n|$. It is a morphism since $\mathbb{R}^{(\mathbb{N})}$ carries the finest structure. So it remains to show that it is a Pre-embedding. We have to show that any bounded $B \subseteq E_1$ is contained in a subspace generated by finitely many x^n. Otherwise there would exist a strictly increasing sequence (n_k) and $b^k = \sum_{n \leqslant n_k} t_n^k x^n \in B$ with $t_{n_k}^k \neq 0$. Define an index j recursively by $j(n) := n |t_n^k|^{-1} \cdot \operatorname{sign}(\sum_{m < n} t_m^k j(m))$ if $n = n_k$ and $j(n) := 0$ if $n \neq n_k$ for all k. Then the absolute value of the jth coordinate of b^k evaluates as follows:

$$|(b^k)_j| = |\textstyle\sum_{n \leq n_k} t_n^k j(n)| = |\sum_{n < n_k} t_n^k j(n) + t_{n_k}^k j(n_k)| =$$
$$|\textstyle\sum_{n < n_k} t_n^k j(n)| + |t_{n_k}^k j(n_k)| \geq |t_{n_k}^k j(n_k)| \geq n_k.$$

Hence the jth coordinates of $\{b^k; k \in \mathbb{N}\}$ are unbounded with respect to $k \in \mathbb{N}$ and thus B is unbounded.

(iv) We cannot apply lemma (6.2.5) since every double sequence has countable support and hence is contained in the dual $\mathbb{R}^{(A)}$ of a Fréchet Schwartz space $\mathbb{R}^A$ for some countable subset $A \subseteq J$. It is enough to show (iv) for $\mathbb{R}^{(J)}$, where $J = \mathbb{N} \cup c_0$. Let $A := \{ j_n(e_n + e_j); n \in \mathbb{N}, j \in c_0, j_n \neq 0 \text{ for all } n\}$, where e_n and e_j denote the unit vectors in the corresponding summand. The set A is M-closed, since its intersection with finite subsums is finite. Admit there exists a symmetric M-open 0-neighborhood U with $U+U\subseteq E\setminus A$. Then for every n there exists a $j_n\neq 0$ with $j_n e_n\in U$ and we may assume that $n\mapsto j_n$ converges to 0 and hence defines an element $j\in c_0$. Furthermore, there has to be an $N\in\mathbb{N}$ with $j_N e_j\in U$, thus $j_N(e_N+e_j)\in(U+U)\cap A$, in contradiction to $U+U\subseteq E\setminus A$. □

Remark. A nice and simple example where one either uses (i) or (ii) is $\mathbb{R}^{\mathbb{N}}\oplus\mathbb{R}^{(\mathbb{N})}$. The locally convex topology on both factors coincides with their Mackey-closure topology (the first being a Fréchet (Schwartz) space, cf. (i) of (6.1.4), the second as dual of the first, cf. (ii) of (6.1.4); but the Mackey closure topology on their product is not even a vector space topology.

Although the Mackey closure topology on a convenient vector space is always functionally separated, hence Hausdorff, it is not always completely regular as the following example shows.

6.2.9 Theorem. *The Mackey closure topology of* $\mathbb{R}^J$ is not completely regular if the cardinality of J is at least $2^{\aleph_0}$.

Proof. It is enough to show this for an index set J of cardinality $2^{\aleph_0}$, since the corresponding product is a complemented subspace in every product with larger index set. We prove the theorem by showing that every function $f\colon\mathbb{R}^J\to\mathbb{R}$ which is continuous for the Mackey-closure topology is also continuous with respect to the locally convex topology. Hence the completely regular topology associated to the Mackey-closure topology is the locally convex topology of E. That these two topologies are different was shown in (6.1.2). We use the following theorem of [Mazur, 1952]: Let $E_0:=\{x\in\mathbb{R}^J;\ \operatorname{supp}(x)\text{ is countable}\}$ and $f\colon E_0\to\mathbb{R}$ be sequentially continuous. Then there is some countable subset $A\subseteq J$ such that $f(x)=f(x_A)$, where in this proof x_A is defined as $x_A(j):=x(j)$ for $j\in A$ and $x_A(j)=0$ for $j\notin A$. Every sequence which is converging in the locally convex topology of E_0 is contained in a metrizable complemented subspace $\mathbb{R}^A$ for some countable A and therefore is even M-convergent. Thus this theorem of Mazur remains true if f is assumed to be continuous for the M-closure topology. This generalization follows also from the fact that $\tau_M E_0=E_0$, cf. (6.1.5). Now let $f\colon\mathbb{R}^J\to\mathbb{R}$ be continuous for the Mackey closure topology. Then $f|_{E_0}\colon E_0\to\mathbb{R}$ is continuous for the Mackey closure topology and hence there exists a countable set $A_0\subseteq J$ such that $f(x)=f(x_{A_0})$ for any $x\in E_0$. We want to show that the same is true for arbitrary $x\in\mathbb{R}^J$. In order to show this we consider for $x\in\mathbb{R}^J$ the map $\varphi_x\colon 2^J\to\mathbb{R}$ defined by $\varphi_x(A):=f(x_A)-f(x_{A\cap A_0})$ for any $A\subseteq J$, i.e. $A\in 2^J$. For countable A one has $x_A\in E_0$, hence $\varphi_x(A)=0$. Furthermore, φ_x is sequentially continuous where one considers on 2^J the product topology of the discrete factors 2. In order to see this consider a converging sequence of subsets $A_n\to A$, i.e. for every $j\in J$ one has for the characteristic functions $\chi_{A_n}(j)=\chi_A(j)$ for n sufficiently large. Then $\{n(x_{A_n}-x_A);n\in\mathbb{N}\}$ is bounded in $\mathbb{R}^J$ since for fixed $j\in J$ the jth coordinate equals 0 for n sufficiently large. Thus x_{A_n} converges Mackey to x_A and since f is continuous for the Mackey closure topology $\varphi_x(A_n)\to\varphi_x(A)$.

Now we can apply another theorem of [Mazur, 1952]: Any function $f\colon 2^J\to\mathbb{R}$

that is sequentially continuous and is zero on all countable subsets of J is identically 0 provided the cardinality of J is smaller than the first inaccessible cardinal. Thus we conclude that $0=\varphi_x(J)=f(x)-f(x_{A_0})$ for all $x\in\mathbb{R}^J$. Hence f factors over the metrizable space $\mathbb{R}^{A_0}$ and is therefore continuous for the locally convex topology. □

6.3 The Mackey closure of subsets

In (5.1.18) we proved that the completion of the subspace $\ell_c^\infty(X)$ of $\ell^\infty(X,\mathbb{R})$ formed by the functions with finite support is $c_0(X)$. This was done by showing that the two-fold Mackey adherence of $\ell_c^\infty(X)$ in $\ell^\infty(X,\mathbb{R})$ gives $c_0(X)$. Now we give an example where the Mackey adherence of $\ell_c^\infty(X)$ is not all of $c_0(X)$.

6.3.1 Example. The Mackey adherence does not yield the Mackey-closure (quoted in (2.2.23) and in (5.1.18)).

Consider $X:=\mathbb{N}\times\mathbb{N}$ with all the sets $\{(n,k);\, n\le\mu(k)\}$ (for $\mu\colon\mathbb{N}\to\mathbb{N}$) as basis of a bornology.

First we show that this bornology is an ℓ^∞-structure: Consider $f^p\colon X\to\mathbb{N}\subseteq\mathbb{R}$ defined by $f^p(n,p):=n$ and $f^p(n,k)=0$ for $k\ne p$. The function f^p is bornological since $\{n;\, n\le\mu(p)\}$ is finite. Let now $B\subseteq X$ be such that $f^p(B)$ is bounded for all p. Define $\mu(k):=\max f^k(B)$. Then $B\subseteq\{(n,k);\, n\le\mu(k)\}$, since $(n,k)\in B$ implies that $n=f^k(n,k)\le\max f^k(B)=\mu(k)$.

We show next that the function $f\colon X\to\mathbb{R}$ defined by $f(n,k):=1/k$ belongs to c_0X. Since any bounded $B\subseteq X$ is contained in $\{(n,k);\, n\le\mu(k)\}$ for some $\mu\colon\mathbb{N}\to\mathbb{N}$ we obtain

$$\left\{(n,k)\in B;\, f(n,k)\ge\frac{1}{p}\right\}\subseteq\left\{(n,k);\, n\le\mu(k),\ \frac{1}{k}\ge\frac{1}{p}\right\}=\{(n,k);\, n\le\mu(k),\ p\ge k\}$$
$$\subseteq\{(n,k);\, n\le\max\{\mu(j);\, j\le p\},\ k\le p\},$$

and this set is finite.

Now assume that f is the M-limit of some sequence (f_j) of functions with finite support, i.e. $\{\lambda_j(f-f_j);\, j\in\mathbb{N}\}$ is bounded for some sequence (λ_j) converging to ∞. Choose j_k with $\lambda_{j_k}\ge k^2$ and $j_k<j_{k+1}$, and define $\mu(k):=1+\max\{n;\, f_{j_k}(n,k)\ne 0\}$. Since $B:=\{(n,k);\, n\le\mu(k)\}$ is bounded the image $\{\lambda_j(f-f_j)(n,k);\, (n,k)\in B, j,n,k\in\mathbb{N}\}$ has to be bounded. But

$$\{\lambda_j(f-f_j)(n,k);\, n\le\mu(k)\}\supseteq\{\lambda_j(f-f_j)(n,k);\, n\le\mu(k), f_j(n,k)=0\}=$$
$$\left\{\lambda_j\cdot\frac{1}{k};\, n\le\mu(k), f_j(n,k)=0\right\}\supseteq\left\{\lambda_{j_k}\cdot\frac{1}{k};\, n=\mu(k),\, f_{j_k}(n,k)=0\right\}=$$
$$\left\{\lambda_{j_k}\cdot\frac{1}{k};\, f_{j_k}(\mu(k),k)=0\right\}=\left\{\lambda_{j_k}\cdot\frac{1}{k};\, k\in\mathbb{N}\right\}.$$

Since this set is unbounded one has a contradiction.

This example can be used to give another important

6.3.2 Example. An M-dense embedding into a convenient vector space need not be the completion (quoted in (2.6.5)).

Let F be a preconvenient vector space whose M-adherence in its completion E is not all of E, cf. (6.3.1). Choose a $y \in E$ that is not contained in the M-adherence of F and let F_1 be the subspace of E generated by $F \cup \{y\}$. We claim that $F_1 \subseteq E$ cannot be the completion although F_1 is Mackey dense in the convenient vector space E. In order to see this we consider the linear map $\ell : F_1 \to \mathbb{R}$ characterized by $\ell(F) = 0$ and $\ell(y) = 1$. Clearly ℓ is well defined.

$\ell : F_1 \to \mathbb{R}$ is bornological: For any bounded $B \subseteq F_1$ there exists an N such that $B \subseteq F + [-N, N]\, y$. Otherwise $b_n = x_n + t_n y \in B$ exist with $t_n \to \infty$. This would imply that $b_n = t_n((x_n/t_n) + y)$ and thus $-(x_n/t_n)$ would converge Mackey to y; contradiction.

Now assume that a bornological extension $\bar{\ell}$ to E exists. Then $F \subseteq \ker(\bar{\ell})$ and $\ker(\bar{\ell})$ is M-closed, which is a contradiction to the Mackey denseness of F in E. So $F_1 \subseteq E$ does not have the universal property of a completion.

Another consequence is that the trace topology on F_1 inherited from the locally convex topology of E cannot be bornological, since by the Hahn–Banach theorem continuous linear functionals can be extended.

Furthermore, the extension of the inclusion $\iota : F \oplus \mathbb{R} \cong F_1 \to E$ to the completion is given by $(x, t) \in E \oplus \mathbb{R} \cong \tilde{F} \oplus \mathbb{R} \cong \tilde{F}_1 \mapsto x + ty \in E$ and has as kernel the linear subspace generated by $(y, -1)$. Hence the extension of a Pre-embedding to the completions need not be an embedding anymore, in particular the inclusion functor does not preserve injectivity of morphisms.

By the closed embedding lemma (2.6.4) the trace of the Mackey closure topology on any M-closed subspace is the Mackey closure topology of this subspace. Now we give an example which shows that M-closedness of the subspace is essential for this result.

6.3.3 Example. Trace of Mackey closure topology is not Mackey closure topology (quoted in (2.6.4)).

Consider $E = \mathbb{R}^{\mathbb{N}} \sqcap \mathbb{R}^{(\mathbb{N})}$,

$$A := \left\{ a_{nk} := \left(\frac{1}{n} \chi_{\{1, \ldots, k\}}, \frac{1}{k} \chi_{\{n\}} \right);\ n, k \in \mathbb{N} \right\} \subseteq E.$$

Let F be the linear subspace of E generated by A. We show that the closure of A with respect to the M-closure topology of F is strictly smaller than that with respect to the trace topology of the M-closure topology of E.

A is closed in the M-closure topology of F: Assume that a sequence (a_{n_j, k_j}) is M-converging to (x, y). Then the second component of a_{n_j, k_j} has to be bounded. Thus $j \mapsto n_j$ has to be bounded and may be assumed to have constant value n_∞. If $j \mapsto k_j$ were unbounded, then

$$(x, y) = \left(\frac{1}{n_\infty} \chi_{\mathbb{N}}, 0 \right),$$

which is not an element of F. Thus $j \mapsto k_j$ has to be bounded too and may be assumed to have constant value k_∞. Thus $(x, y) = a_{n_\infty, k_\infty} \in A$.

A is not closed in the trace topology since $(0,0)$ is contained in the closure of A with respect to the M-closure topology of E: For $k \to \infty$ and fixed n the sequence $a_{n,k}$ is M-converging to $((1/n)\chi_{\mathbb{N}}, 0)$ and $(1/n)\chi_{\mathbb{N}}$ is M-converging to 0 for $n \to \infty$.

6.3.4 Let Ω be the first uncountable ordinal, E a preconvenient vector space and $A \subseteq E$. For every ordinal α one can define the αth Mackey adherence M-adh$^\alpha(A)$ of a set A by M-adh$^\alpha(A) := \bigcup_{\beta<\alpha}$M-adh$^\beta(A)$ if α is a limit ordinal and M-adh$^\alpha(A)$ is the Mackey adherence of M-adh$^\beta(A)$ if α is the successor ordinal of β, i.e. $\alpha = \beta + 1$. It can be shown [Kriegl, unpublished] that there exists a preconvenient vector space E for which the αth Mackey adherence in its completion is not the completion for all $\alpha < \Omega$. The Ωth Mackey adherence always coincides with the Mackey closure, hence it has to be the completion.

6.4 Convex functions

We consider first the purely algebraic question of characterizing convex functions.

6.4.1 Proposition. *Let E be a vector space, $f: E \to \mathbb{R}$ a map, and $n \geq 2$. Then the following statements are equivalent:*

(1) *f is convex, i.e. the function $t \mapsto f(x+tv)$ is convex for all $x, v \in E$;*
(2) *$f(\Sigma r_i x_i) \leq \Sigma r_i f(x_i)$ for all reals $r_1, \ldots, r_n$ satisfying $r_i \geq 0$ and $\Sigma r_i = 1$;*
(3) *The set $U_f := \{(x,t) \in E \times \mathbb{R}; f(x) < t\}$ is convex;*
(4) *The set $A_f := \{(x,t) \in E \times \mathbb{R}; f(x) \leq t\}$ is convex.*

Proof. $(1 \Leftrightarrow 2)$ is trivial.

$(2 \Rightarrow 3)$ Let $(x_i, t_i) \in U_f$, i.e. $f(x_i) < t_i$, and $r_i \geq 0$ with $\Sigma r_i = 1$. Let finally $(x,t) := \Sigma r_i(x_i, t_i)$. Then $f(x) = f(\Sigma r_i x_i) \leq \Sigma r_i f(x_i) < \Sigma r_i t_i = t$, i.e. $(x,t) \in U_f$. Thus U_f is convex.

$(3 \Rightarrow 4)$ Let $(x_i, t_i) \in A_f$; $r_i \geq 0$ and $\Sigma r_i = 1$; $\varepsilon > 0$ be arbitrary. Then $(x_i, t_i + \varepsilon) \in U_f$. Hence by assumption $(0, \varepsilon) + \Sigma r_i(x_i, t_i) = \Sigma r_i(x_i, t_i + \varepsilon) \in U_f \subseteq A_f$. Since ε was arbitrary we conclude that $\Sigma r_i(x_i, t_i) \in A_f$, i.e. A_f is convex.

$(4 \Rightarrow 2)$ Let $x_i \in E$ and $r_i \geq 0$ with $\Sigma r_i = 1$. Then $(x_i, f(x_i)) \in A_f$, hence by assumption $(\Sigma r_i x_i, \Sigma r_i f(x_i)) \in A_f$, i.e. $f(\Sigma r_i x_i) \leq \Sigma r_i f(x_i)$. □

Next we give a characterization of continuity of convex functions which will be applied using the Mackey-closure topology or the locally convex topology.

6.4.2 Proposition. *Let E be a vector space with a topology such that for all $x_0 \in E$ the map $x \mapsto x_0 - x$ is continuous from E to E, and $f: E \to \mathbb{R}$ be a convex function. Then the following statements are equivalent:*

(1) *$f: E \to \mathbb{R}$ is continuous;*
(2) *The set $U_f := \{(x,t) \in E \times \mathbb{R}; f(x) < t\}$ is open in $E \times \mathbb{R}$;*
(3) *The set $\{x \in E; f(x) < t\}$ is open in E for all $t \in \mathbb{R}$.*

Proof. (1 ⇒ 2) Since $f: E \to \mathbb{R}$ is continuous, so is $f \times \mathrm{id}: E \times \mathbb{R} \to \mathbb{R} \times \mathbb{R}$. Thus $U_f := (f \times \mathrm{id})^{-1}(U_0)$ is open as inverse image of the open set $U_0 := \{(t, s); t < s\}$.

(2 ⇒ 3) is obvious, since $\{x; f(x) < t\} := \mathrm{incl}_t^{-1} U_f$, where incl_t denotes the continuous map $\mathrm{incl}_t: E \to E \times \mathbb{R}$ defined by $x \mapsto (x, t)$.

(3 ⇒ 1) One has to show that $f^{-1}(U)$ is open for all U in a basis of the topology on $\mathbb{R}$. Such a basis is given by all open intervals $U :=]s, t[$. For such U one has $f^{-1}(U) = \{x; f(x) < t) \cap \{x; f(x) > s\}$. The first set is open by assumption and the second set $V := \{x; f(x) > s\}$ is open, since it is equal to $\bigcup_{x \in V}(2x - \{y \in E; f(y) < 2f(x) - s\})$. Using $x = 2x - x$ shows that V is contained in this union. Conversely let z be an element of the union, i.e. $z = 2x - y$ with $f(x) > s$ and $f(y) < 2f(x) - s$. By convexity of f the equation $x = (y+z)/2$ yields $2f(x) \le f(y) + f(z)$; hence $f(z) \ge 2f(x) - f(y) = 2f(x) - s + s - f(y) > f(y) + s - f(y) = s$, thus $z \in V$. □

Now the announced characterization:

6.4.3 Theorem. *Let E be a convenient vector space, $f: E \to \mathbb{R}$ a convex function. Then the following statements are equivalent:*

(1) *f is $\mathscr{Lip}^0$;*
(2) *f is continuous for the locally convex topology;*
(3) *f is continuous for the Mackey closure topology;*
(4) *f is continuous for the Mackey convergence;*
(5) *f is $\mathscr{Lip}^{-1}$, i.e. is bounded on M-convergent sequences.*

Proof. (1 ⇒ 3) Is obvious, since the Mackey closure topology is the final one induced by the smooth (or $\mathscr{Lip}^0$) curves.

(3 ⇔ 2) This follows from (6.4.2), since for convex sets (like $\{x; f(x) < r\}$) it is equivalent to be open in the Mackey closure topology or open in the locally convex topology by (6.2.2).

(3 ⇔ 4) is true for any function $f: E \to \mathbb{R}$. In fact, the Mackey closure topology is the topologification τ of the Mackey convergence structure, and since $\tau\mathbb{R} = \mathbb{R}$ the equivalence follows from the adjunction (2.2.6).

(3 ⇒ 5) Is obvious, since the Mackey converging sequences are relatively compact in the Mackey-closure topology (they converge in this topology).

(5 ⇒ 1) Let us first consider this statement for $E = \mathbb{R}$. For $k \in \mathbb{N}$ one has

$$\max\left\{\left|\frac{f(t)-f(s)}{t-s}\right|; |t|, |s| \le k, t \ne s\right\} \le \frac{2}{k} \cdot \max\{|f(r)|; |r| \le 2k\},$$

since the convexity of f implies

$$\frac{f(2k)-f(k)}{2k-k} \ge \frac{f(t)-f(s)}{t-s} \ge \frac{f(-2k)-f(-k)}{-2k-(-k)}$$

and thus

$$\left|\frac{f(t)-f(s)}{t-s}\right| \le \frac{1}{k} \cdot \max\{|f(2k)-f(k)|, |f(-2k)-f(-k)|\} \le \frac{2}{k} \cdot \max\{|f(r)|; |r| \le 2k\}.$$

Now the general case. For any smooth $c:\mathbb{R}\to E$ we have to show that $f\circ c$ is locally Lipschitz. Let again $k\in\mathbb{N}$. By the 1-dimensional situation applied to the map

$$f_{s,t}:\tau\mapsto f\left(c(s)+\tau\frac{c(t)-c(s)}{t-s}\right)$$

we have

$$\left|\frac{f(c(t))-f(c(s))}{t-s}\right|=\left|\frac{f_{s,t}(t-s)-f_{s,t}(0)}{t-s}\right|\leq\frac{2}{k}\max\{|f_{s,t}(r)|;|r|\leq 2k\}$$
$$=\frac{2}{k}\max\left\{\left|f\left(c(s)+r\frac{c(t)-c(s)}{t-s}\right)\right|;|r|\leq 2k\right\}\quad\text{for all }|t-s|\leq k.$$

The map

$$(t,s,r)\mapsto c(s)+r\frac{c(t)-c(s)}{t-s}$$

has a smooth extension to $\mathbb{R}^3$, cf. (4.1.13). The image of $[-2k,2k]^3$ under this extension is bornologically compact, i.e. compact in some Banach space $E_{\mathbf{B}}$. Thus every sequence in this image has a Mackey converging subsequence and consequently f is bounded on this image. This shows that the difference quotient

$$\left|\frac{f(c(s))-f(c(t))}{t-s}\right|$$

is bounded for $|t|,|s|\leq k/2$ and $t\neq s$. Since k was arbitrary the proof that $f\circ c$ is Lipschitz is complete. □

We now consider the special case of seminorms.

6.4.4 Corollary. *For any seminorm p on a convenient vector space E the following statements are equivalent*:

(1) *p is* $\mathscr{Lip}^0$;
(2) *p is continuous for the locally convex topology*;
(3) *p is continuous for the Mackey closure topology*;
(4) *p is continuous for the Mackey convergence*;
(5) *p is* $\mathscr{Lip}^{-1}$;
(6) *p is bornological.*

Proof. $(1\Leftrightarrow 2\Leftrightarrow 3\Leftrightarrow 4\Leftrightarrow 5)$ follows from the theorem (6.4.3) since any seminorm is convex.

$(6\Rightarrow 5)$ is trivial.

$(5\Rightarrow 6)$ We show that this holds for any positively homogeneous function p. Assume (5) is true but not (6). Then there is a bounded set B such that $p(B)$ is unbounded. Choose $b_n\in B$ with $|p(b_n)|\geq n^2$. Then $\{p(b_n/n);n\in\mathbb{N}\}$ is unbounded but (b_n/n) is obviously a Mackey 0-sequence. □

In the following proposition we characterize those maps which appear as derivative of a convex $\mathcal{Lip}^1$-function. For normed spaces this can be found in [Shigeta, 1986].

To formulate the result we need the

6.4.5 Definition. Let $f\colon E \supseteq U \to E'$ be given.

f is called *monotonic* iff $(f(x)-f(y))(x-y) \geq 0$ for all $x, y \in U$.

f is called *cyclically monotonic* iff for any $x_0, x_1, \ldots, x_n = x_0 \in U$ one has $\sum_{j=1}^n f(x_j)(x_j - x_{j-1}) \geq 0$.

A linear map $f\colon E \to E'$ is called *positive* iff it is monotonic.

Obviously every cyclically monotonic function is monotonic; and a linear map $f\colon E \to E'$ is positive iff $f(x)(x) \geq 0$ for all $x \in E$.

6.4.6 Proposition. *Let $U \subseteq E$ be a convex* M*-open subset of a convenient vector space E and $g\colon U \to E'$ a $\mathcal{Lip}^0$-map. Then the following statements are equivalent:*

(1) *g is cyclically monotonic;*
(2) *A convex $\mathcal{Lip}^1$-function $f\colon U \to \mathbb{R}$ exists with $g = f'$.*

Proof. $(2 \Rightarrow 1)$ Let $x \in U$ and $v \in E$ with $x + v \in U$. Using convexity of f we obtain for $t > 0$:

$$\frac{f(x+tv)-f(x)}{t} = \frac{f((1-t)x+t(x+v))-f(x)}{t} \leq \frac{(1-t)f(x)+tf(x+v)-f(x)}{t}$$

$$= f(x+v) - f(x).$$

Thus $f'(x)(v) \leq f(x+v) - f(x)$. Replacing v by $-v$ we obtain $g(x)(v) = f'(x)(v) \geq f(x) - f(x-v)$. Thus for any finite cyclic sequence $x_0, x_1, \ldots, x_n = x_0 \in U$ one has $\sum_{j=1}^n g(x_j)(x_j - x_{j-1}) \geq \sum_{j=1}^n (f(x_j) - f(x_{j-1})) = 0$, i.e. g is cyclically monotonic.

$(1 \Rightarrow 2)$ We may assume that $0 \in U$. We will show that f defined by $f(x) := \int_0^1 g(tx)(x)\,dt$ for $x \in U$ is the desired convex function.

By lemma (4.3.29) one has that f is $\mathcal{Lip}^1$ and $f' = g$ provided $\int_0^1 g(t(x+v))(x+v)\,dt - \int_0^1 g(tx)(x)\,dt = \int_0^1 g(x+tv)(v)\,dt$ for all $x \in U$ and $x + v \in U$. In order to verify this equation we partition each side of the triangle formed by $0, x$ and $x + v$ into n equidistant parts. Applying cyclic monotonicity of g to the $3n$ points

$$0 = x_0, \ldots, x_i := \frac{i}{n}x, \ldots, x_n = x = x + v_0, \ldots, x + v_i := x + \frac{i}{n}v, \ldots, x + v_n$$

$$= x_n + v_n, \ldots, x_i + v_i, \ldots, x_0 + v_0 = 0$$

one obtains

$$\sum_{j=1}^{n} g(v_j)\left(\frac{v}{n}\right) + \sum_{j=1}^{n} g(x+v_j)\left(\frac{v}{n}\right) \geq \sum_{j=0}^{n-1} g(x_j+v_j)\left(\frac{x+v}{n}\right)$$

and for the reverse ordering

$$\sum_{j=0}^{n-1} g(v_j)\left(\frac{v}{n}\right)+\sum_{j=0}^{n-1} g(x+v_j)\left(\frac{v}{n}\right)\leq \sum_{j=1}^{n} g(x_j+v_j)\left(\frac{x+v}{n}\right).$$

Since these Riemann-sums converge (Mackey) to the corresponding integrals by (4.1.4) the equation is proved.

Remains to show that f is convex, i.e. that the $\mathscr{Lip}^1$-function $t\mapsto f(x+tv)$ is convex for all $x\in U$, $v\in E$. By the classic calculus for real functions this is equivalent with $t\mapsto f'(x+tv)(v)$ being monotonic, i.e. $g(x+tv)(v)\geq g(x+sv)(v)$ for all $t\geq s$. This follows from the cyclic monotonicity applied to $x+tv$ and $x+sv$. □

Convex functions which are even $\mathscr{Lip}^2$ can be further characterized by the

6.4.7 Proposition. *Let $U\subseteq E$ be a convex* M*-open subset of a convenient vector space E and $f\colon U\to\mathbb{R}$ a $\mathscr{Lip}^2$-map. Then the following statements are equivalent*:

(1) *f is convex;*
(2) *f' is cyclically monotonic;*
(3) *$f''(x)$ is cyclically monotonic for all $x\in U$;*
(4) *$f''(x)$ is positive for all $x\in U$.*

Proof. ($1\Leftrightarrow 2$) by (6.4.6).

($2\Rightarrow 3$) Let $x_0, x_1, \ldots, x_n=x_0\in U$ be arbitrary. For t sufficiently small $x+tx_i\in U$ and by the cyclic monotonicity of $g:=f'$ we have

$$\sum_{j=1}^{n} (g(x+tx_j)-g(x))(t(x_j-x_{j-1}))=\sum_{j=1}^{n} g(x+tx_j)(t(x_j-x_{j-1}))\geq 0.$$

Dividing by $t^2>0$ we obtain

$$\sum_{j=1}^{n}\frac{g(x+tx_j)-g(x)}{t}(x_j-x_{j-1})\geq 0$$

and hence $\sum_{j=1}^{n} g'(x)(x_j)(x_j-x_{j-1})\geq 0$, i.e. $f''(x)=g'(x)$ is cyclically monotonic.

($3\Rightarrow 4$) since $f''(x)\colon E\to E'$ is linear.

($4\Rightarrow 1$) f is convex iff the $\mathscr{Lip}^2$-map $c\colon t\mapsto f(x+tv)$ is convex for all $x\in U$, $v\in E$. By the classical calculus for real functions this is equivalent with $f''(x+tv)(v)(v)=c''(0)\geq 0$, i.e. with $f''(y)$ being positive for all $y\in U$. □

7 PERMANENCE PROPERTIES AND COUNTER-EXAMPLES

In this chapter we study the permanence properties of most of the treated important functors with respect to limits, colimits, initial and final morphisms. Further results on final morphisms will be given as well as some additional counter-examples.

In section 7.1 we consider the subspace F formed by the infinitely flat functions of the (nuclear) Fréchet space $E := C^\infty(\mathbb{R}, \mathbb{R})$. We construct a smooth function on F that has no smooth extension to E and a smooth curve $\mathbb{R} \to F'$ that has not even locally a smooth lifting along $E' \to F'$. These results are based on Borel's theorem which tells us that $\mathbb{R}^{\mathbb{N}}$ is isomorphic to the quotient E/F and the fact that this quotient map $E \to \mathbb{R}^{\mathbb{N}}$ has no continuous right inverse. Also a result of [Seeley, 1964] is used which says that, in contrast to F, the subspace $\{f \in C^\infty(\mathbb{R}, \mathbb{R});\ f(t) = 0 \text{ for } t \leq 0\}$ of E is complemented. Besides other consequences we show that $\underline{C^\infty}$ is not locally cartesian closed.

In section 7.2 we determine which among the functors like L, ℓ^∞, $\mathscr{Lip}^k$, $\tilde{\otimes}$, $\bar{\omega}$, λ and η preserve limits or colimits and we give counter-examples for those preservation properties that do not hold.

Final morphisms in $\underline{\text{Pre}}$, $\underline{\text{Con}}$ and $\underline{\mathcal{M}}$ for any $\mathcal{M}$ are characterized in section 7.3. Those of the functors that preserve final or initial morphisms are determined and counter-examples are given for the permanence properties that are not valid.

In section 7.4 we collect various counter-examples whose existence were mentioned throughout the other chapters. We will furthermore show that several naturally constructed function spaces and tensor products are not isomorphic in general, although one has morphisms which are isomorphisms in case of finite-dimensional spaces.

7.1 Extension and lifting properties

If $q: E \to F$ is a quotient map of convenient vector spaces one might expect that for every smooth curve $c: \mathbb{R} \to F$ there exists (at least locally) a smooth lifting, i.e. a smooth curve $\bar{c}: \mathbb{R} \to E$ with $q \circ \bar{c} = c$. And if $\iota: F \to E$ is an embedding of a convenient subspace one might expect that for every smooth function $f: F \to \mathbb{R}$ there exists a smooth extension to E. In this section we give examples showing that both properties fail. As convenient vector spaces we choose spaces of smooth real functions and their duals. We start with some lemmas.

7.1.1 Lemma. *Let $\iota: F \to E$ be the inclusion of a $\underline{\text{Con}}$-subspace, and suppose that the locally convex topology of F is the trace topology of that of E. Then $\iota^*: E' \to F'$ is a surjective final $\underline{\text{CBS}}$-morphism and hence a $\underline{\text{Con}}$-quotient map.*

Proof. Any $\ell \in F'$ is continuous with respect to the locally convex topology of E, hence by the Hahn–Banach theorem [Jarchow, 1981, p. 127] has a continuous linear extension $\bar{\ell}: E \to \mathbb{R}$. This shows that ι^* is surjective. We next show that the bornology of F' is the final convex vector bornology induced by ι^*: $E' \to F'$, i.e. equals $\mathscr{B} := \{\iota^*(B);\ B \subseteq E' \text{ bounded}\}$, cf. (3.1.1). Since ι^* is a $\underline{\text{Con}}$-morphism, $A \in \mathscr{B}$ implies $A \subseteq F'$ bounded. Conversely, let now $A \subseteq F'$ be bounded. By (ii) in (5.4.3) there exists a 0-neighborhood V in the locally convex topology of F with $A \subseteq V^0$, the polar of V. By the hypothesis on the locally convex topologies, V contains $W \cap F$ for some absolutely convex 0-neighborhood W in the locally convex topology of E. Since the polar W^0 of W is bounded in E' and $A \subseteq V^0 \subseteq (W \cap F)^0$, the assertion $A \in \mathscr{B}$ follows if we show that $(W \cap F)^0 \subseteq \iota^*(W^0)$. So let $\ell \in (W \cap F)^0$, i.e. $\ell: F \to \mathbb{R}$ linear and $|\ell(x)| \leq 1$ for all $x \in W \cap F$. Consider the Minkowski seminorm $p: E \to \mathbb{R}$ associated to W, i.e. $p(y) := \inf\{t > 0;\ y \in tW\}$. Then one has $|\ell(x)| \leq p(x)$ for all $x \in F$. By the Hahn–Banach theorem [Jarchow, 1981, p. 126] or [Horváth, 1966, p. 176] there exists a linear extension $\bar{\ell}: E \to \mathbb{R}$ with $|\bar{\ell}(y)| \leq p(y)$ for all $y \in Y$. Then $\bar{\ell} \in W^0$, since $y \in W \Rightarrow p(y) \leq 1 \Rightarrow |\bar{\ell}(y)| \leq 1$, and thus $\ell = \iota^*(\bar{\ell}) \in \iota^*(W^0)$.

So we have proved that $\iota^*: E' \to F'$ is final for the bornologies. According to (3.2.4) the lemma follows. □

7.1.2 Proposition. (*Borel's Theorem.*) *Let $E := C^\infty(\mathbb{R}, \mathbb{R})$ and $F := \{f \in E;\ f$ infinitely flat at zero$\}$, $\iota: F \to E$ the inclusion, and $q: E \to \mathbb{R}^{\mathbb{N}_0}$ defined by $c \mapsto (c^{(n)}(0))_n$. Then q is a $\underline{\text{Con}}$-quotient-map with kernel F and $\iota^*: E' \to F'$ is also a $\underline{\text{Con}}$-quotient-map.*

Proof. By (4.2.9) q is a $\underline{\text{Con}}$-morphism. The kernel F of q is a $\underline{\text{Con}}$-subspace of E and ι^* is a quotient by (7.1.1); in fact, the condition on the locally convex topology of F is satisfied since that of E and hence its trace on F are metrizable, thus bornological, cf. (iv) of (2.1.20). It remains to show that q is a $\underline{\text{Con}}$-quotient. So let $B \subseteq \mathbb{R}^{\mathbb{N}_0}$ be bounded. We use an $h \in C^\infty(\mathbb{R}, \mathbb{R})$ with compact support and

$h(t)=t$ for t in a 0-neighborhood of $\mathbb{R}$. Set

$$\mu_n := 2\cdot\sup\left\{1+\left|\frac{x_n}{n!}(h^n)^{(j)}(t)\right|; x\in B, j\leq n, t\in\mathbb{R}\right\}<\infty.$$

We show that for every $x\in B$ the function $f_x\colon \mathbb{R}\to\mathbb{R}$ defined by

$$f_x(t):=\sum_{n=0}^{\infty}\frac{x_n}{n!}\left(\frac{h(t\mu_n)}{\mu_n}\right)^n$$

is smooth. The summands are smooth and for each t only finitely many summands are non-zero. The jth derivative at t of the nth summand is $(x_n/n!)\,(\mu_n)^{j-n}\cdot(h^n)^{(j)}(t\mu_n)$ with absolute value less than $\frac{1}{2}\,\mu_n^{j-n+1}$ which is less than 2^{j-n} for $n>j$. Thus all derivatives of the series converge uniformly and hence f_x is smooth with $|f_x^{(j)}|\leq\frac{1}{2}\sum_{n=0}^{j}\mu_n^{j-n+1}+1$. Using that $h(t)=t$ for small t one obtains that $f_x^{(j)}(0)=x_j$, i.e. $q(f_x)=x$. Furthermore $A:=\{f_x; x\in B\}\subseteq C^\infty(\mathbb{R},\mathbb{R})$ is bounded with $q(A)=B$. □

7.1.3 Corollary *Let $\iota\colon F\to E$ and $q\colon E\to\mathbb{R}^{\mathbb{N}}$ be as in the previous proposition (7.1.2). For every Con-morphism $\mathfrak{s}\colon\mathbb{R}^{\mathbb{N}}\to E$ the composite $q\circ\mathfrak{s}$ factors over $\mathrm{pr}_N\colon\mathbb{R}^{\mathbb{N}}\to\mathbb{R}^N$ for some $N\in\mathbb{N}$, and there exists no Con-morphism $\rho\colon E\to F$ with $\rho\circ\iota=\mathrm{id}_F$.*

Proof. Let $\mathfrak{s}$: $\mathbb{R}^{\mathbb{N}}\to E$ be an arbitrary Con-morphism. The set $U:=\{g\in E; |g(t)|\leq 1 \text{ for } |t|\leq 1\}$ is a 0-neighborhood in the locally convex topology of E. So there has to exist an $N\in\mathbb{N}$ such that $\mathfrak{s}(V)\subseteq U$ with $V:=\{x\in\mathbb{R}^{\mathbb{N}}; |x_n|<(1/N) \text{ for all } n\leq N\}$. We show that $q\circ\mathfrak{s}$ factors over $\mathbb{R}^N$. So let $x\in\mathbb{R}^{\mathbb{N}}$ with $x_n=0$ for all $n\leq N$. Then $k\cdot x\in V$ for all $k\in\mathbb{N}$, hence $k\cdot\mathfrak{s}(x)\in U$, i.e. $|\mathfrak{s}(x)(t)|\leq 1/k$ for all $|t|\leq 1$ and $k\in\mathbb{N}$. Hence $\mathfrak{s}(x)(t)=0$ for $|t|\leq 1$ and therefore $q(\mathfrak{s}(x))=0$.

Suppose now that there exists a Con-morphism $\rho\colon E\to F$ with $\rho\circ\iota=\mathrm{id}_F$. Define $\mathfrak{s}(q(x)):=x-\iota\rho x$. This definition makes sense, since q is surjective and $q(x)=q(x')$ implies $x-x'\in F$ and thus $x-x'=\rho(x-x')$. Moreover $\mathfrak{s}$ is a Con-morphism, since q is a final Con-morphism; and $(q\circ\mathfrak{s})(q(x))=q(x)-q(\iota(\rho(x)))$ $=q(x)-0$. □

7.1.4 Proposition. *The subspace $\{f\in C^\infty(\mathbb{R},\mathbb{R}); f(t)=0 \text{ for } t\leq 0\}$ of $C^\infty(\mathbb{R},\mathbb{R})$ is a direct summand.*

Proof. [Seeley, 1964] We claim that the following map is a Con-morphism being left inverse to the inclusion: $\mathfrak{s}(g)(t):=g(t)-\sum_{k\in\mathbb{N}}a_k h(-t2^k)g(-t2^k)$ for $t>0$ and $\mathfrak{s}(g)(t)=0$ for $t\leq 0$. Where $h\colon\mathbb{R}\to\mathbb{R}$ is a smooth function with compact support satisfying $h(t)=1$ for $t\in[-1,1]$ and (a_k) is a solution of the infinite system of linear equations $\sum_{k\in\mathbb{N}}a_k(-2^k)^n=1$ $(n\in\mathbb{N})$ (the series is assumed to converge absolutely). The existence of such a solution is shown in [Seeley, 1964] by taking the limit of solutions of the finite subsystems. Let us first show that $\mathfrak{s}(g)$ is smooth. For $t>0$ the series is locally around t finite, since $-t2^k$ lies outside the support of h for k sufficiently large. Its derivative $(\mathfrak{s}g)^{(n)}(t)$ is $g^{(n)}(t) -\sum_{k\in\mathbb{N}} a_k(-2^k)^n\sum_{j=0}^n h^{(j)}(-t2^k)g^{(n-j)}(-t2^k)$ and this converges for $t\to 0$ towards $g^{(n)}(0)-\sum_{k\in\mathbb{N}} a_k(-2^k)^n g^{(n)}(0)=0$. Thus $\mathfrak{s}(g)$ is infinitely flat at 0 and is smooth on $\mathbb{R}$. It remains to show that $g\mapsto\mathfrak{s}(g)$ is a $\underline{\text{Con}}$-morphism. Since the $\underline{\text{Con}}$-structure of $C^\infty(\mathbb{R},\mathbb{R})$ is the initial one induced by the evaluations ev_t by (4.2.11), it is enough to show that $g\mapsto(\mathfrak{s}g)(t)$ is a morphism. For $t\leq 0$ this map is 0 and hence a morphism. For $t>0$ it is a finite linear combination of evaluations and thus a morphism. □

7.1.5 Proposition. *Let $E=C^\infty(\mathbb{R},\mathbb{R})$ and $F:=\{f\in E; f$ is infinitely flat at $0\}$, $\iota^*: E'\to F'$ the $\underline{\text{Con}}$-quotient map of (7.1.2). The curve $c:\mathbb{R}\to F'$ defined by $c(t):=\mathrm{ev}_t$ for $t\geq 0$ and $c(t)=0$ for $t<0$ is smooth but on no neighborhood of 0 there exists a smooth lifting.*

Proof. Let us first verify that c is smooth. Since by (3.6.5) the smooth structure of F' is the initial one induced by the evaluation maps, it is enough to show that $\mathrm{ev}_f\circ c:\mathbb{R}\to\mathbb{R}$ is smooth for all $f\in F$. Since $(\mathrm{ev}_f\circ c)(t)=f(t)$ for $t\geq 0$ and $(\mathrm{ev}_f\circ c)(t)=0$ for $t\leq 0$ this obviously holds.

Assume that there exists a smooth lifting of c, i.e. a smooth $e:\mathbb{R}\to E'$ with $\iota^*\circ e=c$. By exchanging the variables, c corresponds to a morphism $\tilde{c}: F\to E$ and e corresponds to a morphism $\tilde{e}: E\to E$ with $\tilde{e}\circ\iota=\tilde{c}$. The curve c was chosen in such a way that $\tilde{c}(f)(t)=f(t)$ for $t\geq 0$ and $\tilde{c}(f)(t)=0$ for $t\leq 0$.

We show now that such an extension $\tilde{e}$ of $\tilde{c}$ cannot exist. In (7.1.4) we have shown the existence of a retraction $\mathfrak{s}$ to the embedding of the subspace $F_+:=\{f\in F; f(t)=0 \text{ for } t\leq 0\}$ of E. For $f\in F$ one has $\mathfrak{s}(\tilde{e}(f))=\mathfrak{s}(\tilde{c}(f))=\tilde{c}(f)$ since $\tilde{c}(E)\subseteq F_+$. Now let $\Psi: E\to E$, $\Psi(f)(t):=f(-t)$ be the reflection at 0. Then $\Psi(F)\subseteq F$ and $f=\tilde{c}(f)+\Psi(\tilde{c}(\Psi(f)))$ for $f\in F$. We claim that $\rho:=\mathfrak{s}\circ\tilde{e}+\Psi\circ\mathfrak{s}\circ\tilde{e}\circ\Psi: E\to F$ is a retraction to the inclusion, and this is a contradiction with (7.1.3). In fact $\rho(f)=(\mathfrak{s}\circ\tilde{e})(f)+(\Psi\circ\mathfrak{s}\circ\tilde{e}\circ\Psi)(f)=\tilde{c}(f)+\Psi(\tilde{c}(\Psi(f)))=f$ for all $f\in F$. So we have proved that c has no global smooth lifting.

Assume now that $c|_{]-\varepsilon,\varepsilon[}$ has a smooth lifting $e_0:]-\varepsilon,\varepsilon[\to E'$. Trivially $c|_{\mathbb{R}\setminus\{0\}}$ has a smooth lifting e_1 defined by the same formula as c. Take now a smooth partition $\{f_0, f_1\}$ of the unity subordinated to the open covering $\{]-\varepsilon,\varepsilon[, \mathbb{R}\setminus\{0\}\}$ of $\mathbb{R}$, i.e. $f_0+f_1=1$ with $\mathrm{supp}(f_0)\subseteq]-\varepsilon,\varepsilon[$ and $0\notin\mathrm{supp}(f_1)$. Then $f_0e_0+f_1e_1$ gives a global smooth lifting of c, in contradiction with the preceding considerations. □

7.1.6 Corollary. *The category of smooth spaces is not locally cartesian closed.*

Proof. If $\underline{C^\infty}$ were locally cartesian closed, then pullbacks would commute with coequalizers, cf. (8.6.5). This is not the case as the following example shows: Consider $p_1: E_1\to F_1$, where $E_1:=(E')^{\mathbb{R}}$, $F_1:=(F')^{\mathbb{R}}$, $p_1:=\prod_{\mathbb{R}}\iota^*$ with $\iota: F\to E$ the inclusion considered in (7.1.2).

We show that p_1 is a cokernel by verifying that it is surjective and final with respect to $\underline{\text{Born}}\to\underline{\text{Set}}$, cf. (7.3.4). Surjectivity is trivial. For the finality consider a bounded $B \subseteq E_1$, i.e. $B \subseteq \Pi_{t \in \mathbb{R}} B_t$ with $B_t \subseteq F'$ bounded. Using the finality of $\iota^*: E' \to F'$ (cf. (7.1.1)), we obtain a bounded $A_t \subseteq E'$ with $\iota^*(A_t) \supseteq B_t$. The set $A := \Pi_{t \in \mathbb{R}} A_t$ is bounded in E_1 and $p_1(A) \supseteq B$.

The smooth curve $c_1: \mathbb{R} \to F_1$ defined by $\mathrm{pr}_t \circ c_1 : s \mapsto c(t+s)$ has nowhere a local smooth lifting and we now consider in $\underline{C^\infty}$ the pullback $p := (c_1)^*(p_1): \mathbb{R} \sqcap_{F_1} E_1 \to \mathbb{R}$ of p_1 along c_1.

We claim that for every function $f: \mathbb{R} \to \mathbb{R}$ the composite $f \circ p$ is smooth. For this it is enough to show that $p \circ c$ is constant for every smooth curve $c: \mathbb{R} \to \mathrm{dom}(p)$. Assume that $p \circ c$ is not constant. Then at some point $(p \circ c)^\cdot$ is not zero, hence $p \circ c$ is a local diffeomorphism. But then $\mathrm{pr}_2 \circ c \circ (p \circ c)^{-1}$ would be a local smooth lift of c_1. Since we have shown that the final smooth structure on $\mathbb{R}$ induced by p is the discrete one, it follows that p is not final and thus not a coequalizer. □

7.1.7 Proposition. *Let $E := C^\infty(\mathbb{R}, \mathbb{R})$ and $F := \{f \in E; f$ infinitely flat at $0\}$. The function $\varphi: F \to \mathbb{R}$ defined by $\varphi(f) := f(f(1))$ if $f(1) \geq 0$ and $\varphi(f) := 0$ if $f(1) < 0$ is smooth but has no smooth extension to E.*

Proof. We first show that φ is smooth. Using the morphism $\tilde{c}: F \to E$ associated to the smooth curve $c: \mathbb{R} \to F'$ of (7.1.5) we can write φ as the composite $\mathrm{ev} \circ (\tilde{c}, \mathrm{ev}_1)$ of smooth maps.

Assume now that a smooth extension $\psi: E \to \mathbb{R}$ of φ exists. Using a fixed smooth function $h: \mathbb{R} \to [0, 1]$ with $h(t) = 0$ for $t \leq 0$ and $h(t) = 1$ for $t \geq 1$, we then define a map $\sigma: E \to E$ as follows: $(\sigma g)(t) := \psi(g + (t - g(1))h) - (t - g(1))h(t)$. Obviously $\sigma g \in E$ for any $g \in E$, and using cartesian closedness of the category of smooth spaces (cf. (1.4.3)) one easily verifies that σ is a smooth map. For $f \in F$ one has, using that $(f + (t - f(1))h)(1) = t$, the equations $(\sigma f)(t) = (f + (t - f(1))h)(t) - (t - f(1))h(t) = f(t)$ for $t \geq 0$ and $(\sigma f)(t) = 0 - (t - f(1))h(t) = 0$ for $t \leq 0$. This means $\sigma f = \tilde{c} f$ for $f \in F$. So one has $\tilde{c} = \sigma \circ \iota$ with σ smooth. Differentiation gives $\tilde{c} = \tilde{c}'(0) = \sigma'(0) \circ \iota$, and $\sigma'(0)$ is a $\underline{\mathrm{Con}}$-morphism $E \to E$. But in the proof of (7.1.5) it was shown that such an extension of $\tilde{c}$ does not exist. □

Remark. For the smooth map of the previous proposition there does not even exist a smooth extension to a neighborhood of F in the Fréchet space E, since such a local extension could be multiplied with a smooth function $E \to \mathbb{R}$ being 1 on F and having support inside the neighborhood (E has as nuclear Fréchet space smooth partitions of unity [Michor, 1983]) to obtain a global extension.

7.1.8 Example. The structure curves of a $\underline{C^\infty}$-quotient need not be liftable as structure curves and the structure functions on a $\underline{C^\infty}$-subspace need not be extendable as structure functions (quoted after (1.1.4)).

We use the previous construction and note that $\iota: F \to E$ is an initial $\underline{C^\infty}$-morphism, since it is an initial $\underline{\mathrm{Pre}}$-morphism. Thus we only have to show that $q := \iota^*: E' \to F'$ is a final $\underline{C^\infty}$-morphism. So let $f: F' \to \mathbb{R}$ be a function such that $f \circ q$ is smooth. Clearly f is ∞-times weakly differentiable and in order to verify that f is smooth it remains to show that the differentials of f are $\mathcal{Lip}^{-1}$, cf. (4.3.30). This is the case since Mackey-convergent sequences lift which can be seen as follows: Boundedness of $\{t_n(x_n - x); n \in \mathbb{N}\}$ implies that a bounded $B \subseteq F'$ exists with $t_n(x_n - x) \in q(B)$. Choose y with $q(y) = x$ and y_n with $t_n(y_n - y) \in B$ and $q(t_n(y_n - y)) = t_n(x_n - x)$. Then $y_n \to y$ is an M-convergent lifting of $x_n \to x$.

7.2 Preservation of categorical limits

In this section we want to discuss the inheritance properties of the basic functors with respect to limits and colimits.
The functors we consider are:

L:	$\underline{\text{Con}}^{op} \times \underline{\text{Con}} \longrightarrow \underline{\text{Con}}$	cf. (3.6.3)
ℓ^∞:	$(\underline{\ell^\infty})^{op} \times \underline{\text{Con}} \longrightarrow \underline{\text{Con}}$	cf. (3.6.1)
$\mathcal{Lip}^k$:	$(\underline{\mathcal{Lip}^k})^{op} \times \underline{\text{Con}} \longrightarrow \underline{\text{Con}}$	cf. (4.4.3)
$\tilde{\otimes}$:	$\underline{\text{Con}} \times \underline{\text{Con}} \longrightarrow \underline{\text{Con}}$	cf. (3.8.4)
$\tilde{\omega}$:	$\underline{\text{Pre}} \longrightarrow \underline{\text{Con}}$	cf. (2.6.5)
ℓ^1:	$\underline{\ell^\infty} \longrightarrow \underline{\text{Con}}$	cf. (5.1.23)
λ:	$\underline{\mathcal{Lip}^k} \longrightarrow \underline{\text{Con}}$	cf. (5.1.1)

7.2.1 Theorem. *The following functors are left adjoints (hence preserve colimits) and have the stated additional properties*:

(*i*) $\tilde{\omega}$: *Pre*$\to$*Con, it also preserves countable products.*
(*ii*) $\ell^1 : \ell^\infty \to$*Con, it also carries finite products to tensor products in Con.*
(*iii*) λ: $\mathcal{Lip}^k \to$*Con, it also carries finite products to tensor products in Con.*
(*iv*) $(\_)\tilde{\otimes} E$: *Con*$\to$*Con.*

Proof. That these functors have right adjoints was proved in (2.6.5) for (i), in (5.1.1) together with (5.1.23) for (ii), in (5.1.1) for (iii), in (3.8.4) for (iv).

In order to show that $\tilde{\omega}$ preserves countable products we use the linear extension lemma (2.6.6). By (3.3.6) every linear morphism on a countable product depends only on finitely many factors and thus extends to the product of the completions. Since the Mackey closure of a product of finitely many subsets is the product of the Mackey closures of these sets, and since the set of points with only finitely many non-zero coordinates is Mackey dense in the countable product, the countable product of preconvenient vector spaces is Mackey dense in the product of their completions.

That the two functors in (ii) and (iii) carry the product to the tensor product was shown in (5.2.4). □

We give some examples which show that none of the functors mentioned in the above theorem preserves all limits:

7.2.2 Examples (0) A particular important morphism $m: (\mathbb{R}^{\mathbb{N}})^{(\mathbb{N})} \to (\mathbb{R}^{(\mathbb{N})})^{\mathbb{N}}$ is characterized by $\mathrm{pr}_n \circ m \circ \mathrm{in}_k := \mathrm{in}_k \circ \mathrm{pr}_n$. Its image is not closed since it contains the dense subspace formed by the double sequences with finite support and since m is not surjective. It is, however, initial, since the initial Con-embedding $(\mathbb{R}^{\mathbb{N}})^{(\mathbb{N})} \to (\mathbb{R}^{\mathbb{N}})^{\mathbb{N}}$ (cf. (3.4.4)) factors over it.

Moreover no Con-isomorphism $(\mathbb{R}^{\mathbb{N}})^{(\mathbb{N})} \to (\mathbb{R}^{(\mathbb{N})})^{\mathbb{N}}$ exists, cf. [Jarchow, 1981, p. 333].

(i) $\bar{\omega}$: Pre$\to$Con.

$\bar{\omega}$ does not preserve uncountable products: Let Ω be the first uncountable ordinal, i.e. the set of all countable ordinals. It can be shown [Kriegl, unpublished] that there exists a preconvenient vector space E for which the αth Mackey adherence M-adh$^{\alpha}(E)$, cf. (6.3.4), in the completion $\tilde{E}$ is not the completion if $\alpha < \Omega$. We prove that the product E^{Ω} is not Mackey dense in $\tilde{E}^{\Omega}$. Choose for every $\alpha < \Omega$ an element $x_\alpha \in \tilde{E} \setminus$ M-Adh$^{\alpha}(E)$. Let $x := (x_\alpha)$ be the corresponding point in the product $\tilde{E}^{\Omega}$. Assume that x is in the Mackey closure of E^{Ω}; then there is an $\alpha < \Omega$ with $x \in$ M-adh$^{\alpha}(E^{\Omega})$. For the image with respect to the projection on the αth factor one has: $x_\alpha \in \mathrm{pr}_\alpha(\text{M-adh}^{\alpha}(E^{\Omega})) \subseteq \text{M-adh}^{\alpha}(E)$, contradiction.

$\bar{\omega}$ does not preserve equalizers (kernels): [Jarchow, 1981, pp. 98] gave an example of a vector space E with two metrizable locally convex topologies, one being non-complete, the other one being coarser and complete. In this case completeness in the locally convex sense is equivalent to completeness as preconvenient vector spaces, cf. the remark after (2.6.2). Then the extension of the identity to the completion has obviously a non-trivial kernel, but the completion of the kernel of the identity is $\{0\}$.

(ii) λ: $\mathscr{Lip}^k \to$ Con and $\ell^1 \cong \lambda: \ell^\infty \to$ Con.

λ does not preserve finite products: For any finite discrete $\mathscr{M}$-space X ($\mathscr{M} = \ell^\infty$ or $\mathscr{M} = \mathscr{Lip}^k$) one has $\lambda(X) = \mathbb{R}^X$ and thus $\lambda(X \sqcap X) = \mathbb{R}^{X \sqcap X}$ is unequal to $\lambda(X) \sqcap \lambda(X) = \mathbb{R}^X \sqcap \mathbb{R}^X = \mathbb{R}^{X \sqcup X}$ provided that X has at least three points.

λ does not preserve equalizers: For any finite discrete space X the diagonal map $\Delta: X \to X \sqcap X$ is the equalizer of the two projections $\mathrm{pr}_1, \mathrm{pr}_2: X \sqcap X \to X$. Applying the functor λ gives $\lambda(\Delta)$: $\mathbb{R}^X \to \mathbb{R}^{X \sqcap X}$ which cannot be the equalizer of $\lambda(\mathrm{pr}_j)$: $\mathbb{R}^{X \sqcap X} \to \mathbb{R}^X$ for reasons of dimension provided X that has at least three points.

(iii) $(\_)\tilde{\otimes} E$: Con$\to$Con.

$(\_)\tilde{\otimes}\mathbb{R}^{(\mathbb{N})}$ does not preserve countable products:

$$\mathbb{R}^{\mathbb{N}} \tilde{\otimes} \mathbb{R}^{(\mathbb{N})} \cong (\mathbb{R}^{\mathbb{N}} \tilde{\otimes} \mathbb{R})^{(\mathbb{N})} \cong (\mathbb{R}^{\mathbb{N}})^{(\mathbb{N})} \ncong (\mathbb{R}^{(\mathbb{N})})^{\mathbb{N}} \cong (\mathbb{R} \tilde{\otimes} \mathbb{R}^{(\mathbb{N})})^{\mathbb{N}}, \text{ cf. (0).}$$

$(\_)\tilde{\otimes}\ell^2$ does not preserve equalizers (kernels): Clearly $\ell^2 \to C(K, \mathbb{R})$, $x \mapsto \langle x, \_ \rangle$ is an isometric embedding of Banach spaces where $K := \{x \in \ell^2; \|x\| \leq 1\}$ is a compact space considered with the weak topology, cf. [Jarchow, 1981, p. 158]. In [Jarchow, 1981, p. 327] it is shown that applying the projective tensor product $(\_)\otimes\ell^2$ yields $\ell^2 \otimes \ell^2 \to C(K, \mathbb{R}) \otimes \ell^2$ which is not an embedding. Since all spaces are metrizable the projective tensor product coincides with the tensor product in Pre (considered as locally convex space). The extension of this morphism to the completion is not initial in Pre and hence is not an equalizer.

7.2.3 Theorem. *Let E and F be convenient vector spaces and X an ℓ^∞-space or a C^∞-space. Then the following functors are right adjoints and hence preserve limits (cf. (8.5.1) for the particular case where the domain is an opposite category):*

(i) $L(E,\_)$: Con→Con;
(ii) $\ell^\infty(X,\_)$: Con→Con;
(iii) $C^\infty(X,\_)$: Con→Con;
(iv) $L(\_,F)$: $\mathrm{Con}^{\mathrm{op}}\to\mathrm{Con}$;
(v) $\ell^\infty(\_,F)$: $(\ell^\infty)^{\mathrm{op}}\to\mathrm{Con}$;
(vi) $\mathscr{Lip}^k(\_,F)$: $(\mathscr{Lip}^k)^{\mathrm{op}}\to\mathrm{Con}$.

The functor in (v) preserves in addition co-equalizers.

Proof. The functor $L(E,\_)$ in (i) is right adjoint to $(\_)\tilde{\otimes}E$ by (3.8.4). The functor $L(\_,F)$ in (iv) is right adjoint to $L(\_,F)$: $\mathrm{Con}\to\mathrm{Con}^{\mathrm{op}}$ since $L(E_1,L(E_2,F))\cong L(E_2,L(E_1,F))$ by (3.7.3). The results for the functors in (ii) and (v) follow from those in (i) and (iv) by using that $\ell^\infty(X,F)=L(\ell^1(X),F)$ and that $\ell^1:(\ell^\infty)^{\mathrm{op}}\to\mathrm{Con}^{\mathrm{op}}$ is a right adjoint by (ii) in (7.2.1). The results on the functors in (iii) and (vi) follow from those in (i) and (iv) by using that $\mathscr{Lip}^k(X,F)=L(\lambda(X),F)$ and that λ: $(\mathscr{Lip}^k)^{\mathrm{op}}\to\mathrm{Con}^{\mathrm{op}}$ is a right adjoint by (iii) in (7.2.1).

The functor $\ell^\infty(\_,F)$ preserves co-equalizers since every co-equalizer $q: X\to Y$ in $(\ell^\infty)^{\mathrm{op}}$ is an embedding in ℓ^∞, hence has a left inverse (for $Y\neq\varnothing$). Thus $q^*:\ell^\infty(Y,F)\to\ell^\infty(X,F)$ has a right inverse and is therefore a final surjection and a co-equalizer. □

Let us now give examples that show that none of the functors mentioned in the theorem above preserve all colimits:

7.2.4 Examples. Many of the examples will make use of the following general remark:

(0) *Let $q=m\circ q_1$ be a composite of two Con-morphisms. If m is injective and not final (e.g. the image is not closed in the locally convex topology of the codomain) then q is not a co-equalizer and not final.*

Proof. If the image of m is not closed then m cannot be final since otherwise the image is a complemented subspace by (7.3.3). If $q=m\circ q_1$ were final then by (8.7.2) m would be final. Assume now that q is the co-equalizer of two morphisms f and g. Then $m\circ q_1\circ f=q\circ f=q\circ g=m\circ q_1\circ g$ and since m is injective $q_1\circ f=q_1\circ g$. By the universal property of the co-equalizer there has to be a morphism h with $q_1=h\circ q$. Again from the universal property and the equation $q=m\circ q_1=m\circ h\circ q$ one concludes that $m\circ h=1$, i.e. m would be a retraction and thus final, contradiction. □

(i) $L(E, \_)$: Con $\to$ Con.

$L(\mathbb{R}^{(\mathbb{N})}, \_)$ does not preserve countable coproducts:
$L(\mathbb{R}^{(\mathbb{N})}, \mathbb{R}^{(\mathbb{N})}) \cong L(\mathbb{R}, \mathbb{R}^{(\mathbb{N})})^{\mathbb{N}} \cong (\mathbb{R}^{(\mathbb{N})})^{\mathbb{N}} \not\cong (\mathbb{R}^{\mathbb{N}})^{(\mathbb{N})} \cong L(\mathbb{R}^{(\mathbb{N})}, \mathbb{R})^{(\mathbb{N})}$, cf. (0) in (7.2.2).

$L(\mathbb{R}^{\mathbb{N}}, \_)$ does not preserve co-equalizers (cokernels): Consider the final surjection q: $C^\infty(\mathbb{R}, \mathbb{R}) \to \mathbb{R}^{\mathbb{N}}$, $c \mapsto (c^{(n-1)}(0))_n$ which is by Borel's theorem (7.1.2) the cokernel of the subspace F formed by the functions that are infinitely flat at 0. By (0) the map q_*: $L(\mathbb{R}^{\mathbb{N}}, C^\infty(\mathbb{R}, \mathbb{R})) \to L(\mathbb{R}^{\mathbb{N}}, \mathbb{R}^{\mathbb{N}})$ is neither final nor a quotient map since using (7.1.3) it factors over the initial Con-morphism $(\mathbb{R}^{\mathbb{N}})^{(\mathbb{N})} \to (\mathbb{R}^{(\mathbb{N})})^{\mathbb{N}} \cong L(\mathbb{R}^{\mathbb{N}}, \mathbb{R}^{\mathbb{N}})$ of (0) in (7.2.2).

(ii) $\ell^\infty(X, \_)$: Con $\to$ Con.

$\ell^\infty(X, \_)$ does not preserve countable coproducts in general: This follows from the corresponding example in (i) using that $\ell^\infty(X, \_) \cong (\_)^X \cong L(\mathbb{R}^{(X)}, \_)$ for any discrete space X.

$\ell^\infty(X, \_)$ does not preserve co-equalizers: Let E be a Fréchet Montel space for which a quotient map q: $E \to \ell^1$ exists, cf. [Jarchow, 1981, p. 233]. Then the map $\ell^\infty(\mathbb{N}, E) \cong L(\ell^1, E) \to L(\ell^1, \ell^1) \cong \ell^\infty(\mathbb{N}, \ell^1)$ is neither surjective nor a co-equalizer since its image is contained in the subspace $K(\ell^1, \ell^1)$ formed by the compact operators, i.e. by those $g \in L(\ell^1, \ell^1)$ for which $g(\{x \in \ell^1; \|x\| \le 1\})$ is relatively compact (use that for $h \in L(\ell^1, E)$ the set $h(\{x \in \ell^1; \|x\| \le 1\})$ is bounded and since E is Montel it is relatively compact and so is $q(h(\{x \in \ell^1; \|x\| \le 1\}))$. Since $K(\ell^1, \ell^1)$ is closed in $L(\ell^1, \ell^1)$ by [Jarchow, 1981, p. 371] the map q_* cannot be a cokernel.

(iii) $C^\infty(X, \_)$: Con $\to$ Con.

$C^\infty(\mathbb{R}, \_)$ does not preserve countable coproducts: $C^\infty(\mathbb{R}, \mathbb{R}^{(\mathbb{N})}) \neq C^\infty(\mathbb{R}, \mathbb{R})^{(\mathbb{N})}$ since the curves $c_n(t) := h(t-n)$ with $h \in C^\infty(\mathbb{R}, \mathbb{R})$, $\emptyset \neq \operatorname{supp}(h) \subseteq [0, 1]$ define a curve $c = (c_n) \in C^\infty(\mathbb{R}, \mathbb{R}^{(\mathbb{N})})$ with $c \notin C^\infty(\mathbb{R}, \mathbb{R})^{(\mathbb{N})}$.

$C^\infty(\mathbb{R}^{\mathbb{N}}, \_)$ does not preserve co-equalizers (cokernels): We apply (8.3.8) to (i) by using that $L(E, F)$ is a complemented subspace of $C^\infty(E, F)$, by (4.4.24) for $j = 1$, $k = \infty$.

(iv) $L(\_, F)$: $\text{Con}^{\text{op}} \to$ Con.

$L(\_, \mathbb{R}^{\mathbb{N}})$ does not preserve countable coproducts: $L(\mathbb{R}^{\mathbb{N}}, \mathbb{R}^{\mathbb{N}}) \cong L(\mathbb{R}^{\mathbb{N}}, \mathbb{R})^{\mathbb{N}} \cong (\mathbb{R}^{(\mathbb{N})})^{\mathbb{N}} \not\cong (\mathbb{R}^{\mathbb{N}})^{(\mathbb{N})} \cong L(\mathbb{R}, \mathbb{R}^{\mathbb{N}})^{(\mathbb{N})}$.

$L(\_, \mathbb{R}^{(\mathbb{N})})$ does not preserve co-equalizers (cokernels): This follows from (i) since $L(\_, \mathbb{R}^{(\mathbb{N})}) \cong L(\mathbb{R}^{\mathbb{N}}, (\_)')$ by (3.7.3), $C^\infty(\mathbb{R}, \mathbb{R})$ is reflexive by (5.1.7) and $(\_)'$ preserves by (iv) of (7.2.3) the kernel q: $C^\infty(\mathbb{R}, \mathbb{R}) \to \mathbb{R}^{\mathbb{N}}$ in Con^{op}, cf. (7.1.2).

(v) $\ell^\infty(\_, F)$: $(\ell^\infty)^{\text{op}} \to$ Con.

$\ell^\infty(\_, F)$ does not preserve finite coproducts provided $F \neq \{0\}$: One uses the natural isomorphisms $\ell^\infty(\{*\}^J, F) \cong \ell^\infty(\{*\}, F) \cong F$ and $F^{(J)} \cong \ell^\infty(\{*\}, F)^{(J)}$ and the fact that the summation map $F^{(J)} \to F$ is an isomorphism only if J is single pointed.

(vi) $C^\infty(\_, F)$: $(C^\infty)^{\text{op}} \to$ Con.

$C^\infty(\_, F)$ does not preserve finite coproducts provided $F \neq \{0\}$: One uses the same argument as in (v) with ℓ^∞ replaced by C^∞.

$C^\infty(\_, \mathbb{R}^{(\mathbb{N})})$ does not preserve co-equalizers (cokernels): This follows from (iv) since $C^\infty(X, F) \cong L(\lambda(X), F)$, $\lambda(\mathbb{R}) \cong C^\infty(\mathbb{R}, \mathbb{R})'$, $\lambda(\mathbb{N}) \cong C^\infty(\mathbb{N}, \mathbb{R})' \cong \mathbb{R}^{(\mathbb{N})}$, and the inclusion $\mathbb{N} \to \mathbb{R}$ is an equalizer in $\underline{C^\infty}$.

Now we will show that some other functors which have been considered do not preserve limits. We begin with the functor η: $\underline{\text{Born}} \to \underline{\ell^\infty}$. By (1.2.4) it is left adjoint to the inclusion, it preserves the underlying vector spaces and the morphisms, and the ℓ^∞-structure of ηX is the one generated by $\underline{\text{Born}}(X, \mathbb{R})$.

First we give a useful construction involving certain bornologies $\mathscr{B}$ on $\mathbb{N}$. We shall denote the respective bornological space by $\mathbb{N}_\mathscr{B}$, while $\mathbb{N}$ will, according to (1.1.6), denote the natural numbers with their standard ℓ^∞-structure for which $\mathbb{N}$ itself is bounded, cf. proof of (1.2.8).

7.2.5 Proposition. *Let $\mathscr{B}$ be a bornology on $\mathbb{N}$ such that $\underline{Born}(\mathbb{N}_\mathscr{B}, \mathbb{R}) = \ell^\infty$, i.e. $\eta(\mathbb{N}_\mathscr{B}) = \mathbb{N}$. Then the $\underline{Pre}$-subspace $\{x \in \mathbb{R}^\mathbb{N};\ \text{supp}(x) \in \mathscr{B}\}$ of $\mathbb{R}^\mathbb{N}$ is the inductive limit in $\underline{Pre}$ of the Fréchet spaces $\mathbb{R}^B \subseteq \mathbb{R}^\mathbb{N}$ with $B \in \mathscr{B}$.*

Proof. Each $\mathbb{R}^B$ can be considered as the subspace $\{x \in \mathbb{R}^\mathbb{N};\ \text{supp}(x) \subseteq B\}$ of $E := \{x \in \mathbb{R}^\mathbb{N};\ \text{supp}(x) \in \mathscr{B}\}$.

First we want to show that E is barrelled, i.e. every $\sigma(E', E)$-bounded subset of E' is equicontinuous, cf. [Jarchow, 1981, p. 219]. Since E contains the sequences with finite support, E is a dense subspace of the Fréchet-space $\mathbb{R}^\mathbb{N}$; thus $E' = (\mathbb{R}^\mathbb{N})' = \mathbb{R}^{(\mathbb{N})}$. So let $A \subseteq \mathbb{R}^{(\mathbb{N})}$ be bounded at each $x \in E$ and suppose A is not equicontinuous, i.e. not bounded in $\mathbb{R}^{(\mathbb{N})}$. Since $A(e_n) \subseteq \mathbb{R}$ is bounded for each vector e_n of the natural basis of $\mathbb{R}^{(\mathbb{N})}$ there have to exist $a^n \in A$ with $n \mapsto k_n := \max(\text{supp}(a^n))$ strictly increasing. Since $f: \mathbb{N} \to \mathbb{R}$ defined by $f(k_n) := n$ and $f(k) := 0$ iff $k \notin \{k_n;\ n \in \mathbb{N}\}$ is not globally bounded, some infinite subset B of $\{k_n;\ n \in \mathbb{N}\}$ belongs to $\mathscr{B}$. By passing to a subsequence we may assume that $\{k_n; n \in \mathbb{N}\} \in \mathscr{B}$. Let us choose $x_k \in \mathbb{R}$ with $\Sigma_k x_k (a^n)_k = n$ (x_{k_n} is defined inductively by this equation and $x_k := 0$ iff $k \notin \{k_n;\ n \in \mathbb{N}\}$). Now $x := (x_n)_n \in E$ and $a^n(x) = n$, contradiction.

Now we are going to prove the universal property which characterizes inductive limits. Obviously every linear map m: $E \to F$ into a preconvenient vector space F is uniquely determined by its restriction to $\mathbb{R}^B$ for all $B \in \mathscr{B}$ and it remains to show that m is a morphism provided $m|_{\mathbb{R}^B}$ is a morphism for all $B \in \mathscr{B}$. It is enough to show this for $F = \mathbb{R}$. Consider the linear morphisms $m_B := m|_{\mathbb{R}^B} \circ \text{pr}_B$: $E \subseteq \mathbb{R}^\mathbb{N} \to \mathbb{R}^B \to \mathbb{R}$. Pointwise the net m_B converges to m, since $x \in E$ implies $m_B(x) = m(x)$ for all $B \subseteq \text{supp}(x)$. By the Banach–Steinhaus theorem for barrelled spaces, cf. [Jarchow, 1981, p. 220], the assertion that m is a morphism follows provided we show that $\{m_B(x);\ B \in \mathscr{B}\}$ is bounded for all $x \in E$. Obviously $\{\text{pr}_B(x);\ B \in \mathscr{B}\} \subseteq \mathbb{R}^{\text{supp}(x)}$ is bounded and $m_B(x) = m(\text{pr}_B(x)) = m_{\text{supp}(x)}(\text{pr}_B(x))$. Hence $\{m_B(x);\ B \in \mathscr{B}\} = m_{\text{supp}(x)}(\{\text{pr}_B(x); B \in \mathscr{B}\})$ is bounded. □

7.2.6 Example. For every $\lambda \in c_0$ the bornology $\{B \subseteq \mathbb{N};\ \sum_{n \in B} \lambda(n) < \infty\}$ satisfies the assumption of the previous proposition. The same holds for the bornology given by the subsets $A \subseteq \mathbb{N}$ of density 0, i.e. those A for which $(1/n)$ card $\{k \in A; k \leq n\}$ converges to 0 for $n \to \infty$.

7.2.7 Lemma. *For any set J the following statements are equivalent:*

(1) *η preserves all products with index set J;*

(2) *$\eta(\Pi_{j\in J} \mathbb{N}_{\mathscr{B}_j}) = \Pi_{j\in J} \eta(\mathbb{N}_{\mathscr{B}_j})$ for any family of bornologies $\mathscr{B}_j$ on $\mathbb{N}$;*

(3) *If $\mathscr{B}_j$ $(j \in \mathbb{N})$ are bornologies on $\mathbb{N}$ such that every bornological function $f\colon \mathbb{N}_{\mathscr{B}_j} \to \mathbb{R}$ is globally bounded then an infinite $B \in \bigcap_{j\in J} \mathscr{B}_j$ exists.*

Proof. $(1 \Rightarrow 2)$ is trivial.

$(2 \Rightarrow 3)$ Assume that $\mathscr{B}_j$ $(j \in J)$ satisfying the hypothesis in (3) exist. Consider the product of the spaces $X_j = \mathbb{N}_{\mathscr{B}_j}$, the maps $c_j = \mathrm{id}\colon \mathbb{N} \to X_j$, and $c := (c_j)_j$. The maps c_j are structure curves of $\eta(X_j)$ since by assumption every $f_j \in \underline{\mathrm{Born}}(X_j, \mathbb{R})$ is globally bounded. Hence c is a structure curve of $\Pi_{j\in J} \eta(X_j)$. Now define $f\colon \Pi_j X_j \to \mathbb{R}$ by $f(c(n)) := n$ and $f(x) := 0$ if $x \notin c(\mathbb{N})$. This function is bornological on $\Pi_{j\in J} X_j$ since for every bounded set $B \subseteq \Pi_j X_j$ the set $\{n;\ c(n) \in B\}$ is contained in $\{n;\ n = \mathrm{pr}_j(c(n)) \in \mathrm{pr}_j(B)\} = \mathrm{pr}_j(B) \in \mathscr{B}_j$ for all $j \in J$ and thus is finite, i.e. $f(B) = \{n;\ c(n) \in B\}$ is finite. Hence f is a structure function of $\eta(\Pi_j X_j)$; but not of $\Pi_j \eta(X_j)$ since $(f \circ c)(\mathbb{N}) = \mathbb{N}$ is unbounded in $\mathbb{R}$.

$(1 \Leftarrow 3)$ We give an indirect proof. So we assume that some product $\Pi_{j\in J} X_j$ is not preserved by η and then construct a family of bornologies $\mathscr{B}_j$ on $\mathbb{N}$ that contradicts (3).

Since by assumption the structure of $\eta(\Pi_j X_j)$ is strictly finer than that of $\Pi_j \eta(X_j)$, there exists a structure curve $c = (c_j)$ of $\Pi_j \eta(X_j)$ and a bornological function $f\colon \Pi_j X_j \to \mathbb{R}$ with $f \circ c \notin \ell^\infty$, and by passing to a subsequence we may assume that $|f(c(k))| \geq k$. Let $\mathscr{B}_j$ be the initial bornology on $\mathbb{N}$ induced by c_j, i.e. $B \in \mathscr{B}_j$ iff $c_j(B) \subseteq X_j$ is bounded. Obviously $B \in \mathscr{B}_j$ implies $c_j^{-1}(c_j(B)) \in \mathscr{B}_j$.

Let $f_j\colon \mathbb{N}_{\mathscr{B}_j} \to \mathbb{R}$ be a bornological function. We have to show that $f_j(\mathbb{N})$ is bounded. Consider a new function g defined by $g(n) := \sup\{|f_j(k)|;\ k \in c_j^{-1}(c_j(n))\} \geq |f_j(n)|$. The function g is well-defined, since $c_j^{-1}(c_j(n))$ is bounded in $\mathscr{B}_j$ and it is bornological since $\sup\{|g(b)|;\ b \in B\} \leq \sup\{|f_j(b')|;\ b' \in c_j^{-1}(c_j(B))\}$, which is finite since f_j has to be bounded on the bounded set $c_j^{-1}(c_j(B))$. Since $c_j(n) = c_j(n')$ implies $g(n) = g(n')$ we may define a new map $h\colon X_j \to \mathbb{R}$ by $h(c_j(n)) := g(n)$ and $h(x) := 0$ if $x \notin c_j(\mathbb{N})$. Then h is bornological, since for any bounded $B \subseteq X_j$ one has $h(B) = h(c_j(c_j^{-1}(B))) = g(c_j^{-1}(B))$, hence $h \circ c_j = g \in \ell^\infty$. Since $g(k) \geq |f_j(k)|$ it follows that $f_j(\mathbb{N})$ is bounded.

By (3) there exists an infinite $B \subseteq \mathbb{N}$ with $B \in \mathscr{B}_j$ for all $j \in J$. Then $c_j(B) \subseteq X_j$ is bounded and hence $\Pi_j c_j(B) \subseteq \Pi_j X_j$ is bounded. Thus $f(\Pi_j c_j(B)) \supseteq (f \circ c)(B)$ is bounded. This is a contradiction since $|f(c(k))| \geq k$ for all $k \in \mathbb{N}$. □

7.2.8 Corollary. *For any countable family of bornological spaces X_n $(n \in \mathbb{N})$ one has $\eta(\Pi_n X_n) = \Pi_n \eta(X_n)$.*

Proof. It is enough to verify (3) of (7.2.7) for $J = \mathbb{N}$. Each bornology $\mathscr{B}_n$ satisfies the formally stronger condition: every infinite subset A of $\mathbb{N}$ contains an infinite $A_n \in \mathscr{B}_n$. Otherwise we can define a bornological not globally bounded function $f: \mathbb{N}_{\mathscr{B}_n} \to \mathbb{R}$ by $f(x_i) := i$ and $f(x) := 0$ if $x \notin A$, where $A = \{x_1, x_2, \ldots\}$. Now one easily chooses inductively the infinite sets $A_n \in \mathscr{B}_n$ and points $a_n \in A_n$ with $\{a_1, \ldots, a_n\} \subseteq A_{n+1} \subseteq A_n$ and $a_{n+1} \in A_{n+1} \setminus \{a_1, \ldots, a_n\}$. Then the infinite set $B := \{a_1, a_2, \ldots\} \subseteq A_n$ belongs to $\mathscr{B}_n$ for all $n \in \mathbb{N}$. □

Finally we show that η does not preserve products of cardinality at least the continuum.

7.2.9 Corollary. *For every index set I of cardinality larger or equal to that of the continuum there exists a family $\{\mathscr{B}_j; j \in J\}$ of bornologies on $\mathbb{N}$ such that the structure of $\eta(\Pi_j \mathbb{N}_{\mathscr{B}_j})$ is strictly finer than that of $\Pi_j \eta(\mathbb{N}_{\mathscr{B}_j})$.*

Proof. It is enough to show that there exist $2^{\aleph_0}$ many bornologies on $\mathbb{N}$ for which (3) of (7.2.7) fails. For every $j \in c_0$ let $\mathscr{B}_j$ be the bornology on $\mathbb{N}$ defined by $A \in \mathscr{B}_j$ iff $\sum_{n \in A} |j(n)| < \infty$, cf. (7.2.6). Each of these bornologies is of the required type. Assume that there is some infinite $B = \{b_1, b_2, \ldots\} \subseteq \mathbb{N}$ with $B \in \mathscr{B}_j$ for all $j \in c_0$, i.e. $\sum_{n \in \mathscr{B}} |j(n)| < \infty$ for all j. Define a $j_0 \in c_0$ by $j_0(b_n) := 1/n$ and $j_0(k) = 0$ if $k \notin B$. Then $\sum_{k \in B} |j_0(k)| = \sum_{n \in \mathbb{N}} |j_0(b_n)| = \sum_n 1/n = \infty$, contradiction. □

Now we consider the embedding $\xi\sigma_b$: Pre → LimVS. It can be factorized into Pre → bLimVS → LimVS, where the first functor is a right adjoint and the second one a left adjoint, cf. remark (i) after (2.4.3). We will show now that this composite is neither a left nor a right adjoint (quoted in (2.4.3)).

7.2.10 Example. (i) The embedding Pre → LimVS does not preserve products, hence cannot be a right adjoint.

In order to see this we consider $E = \mathbb{R}^J$ with J equal to the set c_0 of all real 0-sequences. Let $x^n \in E$ be defined by $(x^n)_j := j_n$ and let $\mathscr{F}$ be the filter on E generated by this sequence. It is easy to see that $\mathscr{F}$ is not M-convergent, but $\mathrm{pr}_j(\mathscr{F})$ is convergent in $\mathbb{R}$ and thus M-convergent. Hence $\mathscr{F}$ is convergent in the product of the M-convergence structures.

(ii) The embedding Pre → LimVS does not preserve cokernels, hence cannot be a left adjoint.

In order to see this, consider the subspace E of $\mathbb{R}^{\mathbb{N}}$ formed by those $x \in \mathbb{R}^{\mathbb{N}}$ that have support A of density 0. According to (7.2.5) the map $q: \amalg_A \mathbb{R}^A =: F \to E$ is a Pre-quotient, cf. (3.5.3). Assume that q is final with respect to the Mackey convergence structures. Define another convergence structure on E by $\mathscr{F} \to y$ iff $\mathscr{F} = q(\mathscr{G})$ for some filter $\mathscr{G}$ converging Mackey towards an $x \in G$ with $q(x) = y$. This turns E into a convergence vector space E_1 and makes $q: F \to E_1$ continuous. By the assumed finality of $q: F \to E$ the identity $E \to E_1$ is continuous. Now consider the sequence $n \mapsto (1/n)\chi_{[0,n]}$ which is M-convergent to 0 in E, hence in E_1, and thus has an M-convergent lifting to F. But this is impossible, since the union of the supports of this sequence is $\mathbb{N}$, which does not have density 0.

7.3 Preservation of initiality and finality

We first give some useful characterizations of final morphisms.

7.3.1 Proposition. *For any $\mathcal{M}$-map $g: X \to Y$ (with arbitrary $\mathcal{M}$, cf. section* 1.1) *the following statements are equivalent*:

(1) *g is final;*
(2) *Y is the coproduct of the two $\mathcal{M}$-subspaces $g(X)$ and $Y \setminus g(X)$ of Y, $g: X \to g(X)$ is final and $Y \setminus g(X)$ is discrete, i.e. carries the finest $\mathcal{M}$-structure.*

Proof. (1⇒2) As set Y is the disjoint union (i.e. the coproduct in Set) of $g(X)$ and $Y \setminus g(X)$. Thus the functions f on Y are determined by the restrictions f_0 to $g(X)$ and f_1 to $Y \setminus g(X)$. Since g is final the structure functions on Y are those f for which $f_0 \circ g = f \circ g$ is a morphism, i.e. those where f_0 is a structure function with respect to the final structure on $g(X)$ and where f_1 is arbitrary. Hence the initial structure on $g(X)$ is the final one induced by $g: X \to g(X)$, that of $Y \setminus g(X)$ is discrete, and Y is the coproduct of $g(X)$ and $Y \setminus g(X)$ with their initial structures.

(1⇐2) Let $f: Y \to Z$ be a map for which $f \circ g$ is a morphism. Let again f_0 and f_1 be the restrictions of f to $g(X)$ and $Y \setminus g(X)$. Then $f_0 \circ g = f \circ g$ is a morphism and hence f_0 is a morphism since $g: X \to g(X)$ is final. Furthermore, f_1 is a morphism since $Y \setminus g(X)$ is discrete. Since X is the coproduct of $g(X)$ and $Y \setminus g(X)$ we conclude that f is a morphism. □

The last proposition allows us to restrict some considerations about final morphism to surjective ones.

7.3.2 Proposition. *For any $\mathcal{M}$-map $g: X \to Y$ the following statements are equivalent*:

(1) *g is a final surjection;*
(2) *g is a coequalizer, cf.* (8.3.6).

Proof. (1⇒2) Consider the pullback $\mathrm{pr}_1, \mathrm{pr}_2: X \times_Y X \to X$ of the map $g: X \to Y$ with itself. Then g is the coequalizer of $\mathrm{pr}_1, \mathrm{pr}_2: X \times_Y X \to X$. In fact, let $f: X \to Z$ be a morphism for which $f \circ \mathrm{pr}_1$ coincides with $f \circ \mathrm{pr}_2$ on the pullback $X \times_Y X$, i.e. $f(x_1) = f(x_2)$ provided $g(x_1) = g(x_2)$. Then there exists a unique map $\bar{f}: Y \to Z$ with $\bar{f} \circ g = f$, and since g is final this map is a morphism.

(1⇐2) One only has to recall that a colimit in $\mathcal{M}$ has as underlying set the colimit in Set of the underlying sets and as structure the final one and that any co-equalizer of two maps $g_1, g_2: X \to Y$ in Set is surjective (it is the quotient map from Y to the equivalence classes with respect to the equivalence relation generated by $\{(g_1(x), g_2(x)); x \in X\}$). □

Now we will prove for Pre and Con results analogous to (7.3.1) and (7.3.2).

7.3.3 Proposition. *For any Pre-morphism and any Con-morphism $g: E \to F$ the following statements are equivalent:*

(1) *g is final;*
(2) *F is the coproduct of the subspace $g(E)$ and a discrete subspace and $g: E \to g(E)$ is final.*

Proof. (1⇒2) Let F_1 be an algebraic complement of $g(E)$ in F, i.e. F is algebraically the direct sum of $g(E)$ and F_1. Then any linear map $f: F \to G$ is determined by its restrictions f_0 to $g(E)$ and f_1 to F_1. Since g is final those f are morphisms for which $f_0 \circ g = f \circ g$ is a morphism, i.e. those where f_0 is a morphism with respect to the final structure on $g(E)$ and where f_1 is arbitrary. Hence the subspace structure on $g(E)$ is the final one induced by $g: E \to g(E)$, that of F_1 is discrete, and F is the coproduct of these two subspaces. In particular these subspaces are M-closed and thus are convenient, provided F is convenient.

(1⇐2) Let $f: F \to G$ be a linear map for which $f \circ g$ is a morphism and F the coproduct of $g(E)$ with a discrete subspace F_1. Again we denote with f_0 and f_1 the restrictions of f to $g(E)$ and F_1. Then $f_0 \circ g = f \circ g$ is a morphism and hence f_0 is a morphism since $g: E \to g(E)$ is final. Furthermore, f_1 is a morphism since F_1 is discrete. Since E is the coproduct of $g(E)$ and F_1 we conclude that f is a morphism. □

The analogue to proposition (7.3.2) works only for Pre and not for Con.

7.3.4 Proposition. *For any Pre-morphism $g: E \to F$ the following statements are equivalent:*

(1) *g is a final surjection;*
(2) *g is a coequalizer.*

Proof. (1⇒2) Consider the pullback $\mathrm{pr}_1, \mathrm{pr}_2: E \times_F E \to E$ of the map $g: E \to F$ with itself. Then g is the co-equalizer of $\mathrm{pr}_1, \mathrm{pr}_2: E \times_F E \to E$. In fact, let $f: E \to G$ be a morphism for which $f \circ \mathrm{pr}_1$ coincides with $f \circ \mathrm{pr}_2$ on the pullback $E \times_F E$, i.e. $f(x_1) = f(x_2)$ provided $g(x_1) = g(x_2)$. Thus there exists a unique map $\bar{f}: F \to G$ with $\bar{f} \circ g = f$, and since g is final this map is a morphism.

(2⇐1) One only has to recall that a colimit in Pre has as underlying vector space the colimit in VS of the underlying vector spaces and as structure the final one and that any co-equalizer of two linear maps $g_1, g_2: E \to F$ in VS is surjective (it is the quotient map from F to the equivalence classes with respect to the congruence relation generated by $\{(g_1(x), g_2(x)); x \in E\}$). □

Now we discuss the permanence properties of the basic functors with respect to initial and final morphisms.

7.3.5 Theorem

(i) $\ell^1: \underline{\ell^\infty} \to \underline{Con}$ *preserves initial and final morphisms.*
(ii) $\lambda: \underline{\mathscr{Lip}^k} \to \underline{Con}$ *preserves final morphisms.*
(iii) $L(E, \_): \underline{Con} \to \underline{Con}$ *preserves initial morphisms.*
(iv) $\ell^\infty(X, \_): \underline{Con} \to \underline{Con}$ *preserves initial morphisms.*
(v) $C^\infty(X, \_): \underline{Con} \to \underline{Con}$ *preserves initial morphisms.*
(vi) $L(\_, F): \underline{Con}^{op} \to \underline{Con}$ *carries surjections to embeddings.*
(vii) $\ell^\infty(\_, F): (\underline{\ell^\infty})^{op} \to \underline{Con}$ *carries surjections to embeddings and initial morphisms of* $\underline{\ell^\infty}$ *to final morphisms.*
(viii) $\mathscr{Lip}^k(\_, F): (\underline{\mathscr{Lip}^k})^{op} \to \underline{Con}$ *carries surjections to embeddings.*

Proof. The functor in (i) preserves initial morphisms, since every initial map $(\neq \emptyset)$ in $\underline{\ell^\infty}$ has a right inverse. And maps having a right inverse in $\underline{Con}$ are initial.

That the functors in (i) and (ii) preserve final morphisms follows from (7.3.1) and (7.3.2), since they preserve colimits and discrete spaces. They preserve discrete spaces since every discrete space in the domain category is a coproduct of single points, the functors applied to a single point give $\mathbb{R}$ and a coproduct of factors $\mathbb{R}$ is discrete in $\underline{Con}$ by (3.4.6).

(iii) Let $g: F_1 \to F_2$ be an initial morphism. We have to show that $g_*: L(E, F_1) \to L(E, F_2)$ is initial. So let $h: G \to L(E, F_1)$ be a linear map for which $g_* \circ h$ is a morphism. Since the $\underline{Con}$-structure of $L(E, F_1)$ is the initial one induced by the evaluations $ev_x: L(E, F_1) \to F_1$ it is enough to show that $ev_x \circ h$ is a morphism. This follows since $g \circ ev_x \circ h = ev_x \circ g_* \circ h$ is a morphism and g is initial.

(iv) follows from (iii) since $\ell^\infty(X, \_) \cong L(\ell^1(X), \_)$.

(v) follows from (iii) since $C^\infty(X, \_) \cong L(\lambda(X), \_)$.

(vi) Let $g: E_1 \to E_2$ be surjective. Then $g^*: L(E_2, F) \to L(E_1, F)$ is injective. Let $h: G \to L(E_2, F)$ be a linear map for which $g^* \circ h$ is a morphism. Since the $\underline{Con}$-structure on $L(E_2, F)$ is the initial one induced by the evaluations ev_y: $L(E_2, F) \to F$ it is enough to show that $ev_y \circ h$ is a morphism. This follows since $ev_y \circ h = ev_x \circ g^* \circ h$ provided x is chosen such that $g(x) = y$, which is possible since g is assumed to be surjective.

(vii) and (viii) The statements concerning surjections are proved as in (vi). So let $\iota: X \to Y$ be an initial morphism of $\underline{\ell^\infty}$. It has a left inverse (for $X \neq \emptyset$) hence $\iota^* = \ell^\infty(\iota, F)$ has a right inverse and consequently is final. □

Let us give some examples which show that among the functors considered only those explicitly mentioned in the last theorem preserve initial or final morphisms.

7.3.6 Examples. (0) We will make use of the examples (7.2.2) and (7.2.4) and the fact that a composite $q = m \circ q_1$ of two $\underline{Con}$-morphisms is not final by (8.7.2), provided m is not final (and hence in particular provided the image of m is not closed in the locally convex topology, cf. (0) in (7.2.4)).

(i) $\bar{\omega}$: Pre→Con.

$\bar{\omega}$ does not preserve final morphisms: In (7.2.5) we gave an example of a Pre-quotient $q: F \to E$, cf. (3.5.3), where F is convenient and E is an M-dense Pre-subspace of $\mathbb{R}^{\mathbb{N}}$. Applying $\bar{\omega}$ gives a non-surjective, non-final map $\bar{\omega}(q): F \to E \to \tilde{E}$ ($\chi_{[n,\ldots,\infty)} \notin E$ form a bounded sequence).

$\bar{\omega}$ does not preserve initial morphisms: In (6.3.2) we gave an example of a Pre-subspace $\iota: E \oplus \mathbb{R} \to F$, $(x, t) \mapsto x + ty$ of a convenient vector space $F = \tilde{E}$ whose Mackey-closure is F. Applying $\bar{\omega}$ to ι gives a map $\tilde{\iota}: F \oplus \mathbb{R} \to F$ acting as $(x, t) \mapsto x + ty$. The kernel $\{t(y, -1); t \in \mathbb{R}\}$ of this map is not trivial. Hence $\tilde{\iota}$ is not injective and thus not initial, cf. (3.2.1).

(ii) λ: $\mathscr{Lip}^k$→Con.

λ does not preserve initial morphisms in general: In [Joris, 1982] it is shown that $f: \mathbb{R} \to \mathbb{R}^2$, $t \mapsto (t^2, t^3)$ is initial in C^∞. But $\lambda f: \lambda\mathbb{R} = C^\infty(\mathbb{R}, \mathbb{R})' \to C^\infty(\mathbb{R}^2, \mathbb{R})' = \lambda(\mathbb{R}^2)$ is not even injective ($\mathrm{ev}_0 \circ \mathscr{D} \in C^\infty(\mathbb{R}, \mathbb{R})'$ is mapped onto 0).

(iii) $(\_)\tilde{\otimes} E$: Con→Con.

$(\_)\tilde{\otimes} E$ does not preserve final morphisms provided E is not discrete: Applied to the final morphism $\{0\} \to \mathbb{R}$ it gives the non-final morphism $\{0\} \cong \{0\} \tilde{\otimes} E \to \mathbb{R} \tilde{\otimes} E \cong E$.

$(\_)\tilde{\otimes} \ell^2$ does not preserve initial morphisms: This was already contained in the counter-example (iii) of (7.2.2).

(iv) $L(E, \_)$: Con→Con.

$L(\mathbb{R}^{(\mathbb{N})}, \_)$ does not preserve final morphisms: This was already contained in the counter-example (i) of (7.2.4).

(v) $C^\infty(X, \_)$: Con→Con.

$C^\infty(\mathbb{R}, \_)$ does not preserve final morphisms: By (7.1.5) there exists a quotient map $q: E \to F$ and a smooth curve $c: \mathbb{R} \to F$ with $c(t) = 0$ for $t \leq 0$ not having locally around 0 a smooth lifting to E. Thus $q_*: C^\infty(\mathbb{R}, E) \to C^\infty(\mathbb{R}, F)$ is not onto. In order to prove that it is not final we consider the smooth curves $c_s: \mathbb{R} \to F$ defined by $c_s(t) := c(t - s)$. We claim that $\{c_s;\ s \in \mathbb{R}\}$ is linearly independent in $C^\infty(\mathbb{R}, F)/\text{image}\,(q_*)$. Assume some finite sum $\Sigma \alpha_s c_s$ belongs to image (q_*) with not all α_s equal to 0. Then there exists a smooth curve $e: \mathbb{R} \to E$ with $q \circ e = \Sigma \alpha_s c_s$. Let $s_0 := \min\{s;\ \alpha_s \neq 0\}$. Then $(q \circ e)(t) = \alpha_{s_0} c_{s_0}(t)$ for t near s_0, and hence $t \mapsto (\alpha_{s_0})^{-1} \cdot e(t + s_0)$ is a smooth lifting of c locally around 0, contradiction. If q were final then $C^\infty(\mathbb{R}, F)/\text{image}\,(q_*)$ would be discrete, which is not the case, since the linearly independent family $\{c_s;\ s \in [0, 1]\}$ is bounded in $C^\infty(\mathbb{R}, F)$ and thus in $C^\infty(\mathbb{R}, F)/\text{image}(q_*)$.

(vi) $L(\_, F)$: Con^{op}→Con.

$L(\_, F)$ does not preserve initial morphisms, i.e. does not transform final Con-morphisms into initial Con-morphisms, provided $F \neq \{0\}$: Let $E_1 \subseteq E_2$ be two discrete convenient vector spaces. Then the inclusion $\iota: E_1 \to E_2$ is final in Con. But $L(E, F) = L(\mathbb{R}^{(J)}, F) = F^J$ for every discrete space $E = \mathbb{R}^{(J)}$ and hence $\iota^* = L(\iota, F)$ is not injective and thus not initial.

$L(\_, \mathbb{R}^{(\mathbb{N})})$ does not preserve final morphisms: This was already shown in the counter-example (iv) of (7.2.4). Even the duality functor $(\_)'$ does not preserve final morphisms. Applying it to the initial but non-final morphism $\mathbb{R}^{(\mathbb{N})} \to \mathbb{R}^{\mathbb{N}}$ gives $\mathbb{R}^{(\mathbb{N})} \to \mathbb{R}^{\mathbb{N}}$.

(vii) $\ell^\infty(\_, F): (\ell^\infty)^{\mathrm{op}} \to \mathrm{Con}$.

$\ell^\infty(\_, F)$ does not preserve initial morphisms provided $F \neq \{0\}$: Let $X_1 \subseteq X_2$ be two discrete ℓ^∞-spaces. Then the inclusion $\iota: X_1 \to X_2$ is final in ℓ^∞. But $\ell^\infty(X, F) = F^X$ for every discrete space X and hence $\iota^* = \ell^\infty(\iota, F)$ is not injective and thus not initial.

(viii) $\mathcal{Lip}^k(\_, F): (\mathcal{Lip}^k)^{\mathrm{op}} \to \mathrm{Con}$.

$\mathcal{Lip}^k(\_, F)$ does not preserve initial morphisms provided $F \neq \{0\}$: Let $X_1 \subseteq X_2$ be two discrete $\mathcal{Lip}^k$-spaces. Then the inclusion $\iota: X_1 \to X_2$ is final in $\mathcal{Lip}^k$. But $\mathcal{Lip}^k(X, F) = F^X$ for every discrete space X and hence $\iota^* = \mathcal{Lip}^k(\iota, F)$ is not injective and thus not initial.

$C^\infty(\_, \mathbb{R}^{(\mathbb{N})})$ does not preserve final morphisms: This was already shown in the counter-example (vi) of (7.2.4).

Now we want to give an example which shows that finality is not preserved by the functors Pre→Born, Pre→ℓ^∞, and Pre→C^∞.

7.3.7 Example. The bornology, the ℓ^∞-structure and the $\mathcal{Lip}^k$-structure of a Pre-quotient need not be the final structures.

(i) Recall that the final bornology (final convex bornology) for a surjective map consists of the images of all bounded sets. There exists a Fréchet Montel space which has ℓ^1 as LCS-quotient [Jarchow, 1981, p. 233]. The unit ball cannot have a bounded lift, otherwise this lift would be precompact (since the space is Montel) and thus the unit ball would be compact as closed image of a precompact set.

(ii) Recall that the final linearly generated $\mathcal{M}$-structure ($\mathcal{M} = \ell^\infty$ or $\mathcal{M} = \mathcal{Lip}^k$) is the final Pre-structure.

Consider the subspace E of $\mathbb{R}^{\mathbb{N}}$ consisting of all those sequences whose support has density 0. Then $\coprod_A \mathbb{R}^A \to E$ is a final Pre-morphism by (7.2.5) and (7.2.6), where A runs through all subsets of $\mathbb{N}$ with density 0.

The final $\mathcal{M}$-structure on this space E is compatible ($M = \ell^\infty$, or $M = \mathcal{Lip}^\infty$), i.e. the vector operations are $\mathcal{M}$-morphisms; but it is not the final Pre-structure and thus not linearly generated.

In order to see that the final $\mathcal{M}$-structure on E is compatible one uses that the product of final morphisms is final (by cartesian closedness) and the vector operations on E are induced by those of $\coprod_A \mathbb{R}^A$. Now we show that this structure cannot be the $\mathcal{M}$-structure of the final Pre-structure.

Let $\varphi: E \to \mathbb{R}$ be defined by $\varphi(x) = \sum_n n^n x_1 \cdot \ldots \cdot x_n$ (for $x \in E$ this is a finite sum, since some coordinate of x has to be 0). Then φ is not bounded on the

M-convergent sequence $(1/n)\chi_{[1,n]}$ and thus is not smooth, i.e. it is not a structure-function of the $\mathscr{M}$-structure associated to the final Pre-structure. But for any $A\subseteq\mathbb{N}$ of density 0 the restriction $\varphi|_{\mathbb{R}^A}$ is smooth and bornological, since there exists an $N\in\mathbb{N}\setminus A$ and thus $\varphi(x)=\sum_{n<N}n^n x_1\ldots x_n$ for all $x\in A$. Thus φ is a structure function for the final M-structure.

Remark. The previous example shows at the same time that a compatible smooth structure on a vector space is not necessarily linearly generated.

7.4 Various counter-examples

7.4.1 Example. A convenient vector space E with a point separating subset $\mathscr{S}\subseteq E'$ such that the bornology on E does not have a basis of $\sigma(E,\mathscr{S})$-closed subsets (quoted in (4.1.17)).

Let $E=\ell^1$, ψ: $\mathbb{N}\times\mathbb{N}\to\mathbb{N}$ a bijection and $\mathscr{S}:=\{\ell\in\ell^\infty\cong(\ell^1)'; \lim_{k\to\infty}\ell(\psi(n,k))$ exists and equals $n\cdot\ell(\psi(n,1))$ for all $n\in\mathbb{N}\}$.

Let us show first that $\mathscr{S}$ separates points: Let $x\in\ell^1$, $x\neq0$. Take i_0 minimal with $x_{i_0}\neq0$, $(n_0,k_0):=\psi^{-1}(i_0)$, and define ℓ by $\ell(\psi(n_0,k)):=\operatorname{sgn}(x_{i_0})$ for $k\leq k_0$ and appropriate for $k>k_0$ such that $\lim_{k\to\infty}\ell(\psi(n_0,k))=n_0\cdot\operatorname{sgn}(x_{i_0})$ and $\ell(\psi(n,k)):=0$ for all $n\neq n_0$. Then clearly $\ell\in\mathscr{S}$ and $\langle\ell,x\rangle=\operatorname{sgn}(x_{i_0})x_{i_0}+\sum_{k>k_0}\langle\ell,\psi(n_0,k)x_{\psi(n_0,k)}\rangle>0$.

We show that $B_1:=\bigcap_{\ell\in\mathscr{S}}\ell^{-1}\ell B$ is not bounded, B being the open unit ball in ℓ^1. It is enough to show that $(n/2)e_{\psi(n,1)}\in B_1$. One has $\langle\ell,B\rangle=]-\|\ell\|_\infty,\|\ell\|_\infty[$, because on the one hand $|\langle\ell,x\rangle|\leq\|\ell\|_\infty\|x\|_1$ and on the other $|t|<\|\ell\|_\infty$ implies $|\ell_n|>|t|$ for some n, thus $|\langle\ell,e_n\rangle|=|\ell_n|>|t|$. So we obtain

$$\left|\left\langle\ell,\frac{n}{2}e_{\psi(n,1)}\right\rangle\right|=\left|\frac{n}{2}\ell(\psi(n,1))\right|=|\lim_{k\to\infty}\ell(\psi(n,k))|\frac{1}{2}\leq\frac{1}{2}\|\ell\|_\infty.$$

7.4.2 Example. Pre-embeddings which are not LCS-embeddings (quoted in (3.2.2)).

Since every separated locally convex space E is a projective limit of normed spaces by (2.1.17) one has an LCS-embedding of E into the product of these normed spaces. Now we can take any non-bornological E (e.g. an infinite-dimensional Banach space with the weak topology, cf. [Jarchow, 1981, p. 271]). Then the bornologification $\gamma\beta E$ of E is Pre-embedded into the product, but it is not an LCS-embedding (provided the cardinality of the index set of the product is non-measurable), since $\gamma\beta E$ is different from E.

7.4.3 Example. A convenient vector space E whose locally convex topology is not quasi-complete (quoted in (2.6.2)).

An easy example consists of the Pre-subspace E of $\mathbb{R}^J$ formed by all $x\in\mathbb{R}^J$ with countable support provided J is uncountable. Clearly E is M-closed in $\mathbb{R}^J$, hence convenient. If the locally convex topology of E were quasi-complete, then the closed bounded set $\{x \in E; |x_j| \leq 1 \text{ for all } j\}$ would be complete; but this is not the case since the characteristic functions χ_F ($F \subseteq J$ finite) form a net which converges in $\mathbb{R}^J$ to $\chi_J \notin E$. For a different example see [Jarchow, 1981, p. 71].

7.4.4 Example. An M-complete convex bornological space whose bornology is not complete (quoted in the remark after (2.6.2)).

An example of such a space E is due to [Nel, 1965] and is as follows: the underlying vector space of E is ℓ^∞ and $B \subseteq E$ is bounded iff $B \subseteq \ell^\infty$ is bounded and there exists some finite-dimensional subspace $F \subseteq E$ with $B \subseteq E_0 + F$, where E_0 denotes the subspace formed by the sequence with finite support (one uses that in complete convex bornological spaces the ℓ^1-hull of bounded subsets is bounded, cf. [Hogbe-Nlend, 1977, p. 126]).

7.4.5 Examples. We consider the following natural morphisms, which are isomorphisms if the spaces involved are finite-dimensional vector spaces, resp. manifolds:

(i)	$E' \tilde{\otimes} F \to L(E; F)$	$x_1 \otimes y \mapsto (x \mapsto x_1(x) \cdot y)$
(ii)	$E' \tilde{\otimes} F' \to (E \tilde{\otimes} F)'$	$x_1 \otimes y_1 \mapsto (x \otimes y \mapsto x_1(x) \cdot y_1(y))$
(iii)	$L(E, F)' \to L(F, E'')$	$g_1 \mapsto (y \mapsto (x_1 \mapsto g_1(x_1(\_) \cdot y)))$
(iv)	$C^\infty(X, \mathbb{R}) \tilde{\otimes} F \to C^\infty(X, F)$	$f \otimes y \mapsto (x \mapsto f(x) \cdot y)$
(v)	$C^\infty(X, F)' \to L(C^\infty(X, \mathbb{R}), F')$	$g_1 \mapsto (f \mapsto (y \mapsto g_1(f(\_) \cdot y)))$
(vi)	$C^\infty(X, \mathbb{R}) \tilde{\otimes} C^\infty(Y, \mathbb{R}) \to C^\infty(X \sqcap Y, \mathbb{R})$	$f \otimes g \mapsto ((x, y) \mapsto f(x) \cdot g(y))$
(vii)	$E'' \tilde{\otimes} F' \to L(E, F)'$	$x_1 \otimes y_1 \mapsto (g \mapsto x_1(y_1 \circ g))$
(viii)	$C^\infty(X, \mathbb{R})' \tilde{\otimes} F' \to C^\infty(X, F)'$	$f_1 \otimes y_1 \mapsto (g \mapsto f_1(y_1 \circ g))$
(ix)	$C^\infty(X, \mathbb{R})' \tilde{\otimes} C^\infty(Y, \mathbb{R})' \to C^\infty(X \sqcap Y, \mathbb{R})'$	$f_1 \otimes g_1 \mapsto (h \mapsto f_1(g_1 \circ h^\vee))$
(x)	$\lambda(X) \to C^\infty(X, \mathbb{R})'$	cf. (5.1.1)

(i) $(\mathbb{R}^\mathbb{N})' \tilde{\otimes} \mathbb{R}^\mathbb{N}$ and $L(\mathbb{R}^\mathbb{N}, \mathbb{R}^\mathbb{N})$ are not isomorphic:

$$\mathbb{R}^{(\mathbb{N})} \tilde{\otimes} \mathbb{R}^\mathbb{N} \cong (\mathbb{R} \tilde{\otimes} \mathbb{R}^\mathbb{N})^{(\mathbb{N})} \cong (\mathbb{R}^\mathbb{N})^{(\mathbb{N})} \ncong (\mathbb{R}^{(\mathbb{N})})^\mathbb{N} \cong L(\mathbb{R}^\mathbb{N}, \mathbb{R})^\mathbb{N} \cong L(\mathbb{R}^\mathbb{N}, \mathbb{R}^\mathbb{N}).$$

(ii) $(\mathbb{R}^\mathbb{N})' \tilde{\otimes} (\mathbb{R}^{(\mathbb{N})})'$ and $(\mathbb{R}^\mathbb{N} \tilde{\otimes} \mathbb{R}^{(\mathbb{N})})'$ are not isomorphic:

$$(\mathbb{R}^\mathbb{N})' \tilde{\otimes} (\mathbb{R}^{(\mathbb{N})})' \cong \mathbb{R}^{(\mathbb{N})} \tilde{\otimes} \mathbb{R}^\mathbb{N} \cong (\mathbb{R}^\mathbb{N})^{(\mathbb{N})} \ncong (\mathbb{R}^{(\mathbb{N})})^\mathbb{N} \simeq ((\mathbb{R}^\mathbb{N})^{(\mathbb{N})})' \cong (\mathbb{R}^\mathbb{N} \tilde{\otimes} \mathbb{R}^{(\mathbb{N})})'.$$

One can even achieve E and F to be the same, if one takes both equal to $\mathbb{R}^\mathbb{N} \oplus \mathbb{R}^{(\mathbb{N})}$.

(iii) $L(\mathbb{R}^\mathbb{N}, \mathbb{R}^\mathbb{N})'$ and $L(\mathbb{R}^\mathbb{N}, (\mathbb{R}^\mathbb{N})'')$ are not isomorphic:

$$L(\mathbb{R}^\mathbb{N}, \mathbb{R}^\mathbb{N})' \cong ((\mathbb{R}^{(\mathbb{N})})^\mathbb{N})' \cong (\mathbb{R}^\mathbb{N})^{(\mathbb{N})} \ncong (\mathbb{R}^{(\mathbb{N})})^\mathbb{N} \cong L(\mathbb{R}^\mathbb{N}, \mathbb{R}^\mathbb{N}).$$

(iv) $C^\infty(\mathbb{R}, \mathbb{R}) \tilde{\otimes} \mathbb{R}^{(\mathbb{N})} \to C^\infty(\mathbb{R}, \mathbb{R}^{(\mathbb{N})})$ is not an isomorphism: Using that $C^\infty(\mathbb{R}, E) \to E^\mathbb{N}$, $c \mapsto (c(n))_n$ admits a right inverse $x \mapsto \sum_n x_n h(t-n)$, where $h \in C^\infty(\mathbb{R}, \mathbb{R})$ with $h(0) = 1$ and $\operatorname{supp}(h) \subseteq [-1, 1]$, we conclude that $\mathbb{R}^\mathbb{N}$ is a complemented subspace of $C^\infty(\mathbb{R}, \mathbb{R})$ and that $(\mathbb{R}^{(\mathbb{N})})^\mathbb{N}$ is a complemented subspace of $C^\infty(\mathbb{R}, \mathbb{R}^{(\mathbb{N})})$. Thus this counter-example follows from the one given in (i) using (8.3.8).

(v) $C^\infty(\mathbb{R}, \mathbb{R}^{(\mathbb{N})})' \to L(C^\infty(\mathbb{R}, \mathbb{R}), (\mathbb{R}^{(\mathbb{N})})')$ is not an isomorphism:

As in (iv) one considers $\mathbb{R}^{\mathbb{N}}$ as complemented subspace of $C^\infty(\mathbb{R}, \mathbb{R})$ and $(\mathbb{R}^{(\mathbb{N})})^{\mathbb{N}}$ as complemented subspace of $C^\infty(\mathbb{R}, \mathbb{R}^{(\mathbb{N})})$. Thus this counter-example follows from the one given in (iii) using (8.3.8).

(vi) $C^\infty(\mathbb{R}, \mathbb{R}) \tilde{\otimes} C^\infty(\mathbb{R}^{\mathbb{N}}, \mathbb{R}) \to C^\infty(\mathbb{R} \sqcap \mathbb{R}^{\mathbb{N}}, \mathbb{R})$ is not an isomorphism: One considers $\mathbb{R}^{(\mathbb{N})} \cong L(\mathbb{R}^{\mathbb{N}}, \mathbb{R})$ as complemented subspace of $C^\infty(\mathbb{R}^{\mathbb{N}}, \mathbb{R})$; because of $C^\infty(\mathbb{R} \sqcap \mathbb{R}^{\mathbb{N}}, \mathbb{R}) \cong C^\infty(\mathbb{R}, C^\infty(\mathbb{R}^{\mathbb{N}}, \mathbb{R}))$ this example follows from the one given in (iv) using (8.3.8).

One can even take $X = Y := \mathbb{R}^{\mathbb{N}}$, since $C^\infty(\mathbb{R}^{\mathbb{N}}, E)$ contains $C^\infty(\mathbb{R}, E)$ as complemented subspace.

We do not think that the morphisms in (vii)–(x) are always isomorphisms. We have no counter-examples but we now show that one for (ix) would yield others for (vii), (viii) and (x).

(vii) Let φ denote the natural morphism $E'' \tilde{\otimes} F' \to L(E, F)'$. In the case where E and F are reflexive the dual φ^* is a retraction whose right inverse is ι: $L(E, F) \to L(E, F)''$ composed with the isomorphism $(E'' \tilde{\otimes} F')' \cong (E \tilde{\otimes} F')' \cong L(E, F'; \mathbb{R}) \cong L(E, F'') \cong L(E, F)$. Thus the bidual φ^{**} is a section which implies that φ is initial.

We do not know an example where φ is not surjective. Recall in this connection that denseness of $E \tilde{\otimes} F'$ in $L(E, F)'$ with respect to the locally convex topology is equivalent to the reflexivity of $L(E, F)$, cf. (5.4.18).

(viii) An example showing that this map is not an isomorphism would also be an example for (vii) by setting $E := \lambda(X) \to C^\infty(X, \mathbb{R})'$.

(ix) An example showing that this map is not an isomorphism would also be an example for (viii) by setting $F := C^\infty(Y, \mathbb{R})$.

(x) An example showing that the map in (ix) need not be an isomorphism would also yield an example for (x): If for all X the morphism $\lambda X \to C^\infty(X, \mathbb{R})'$ in (x) were an isomorphism then using the isomorphism $\alpha(X \sqcap Y) \cong \lambda X \tilde{\otimes} \lambda Y$ of (5.2.4) would show that the morphism in (ix) is also an isomorphism.

8 SOME CATEGORICAL NOTIONS AND NOTATIONS

This chapter is by no means an introduction or a survey on category theory. It is intended as a helpful guide for a reader who is not familiar with categorical terms and reasonings. Therefore it only recalls those standard definitions and basic results which are used in this book. No proofs are given; they can either be found in the standard textbooks on category theory or are trivial.

8.1 Categories

8.1.1 Definition. A category $\mathscr{X}$ consists of

(i) a class denoted $|\mathscr{X}|$;
(ii) for each pair $X, Y \in |\mathscr{X}|$ a set denoted $\mathscr{X}(X, Y)$;
(iii) for each triple $X, Y, Z \in |\mathscr{X}|$ a map $\mathscr{X}(Y, Z) \times \mathscr{X}(X, Y) \to \mathscr{X}(X, Z)$ denoted $(g, f) \mapsto g \circ f$;

and the following axioms are supposed to hold:

(a) $(X_1, Y_1) \neq (X_2, Y_2) \Rightarrow \mathscr{X}(X_1, Y_1) \cap \mathscr{X}(X_2, Y_2) = \varnothing$;
(b) $f \in \mathscr{X}(X, Y)$, $g \in \mathscr{X}(Y, Z)$, $h \in \mathscr{X}(Z, U) \Rightarrow h \circ (g \circ f) = (h \circ g) \circ f$;
(c) for each $X \in |\mathscr{X}|$ there exists an element $1_X \in \mathscr{X}(X, X)$ such that $f \circ 1_X = f$ for any $f \in \mathscr{X}(X, Y)$ and $1_X \circ g = g$ for any $g \in \mathscr{X}(Y, X)$.

The elements of $|\mathscr{X}|$ are called the *objects* of $\mathscr{X}$; the elements of $\mathscr{X}(X, Y)$ are called the *$\mathscr{X}$-morphisms* from X to Y or also the morphisms having X as domain and Y as range and one also writes $f: X \to Y$; the maps $(g, f) \mapsto g \circ f$ are called *compositions* of $\mathscr{X}$.

A morphism $f: X \to Y$ of $\mathscr{X}$ is called *isomorphism* iff there exists a morphism $g: Y \to X$ in $\mathscr{X}$ satisfying $f \circ g = 1_Y$ and $g \circ f = 1_X$. Of course g is uniquely determined and denoted by f^{-1}.

If one has $f \circ g = 1_X$ for two $\mathscr{X}$-morphisms $f: X \to Y$ and $g: Y \to X$ then f is called *retraction* (of g) and g is called *section* (of f).

In most examples of this book the objects are sets with additional structures (e.g. algebraic or topological); and the morphisms certain maps between the underlying sets, compatible with the structures (e.g. homomorphic or continuous); and the composition will be the usual composition of maps.

In many important examples such as the category $\underline{\text{Set}}$ of sets, the category $\underline{\text{VS}}$ of (real) vector spaces or the category $\underline{\text{Top}}$ of topological spaces the objects form a proper class and not a set; if they form a set one speaks of a *small category*.

For two categories $\mathscr{X}$ and $\mathscr{Y}$ the *product category* $\mathscr{X} \times \mathscr{Y}$ is defined as follows. Objects of $\mathscr{X} \times \mathscr{Y}$ are the pairs (X, Y) with X an object of $\mathscr{X}$ and Y an object of $\mathscr{Y}$; morphisms of $\mathscr{X} \times \mathscr{Y}$ from (X_1, Y_1) to (X_2, Y_2) are pairs (f, g) with $f: X_1 \to X_2$ an $\mathscr{X}$-morphism and $g: Y_1 \to Y_2$ a $\mathscr{Y}$-morphism; composition consists in composing the components.

A category $\mathscr{X}$ is called *subcategory* of a category $\mathscr{Y}$ iff the objects of $\mathscr{X}$ form a subclass of the objects of $\mathscr{Y}$, and the morphisms of $\mathscr{X}$ are morphisms of $\mathscr{Y}$, more precisely $\mathscr{X}(X_1, X_2) \subseteq \mathscr{Y}(X_1, X_2)$ for all $X_1, X_2 \in |\mathscr{X}|$; and $\mathscr{X}$-morphisms are composed as $\mathscr{Y}$-morphisms. A subcategory $\mathscr{X}$ of $\mathscr{Y}$ is called *full* iff $\mathscr{X}(X_1, X_2) = \mathscr{Y}(X_1, X_2)$ for all $X_1, X_2 \in |\mathscr{X}|$; and it is called *replete* iff any object in $\mathscr{Y}$ that is $\mathscr{Y}$-isomorphic to an object in $\mathscr{X}$ lies also in $\mathscr{X}$.

8.2 Functors and natural transformations

8.2.1 Definition. A *functor* $\alpha: \mathscr{X} \to \mathscr{Y}$ from a category $\mathscr{X}$ to a category $\mathscr{Y}$ consists of

(i) a map $|\mathscr{X}| \to |\mathscr{Y}|$ for the objects denoted by $X \mapsto \alpha X$;
(ii) for all $\mathscr{X}$-objects X_1, X_2 a map $\mathscr{X}(X_1, X_2) \to \mathscr{Y}(\alpha X_1, \alpha X_2)$ for the morphisms denoted by $f \mapsto \alpha f$;

and the following axioms are supposed to hold:

(a) $\alpha(1_X) = 1_{\alpha X}$ for any object X of $\mathscr{X}$;
(b) $\alpha(f \circ g) = \alpha f \circ \alpha g$ whenever $f \circ g$ is defined.

So-called contravariant functors $\mathscr{X} \to \mathscr{Y}$ can be avoided by considering ordinary functors $\mathscr{X}^{\text{op}} \to \mathscr{Y}$, where the category $\mathscr{X}^{\text{op}}$ opposite to a category $\mathscr{X}$ has as objects the same as $\mathscr{X}$; $\mathscr{X}^{\text{op}}(X, Y) := \mathscr{X}(Y, X)$ for all objects X, Y; and $f \circ g$ in $\mathscr{X}^{\text{op}}$ is $g \circ f$ in the sense of $\mathscr{X}$.

For any category $\mathscr{X}$ one has a functor $\mathscr{X}(\_, \_): \mathscr{X}^{\text{op}} \times \mathscr{X} \to \underline{\text{Set}}$, called the *hom-functor* of $\mathscr{X}$. It associates to (X_1, X_2) the set $\mathscr{X}(X_1, X_2)$ of $\mathscr{X}$-morphisms from X_1 to X_2 and to (g, f) the map $\mathscr{X}(g, f): h \mapsto f \circ h \circ g$. If $g = 1_X$ then we write f_* or $\mathscr{X}(X, f)$ instead of $\mathscr{X}(1_X, f)$, i.e. $f_*(h) = f \circ h$. Similarly we write g^* or $\mathscr{X}(g, X)$ instead of $\mathscr{X}(g, 1_X)$, i.e. $g^*(h) = h \circ g$. Then $\mathscr{X}(g, f) = f_* \circ g^* = g^* \circ f_*$.

One has always a functor denoted $\text{Id}_{\mathscr{X}}: \mathscr{X} \to \mathscr{X}$ that acts on objects and morphism as identity.

Functors $\alpha: \mathscr{X} \to \mathscr{Y}$ and $\beta: \mathscr{Y} \to \mathscr{Z}$ can be composed in the obvious way to form a functor $\beta \circ \alpha: \mathscr{X} \to \mathscr{Z}$.

A functor $\alpha: \mathscr{X} \to \mathscr{Y}$ is called *faithful* (resp. *full*) iff for any two objects X_1, X_2 of $\mathscr{X}$ the mapping $f \mapsto \alpha(f)$, $\mathscr{X}(X_1, X_2) \to \mathscr{Y}(\alpha X_1, \alpha X_2)$ is injective (resp. surjective). A full and faithful functor is called an *embedding*. A category $\mathscr{X}$ with a given faithful functor $\mathscr{X} \to \underline{\mathrm{Set}}$ is called a *concrete category*.

8.2.2 Definition. A *natural transformation* $\vartheta: \alpha \to \beta$, where α, $\beta: \mathscr{X} \to \mathscr{Y}$ are two functors, is a family of $\mathscr{Y}$-morphisms $\vartheta_X: \alpha X \to \beta X$ $(X \in |\mathscr{X}|)$ that satisfies the following axiom: For every $\mathscr{X}$-morphism $f: X_1 \to X_2$ one has $\vartheta_{X_2} \circ \alpha f = \beta f \circ \vartheta_{X_1}$.

Two functors α, $\beta: \mathscr{X} \to \mathscr{Y}$ are called *isomorphic* iff there exists a *natural isomorphism*, i.e. a natural transformation ϑ: $\alpha \to \beta$ such that ϑ_X is a $\mathscr{Y}$-isomorphism for all $X \in \mathscr{X}$.

A functor λ: $\mathscr{X} \to \mathscr{Y}$ is called *representable* iff there exists a *representation*, i.e. a natural isomorphism ϑ: $\lambda \to \mathscr{X}(A, \_)$ for some $\mathscr{X}$-object A.

8.3 Limits and colimits

8.3.1 Definition. A *diagram* in a category $\mathscr{X}$ is a functor $\nabla: \mathscr{J} \to \mathscr{X}$, where $\mathscr{J}$ is a small category. Let us write $X_j := \nabla(j)$ for any $j \in |\mathscr{J}|$.

A *source* in $\mathscr{X}$ is a $\mathscr{X}$-object X together with a family of $\mathscr{X}$-morphisms $f_j: X \to X_j$ $(j \in J)$. A *sink* in $\mathscr{X}$ is a $\mathscr{X}$-object X together with a family of $\mathscr{X}$-morphisms $f_j: X_j \to X$ $(j \in J)$.

A *limit* of ∇ is a source $(X_\infty;\ p_j: X_\infty \to X_j\ (j \in |\mathscr{J}|))$ in $\mathscr{X}$ with $p_j = \nabla\varphi \circ p_i$ for any $\mathscr{J}$-morphism $\varphi: i \to j$, having the following *universal property*: for each source $(X, f_j: X \to X_j\ (j \in |\mathscr{J}|))$ in $\mathscr{X}$ with the compatibility property $f_j = \nabla\varphi \circ f_i$ for any $\mathscr{J}$-morphism $\varphi: i \to j$ there exists a unique $\mathscr{X}$-morphism $f: X \to X_\infty$ with $f_j = p_j \circ f$ for all $j \in |\mathscr{J}|$.

The dual notion is the following: A *colimit* of ∇ is a sink $(X_\infty;\ q_j: X_j \to X_\infty\ (j \in |\mathscr{J}|))$ in $\mathscr{X}$ with $q_i = q_j \circ \nabla\varphi$ for any $\mathscr{J}$-morphism $\varphi: i \to j$, having the following *universal property*: for each source $(X;\ g_j: X_j \to X\ (j \in |\mathscr{J}|))$ in $\mathscr{X}$ with the same compatibility property there exists a unique $\mathscr{X}$-morphism $g: X_\infty \to X$ with $g_j = g \circ q_j$ for all $j \in |\mathscr{J}|$.

We mention some important special cases:

8.3.2 Products and coproducts. One takes as $\mathscr{J}$ a *discrete small category*, i.e. a set J of objects, and as only morphisms the identities 1_j for $j \in J$. A diagram $\nabla: \mathscr{J} \to \mathscr{X}$ is uniquely determined by the family X_j $(j \in J)$ of objects. The given definitions of limit and colimit are simplified because the conditions involving $\mathscr{J}$-morphisms become void. A limit of ∇ is called *product* of the family X_j and is denoted by $\mathrm{pr}_j: \Pi_{i \in J} X_i \to X_j$ $(j \in J)$; similarly a colimit is called *coproduct* and is denoted by $\mathrm{in}_j: X_j \to \amalg_{i \in J} X_i$ $(j \in J)$.

8.3.3 Inverse or projective limits. Let J be a set directed by a relation $\succ$, cf. (2.2.1). One defines a small category $\mathscr{I}$: The set of objects is J; $\mathscr{I}(j, i)$ has exactly one element if $j \succ i$ and is empty otherwise; the composition is then unambiguous. A diagram $\mathscr{I} \to \mathscr{X}$ consists of a family of $\mathscr{X}$-objects X_j together with $\mathscr{X}$-morphisms $f_{j,i}: X_j \to X_i$ for $j \succ i$ that satisfy $f_{j,j} = 1_{X_j}$ for all $j \in J$ and $f_{k,i} = f_{j,i} \circ f_{k,j}$ for $k \succ j \succ i$. Such a diagram is usually called an inverse or projective system and a limit of it is called an *inverse* or *projective limit*.

8.3.4 Direct or inductive limits. Let $(J, \succ)$ be a directed set as before. One forms the category $\mathscr{I}$ in the dual way: The set of objects is J and $\mathscr{I}(i, j)$ consists of exactly one element if $j \succ i$ and is empty otherwise. A diagram $\mathscr{I} \to \mathscr{X}$ consists of a family of $\mathscr{X}$-objects X_j together with $\mathscr{X}$-morphisms $f_{i,j}: X_i \to X_j$ for $j \succ i$ that satisfy $f_{j,j} = 1_{X_j}$ for all j and $f_{i,k} = f_{j,k} \circ f_{i,j}$ for $k \succ j \succ i$. Such a diagram is usually called a direct or inductive system and a colimit of it is called a *direct* or *inductive limit*.

8.3.5 Pullbacks and pushouts. One takes as $\mathscr{I}$ a category with three objects, say 0, 1, 2, and with two morphisms besides the identities, namely $1 \to 0$ and $2 \to 0$. A diagram $\nabla: \mathscr{I} \to \mathscr{X}$ consists of three objects X_0, X_1, X_2 and two morphisms $g_1: X_1 \to X_0$ and $g_2: X_2 \to X_0$. A limit of this diagram is determined by two maps $f_1: X_\infty \to X_1$ and $f_2: X_\infty \to X_2$ satisfying $g_1 \circ f_1 = g_2 \circ f_2$. It is called a *pullback* of (g_1, g_2). *Pushout* is the dual notion, i.e. a pushout of two morphisms $g_1: X_0 \to X_1$ and $g_2: X_0 \to X_2$ is a pullback in $\mathscr{X}^{\text{op}}$ of (g_1, g_2).

8.3.6 Equalizers and coequalizers. One takes as $\mathscr{I}$ a category with two objects, say 0 and 1, and with two morphisms $0 \to 1$ besides the identities. A diagram $\nabla: \mathscr{I} \to \mathscr{X}$ consists of two objects X_0, X_1 and two morphisms $g_i: X_0 \to X_1$ $(i = 1, 2)$. A limit of this diagram is determined by a map $f: X_\infty \to X_0$ satisfying $g_1 \circ f = g_2 \circ f$. It is called an *equalizer* of (g_1, g_2).

A coequalizer of (g_1, g_2) is the dual notion, i.e. is an equalizer in $\mathscr{X}^{\text{op}}$ of (g_1, g_2).

8.3.7 Definition. A category $\mathscr{X}$ is called *complete* iff all diagrams in $\mathscr{X}$ have a limit. It is called *cocomplete* iff all diagrams in $\mathscr{X}$ have a colimit.

8.3.8 Lemma. *Suppose there are given morphisms:*

$$\begin{array}{ccc} X & \overset{r}{\underset{s}{\rightleftarrows}} & Y \\ \downarrow f & & \downarrow g \\ X' & \overset{r'}{\underset{s'}{\rightleftarrows}} & Y' \end{array}$$

such that $r \circ s = \text{id}_Y$, $r' \circ s' = \text{id}_{Y'}$, $g \circ r = r' \circ f$, $f \circ s = s' \circ g$.

If f is a coequalizer, a section or a retraction then g has the same property.

8.4 Adjoint functors

8.4.1 Definition. Let ρ and λ be two functors as follows: $\mathscr{X} \underset{\lambda}{\overset{\rho}{\rightleftarrows}} \mathscr{Y}$. Then ρ is called *right adjoint* to λ (and λ *left adjoint* to ρ) if there are bijections between the following sets:

$$\mathscr{X}(\lambda Y, X) \cong \mathscr{Y}(Y, \rho X) \text{ for } X \in \mathscr{X},\ Y \in \mathscr{Y}$$

in such a way that they form a natural transformation between the functors $\mathscr{X}(\lambda\_, \_)$, $\mathscr{Y}(\_, \rho\_): \mathscr{Y}^{\mathrm{op}} \times \mathscr{X} \to \underline{\mathrm{Set}}$. This natural transformation is called an *adjunction* between ρ and λ.

8.4.2 If such an adjunction is given, then to $1_{\lambda Y}$ corresponds a $\mathscr{Y}$-morphism $\eta_Y: Y \to \rho\lambda Y$ and similarly to $1_{\rho X}$ an $\mathscr{X}$-morphism $\varepsilon_X: \lambda\rho X \to X$. These are easily verified to constitute natural transformations $\eta: \mathrm{Id}_{\mathscr{Y}} \to \rho \circ \lambda$ and $\varepsilon: \lambda \circ \rho \to \mathrm{Id}_{\mathscr{X}}$, and to satisfy the identities $\varepsilon_{\lambda Y} \circ \lambda\eta_Y = 1_{\lambda Y}$ and $\rho\varepsilon_X \circ \eta_{\rho X} = 1_{\rho X}$. One calls η the *unit* and ε the *counit* of the adjunction. Conversely, any two natural transformations $\eta: \mathrm{Id}_{\mathscr{Y}} \to \rho \circ \lambda$ and $\varepsilon: \lambda \circ \rho \to \mathrm{Id}_{\mathscr{X}}$ satisfying the above identities yield an adjunction: one defines $\mathscr{X}(\lambda Y, X) \to \mathscr{Y}(Y, \rho X)$ by $f \mapsto \rho f \circ \eta_Y$ and $(Y, \rho X) \to \mathscr{X}(\lambda Y, X)$ by $g \mapsto \varepsilon_X \circ \lambda g$. The assumed identities imply that these are inverse to each other.

In many of our examples the objects of $\mathscr{X}$ and $\mathscr{Y}$ are structured sets and η_Y and ε_X the identity maps of the underlying sets. In these cases the transformations $\mathrm{Id}_{\mathscr{Y}} \to \rho \circ \lambda$ and $\lambda \circ \rho \to \mathrm{Id}_{\mathscr{X}}$ mean that $\rho\lambda$ coarsens and $\lambda\rho$ refines the respective structures.

8.4.3 Often only one functor $\rho: \mathscr{X} \to \mathscr{Y}$ is given and one wants to know whether it has an adjoint. Suppose for every object Y of $\mathscr{Y}$ one has an object λY in $\mathscr{X}$ together with a $\mathscr{Y}$-morphism $\eta_Y: Y \to \rho(\lambda Y)$ such that for any $\mathscr{Y}$-morphism $f: Y \to \rho(X)$ with $X \in |\mathscr{X}|$ there exists a unique $\mathscr{X}$-morphism $\bar{f}: \lambda Y \to X$ such that $f = \rho(\bar{f}) \circ \eta_Y$ (In this situation the pair $(\lambda Y, \eta_Y)$ is called *universal arrow* from Y to ρ). Then λ extends in a unique way to a functor $\lambda: \mathscr{Y} \to \mathscr{X}$ that is left adjoint to ρ and has η as unit of adjunction. The adjunction is given by $f \in \mathscr{Y}(Y, \rho X) \mapsto \bar{f} \in \mathscr{X}(\lambda Y, X)$.

In the dual way one obtains a right adjoint to a given functor $\lambda: \mathscr{Y} \to \mathscr{X}$ if for every $\mathscr{X}$-object X one has a $\mathscr{Y}$-object ρX together with a $\mathscr{X}$-morphism $\varepsilon_X: \lambda(\rho X) \to X$ such that for any $\mathscr{X}$-morphism $g: \lambda Y \to X$ there exists a unique $\mathscr{Y}$-morphism $\bar{g}: Y \to \rho X$ satisfying $g = \varepsilon_X \circ \lambda\bar{g}$.

8.4.4 If the inclusion functor of a subcategory has a left (right) adjoint one speaks of a *reflective* (*coreflective*) subcategory. A left (right) adjoint to an inclusion functor is called *reflector* (*coreflector*).

8.5 Adjoint functors and limits

8.5.1 Proposition. (*i*) *A functor* $\rho: \mathscr{X} \to \mathscr{Y}$ *which has a left adjoint preserves limits; i.e. if* $p_j: X_\infty \to X_j$ $(j \in |\mathscr{I}|)$ *is a limit of a diagram* $\nabla: \mathscr{I} \to \mathscr{X}$, *then* $\rho(p_j): \rho(X_\infty) \to \rho(X_j)$ $(j \in |\mathscr{I}|)$ *is a limit of the diagram* $\rho \circ \nabla: \mathscr{I} \to \mathscr{Y}$.

(*ii*) *Similarly a functor which has a right adjoint preserves colimits.*

Remark. That a functor $\rho: \mathscr{X}^{\mathrm{op}} \to \mathscr{Y}$ preserves limits means that it transforms colimits in $\mathscr{X}$ into limits in $\mathscr{Y}$.

8.5.2 Proposition. *Let* $\iota: \mathscr{X} \to \mathscr{Y}$ *be the inclusion functor of a full replete subcategory.*

(*i*) *Suppose* $\mathscr{X}$ *is a reflective subcategory of* $\mathscr{Y}$. *Then one can choose a left adjoint (reflector)* $\lambda: \mathscr{Y} \to \mathscr{X}$ *such that* $\lambda \circ \iota = \mathrm{Id}_{\mathscr{X}}$. *Let* $\nabla: \mathscr{I} \to \mathscr{X}$ *be a diagram. If the diagram* $\iota \circ \nabla: \mathscr{I} \to \mathscr{Y}$ *has a limit (in* $\mathscr{Y}$*) then this limit belongs to* $\mathscr{X}$ *and is a limit of* ∇ *(in* $\mathscr{X}$*). If the diagram* $\iota \circ \nabla: \mathscr{I} \to \mathscr{Y}$ *has a colimit (in* $\mathscr{Y}$*) then one obtains (according to* (8.5.1)*) a colimit of* ∇ *(in* $\mathscr{X}$*) by applying the functor* λ.

(*ii*) *Suppose* $\mathscr{X}$ *is a coreflective subcategory of* $\mathscr{Y}$. *Then one can choose a right adjoint (coreflector)* $\rho: \mathscr{Y} \to \mathscr{X}$ *such that* $\rho \circ \iota = \mathrm{Id}_{\mathscr{X}}$. *Let* $\nabla: \mathscr{I} \to \mathscr{X}$ *be a diagram. If the diagram* $\iota \circ \nabla: \mathscr{I} \to \mathscr{Y}$ *has a colimit (in* $\mathscr{Y}$*) then this colimit belongs to* $\mathscr{X}$ *and is a colimit of* ∇ *(in* $\mathscr{X}$*). If the diagram* $\iota \circ \nabla: \mathscr{I} \to \mathscr{Y}$ *has a limit (in* $\mathscr{Y}$*) then one obtains (according to* (8.5.1)*) a limit of* ∇ *(in* $\mathscr{X}$*) by applying the functor* ρ.

8.5.3 Corollary. *Let* $\mathscr{X}$ *be a full replete reflective (resp. coreflective) subcategory of a complete and cocomplete category* $\mathscr{Y}$. *Then* $\mathscr{X}$ *is complete and cocomplete. Limits (resp. colimits) in* $\mathscr{X}$ *are obtained by forming them in* $\mathscr{Y}$. *Colimits (resp. limits) in* $\mathscr{X}$ *are obtained by applying the retraction functor to the colimit (resp. limit) in* $\mathscr{Y}$.

8.6 Cartesian closed categories

8.6.1 Definition. A category $\mathscr{X}$ is called *cartesian closed* iff

(i) $\mathscr{X}$ has a terminal object T, i.e. an object such that for any object X there exists exactly one morphism $X \to T$;
(ii) For any pair of objects a product exists, i.e. the functor $\mathscr{X} \to \mathscr{X} \times \mathscr{X}$, $X \mapsto (X, X)$, $f \mapsto (f, f)$ has a right adjoint $\_ \sqcap \_ : \mathscr{X} \times \mathscr{X} \to \mathscr{X}$;
(iii) For every object Y the functor $\_ \sqcap Y$ has a right adjoint.

Remark. Let $\Omega(Y, \_): \mathscr{X} \to \mathscr{X}$ denote a right adjoint to $\_ \sqcap Y$. Then Ω extends in a unique way to a functor $\Omega: \mathscr{X}^{\mathrm{op}} \times \mathscr{X} \to \mathscr{X}$ such that one has bijections $\mathscr{X}(X \sqcap Y, Z) \cong \mathscr{X}(X, \Omega(Y, Z))$ being natural in X, Y, Z.

Unit and counit of the adjunction(s) are of the form $\eta_Z^Y: Z \to \Omega(Y, Z \sqcap Y)$ and $\varepsilon_X^Y: \Omega(Y, X) \sqcap Y \to X$.

8.6.2 Example. The category Set is cartesian closed. Any singleton is a terminal object, the product is the cartesian product, and as Ω one can take the hom-functor Set(_, _). The unit consists of the *insertion* maps $\text{ins}_Z^Y: Z \to \text{Set}(Y, Z \times Y)$, $\text{ins}(z)(y) := (z, y)$ and the counit of the *evaluation* maps ev_X^Y: $\text{Set}(Y, X) \times Y \to X$, $\text{ev}(f, y) := f(y)$.

The adjunction $\text{Set}(X \times Y, Z) \cong \text{Set}(X, \text{Set}(Y, Z))$ is given by the map $f \mapsto f^\vee$, where $f^\vee(x)(y) := f(x, y)$; the inverse map is $g \mapsto g^\wedge$, where $g^\wedge(x, y) = g(x)(y)$.

8.6.3 Remark. For the cartesian closed categories $\mathscr{X}$ in this book the forgetful functor $\mathscr{X} \to \text{Set}$ is represented by $\mathscr{X}(T, \_)$ (T denotes a terminal object). In this situation one can choose Ω, $\sqcap$ and the adjunction in such a way that the underlying set of $\Omega(X, Y)$ is the set of morphisms $\mathscr{X}(X, Y)$ and that of $X \sqcap Y$ is the cartesian product of the underlying sets of X and Y. Unit and counit consist of the insertion and evaluation maps, and the adjunction is given $f \mapsto f^\vee$ and $g \mapsto g^\wedge$ where $f^\vee$ and $g^\wedge$ are defined by the same formulas as above.

8.6.4 Remark. In any cartesian closed category $\mathscr{X}$ one has a natural isomorphism $\Omega(X \sqcap Y, Z) \cong \Omega(X, \Omega(Y, Z))$, which can be written as *exponential law* $Z^{(X \sqcap Y)} \cong (Z^Y)^X$ if one uses the notation Z^Y for $\Omega(Z, Y)$. Other exponential laws that are valid are $X^T \cong X$, $(X \sqcap Y)^Z \cong X^Z \sqcap Y^Z$ and $X \sqcap T \cong X$. One has always a natural morphism $Z^Y \sqcap Y^X \to Z^X$; in the concrete situation above this is the composition map.

8.6.5 A category $\mathscr{X}$ is called *locally cartesian* closed iff the comma-categories $(\mathscr{X}, X)$ are cartesian closed for all $\mathscr{X}$-objects X, where the *comma-category* $(\mathscr{X}, X)$ has as objects all $\mathscr{X}$-morphism with range X and as morphisms $\varphi: f \to g$ those $\mathscr{X}$-morphisms $\text{dom}(f) \to \text{dom}(g)$ for which $g \circ \varphi = f$. The product in $(\mathscr{X}, X)$ of two $(\mathscr{X}, X)$-objects f and g is just the pullback in $\mathscr{X}$ of f and g. Thus locally cartesian closedness of $\mathscr{X}$ implies that pullbacks in $\mathscr{X}$ preserve colimits. And since for any terminal object T of $\mathscr{X}$ the comma-category $(\mathscr{X}, T)$ is isomorphic to $\mathscr{X}$ the locally cartesian closedness implies cartesian closedness provided a terminal object exists.

8.7 Initial sources and final sinks

For this section we suppose that a faithful functor $\varphi: \mathscr{X} \to \mathscr{Y}$ is given, i.e. a functor such that for all $\mathscr{X}$-objects X, Y the map $\mathscr{X}(X, Y) \to \mathscr{Y}(\varphi X, \varphi Y)$ is injective.

8.7.1 Definition. (i) An *initial source* with respect to φ is a source formed by an $\mathscr{X}$-object X and a family $f_j: X \to X_j$ $(j \in J)$ of $\mathscr{X}$-morphisms with the property: If Z is any $\mathscr{X}$-object and $g: \varphi Z \to \varphi X$ is any $\mathscr{Y}$-morphism then $\varphi f_j \circ g \in \varphi \mathscr{X}(Z, X_j)$ for all $j \in J$ implies $g \in \varphi \mathscr{X}(Z, X)$. An *initial morphism* with respect to φ is an initial source with a singleton as index set J.

(ii) The dual notion is the following. A *final sink* with respect to φ is a sink formed by an $\mathscr{X}$-object X and a family $f_j: X_j \to X$ of $\mathscr{X}$-morphisms with the property: if Z is any $\mathscr{X}$-object and $g: \varphi X \to \varphi Z$ any $\mathscr{Y}$-morphism then $g \circ \varphi f_j \in \varphi \mathscr{X}(X_j, Z)$ for all $j \in J$ implies $g \in \varphi \mathscr{X}(X, Z)$. A *final morphism* with respect to φ is a final sink with a singleton as index set J.

(iii) An initial source $f_j: X \to X_j$ (resp. final sink $f_j: X_j \to X$), $j \in J$, with respect to φ is called *initial source* (*resp. final sink*) *over a family* of $\mathscr{Y}$-morphisms $g_j: Y \to \varphi X_j$ (resp. $g_j: \varphi X_j \to Y$) iff $\varphi X = Y$ and $\varphi f_j = g_j$ for all $j \in J$. We say that $\mathscr{X}$ *has initial sources* (*resp. final sinks*) with respect to φ iff over every source $g_j: Y \to \varphi X_j$ (resp. sink $g_j: \varphi X_j \to Y$), $j \in J$, in $\mathscr{Y}$ there exists an initial source (resp. a final sink) in $\mathscr{X}$.

Remark. We only use the special case where the objects in $\mathscr{X}$ can be considered as objects of $\mathscr{Y}$ together with some additional structure. Then taking an initial source over a family $g_j: Y \to Y_j$ $(j \in J)$ amounts to supply Y with an additional structure and we will call this structure the *initial structure induced by the morphisms* g_j $(j \in J)$. The additional structure corresponding to a final sink over a family will be called *final structure induced by that family.*

Remark. If one has a family $f_j: X \to X_j$ $(j \in J)$ of $\mathscr{X}$-morphisms and the product $\mathrm{pr}_j: \Pi_{i \in J} X_i \to X_j$ exists, then the f_j $(j \in J)$ form an initial source if and only if the corresponding morphism $f: X \to \Pi_{i \in J} X_i$ is an initial morphism.

The dual result is: if one has a family $f_j: X_j \to X$ $(j \in J)$ of $\mathscr{X}$-morphisms and the coproduct $\mathrm{in}_j: X_j \to \amalg_{i \in J} X_i$ exists, then the f_j $(j \in J)$ form a final sink if and only if the corresponding morphism $f: \amalg_{i \in J} X_i \to X$ is a final morphism.

8.7.2 Proposition. (i) *The composition $f \circ g$ of initial $\mathscr{X}$-morphisms is initial. Conversely, if the composition $f \circ g$ of two $\mathscr{X}$-morphisms is initial then g is also initial.*

(ii) *The composition $f \circ g$ of final $\mathscr{X}$-morphisms is final. Conversely, if the composition $f \circ g$ of two $\mathscr{X}$-morphisms is final then f is also final.*

8.7.3 Proposition. (i) *If $\mathscr{X}$ has initial sources with respect to $\varphi: \mathscr{X} \to \mathscr{Y}$ then φ has a right adjoint. If in addition $\mathscr{Y}$ is complete, then $\mathscr{X}$ is also complete, and a limit of a diagram $\nabla: \mathscr{I} \to \mathscr{X}$ is obtained by taking an initial source over a limit $g_j: Y_\infty \to \varphi X_j$ $(j \in |\mathscr{I}|)$ of $\varphi \circ \nabla: \mathscr{I} \to \mathscr{Y}$.*

(ii) *If $\mathscr{X}$ has final sinks with respect to $\varphi: \mathscr{X} \to \mathscr{Y}$ then φ has a left adjoint. If in addition $\mathscr{Y}$ is cocomplete, then also $\mathscr{X}$ is cocomplete, and a colimit of a diagram $\nabla: \mathscr{I} \to \mathscr{X}$ is obtained by taking a final sink over a colimit $g_j: \varphi X_j \to Y_\infty$ $(j \in |\mathscr{I}|)$ of $\varphi \circ \nabla: \mathscr{I} \to \mathscr{Y}$.*

Remark. If $\varphi^{-1}(Y)$ is a set for every $\mathscr{Y}$-object Y then $\mathscr{X}$ has initial sources iff it has final sinks. For this as well as for the existence of the adjoints one uses the special case $J = \varnothing$.

8.7.4 Proposition. *Suppose one has functors according to the diagram:*

$$\begin{array}{ccc} \mathscr{X} & \underset{\lambda}{\overset{\rho}{\rightleftarrows}} & \mathscr{Y} \\ \downarrow{\scriptstyle\varphi} & & \downarrow{\scriptstyle\psi} \\ \mathscr{X}' & \underset{\lambda'}{\overset{\rho'}{\rightleftarrows}} & \mathscr{Y}' \end{array}$$

such that the diagram commutes (i.e. $\varphi\lambda = \lambda'\psi$ and $\rho'\varphi = \psi\rho$); that φ and ψ are faithful; that λ is left adjoint to ρ with unit η and co-unit ε; that λ' is left adjoint to ρ' with unit η' and co-unit ε'; and that $\varphi(\varepsilon_X) = \varepsilon'_{\varphi(X)}$ for all $\mathscr{X}$-objects X and $\psi(\eta_Y) = \eta'_{\psi(Y)}$ for all $\mathscr{Y}$-objects Y.

Then ρ transforms initial sources (with respect to φ) in $\mathscr{X}$ into initial sources (with respect to ψ) in $\mathscr{Y}$; and λ transforms final sinks (with respect to ψ) in $\mathscr{Y}$ into final sinks (with respect to φ) in $\mathscr{X}$.

8.7.5 The above proposition is often used in the following special situation: $\mathscr{X}' = \mathscr{Y}'$, $\rho' = \lambda'$ the identity functor, and the unit η' and co-unit ε' are the identity transformations.

8.8 Embeddings and quotient maps

For this section we suppose that a faithful functor $\varphi: \mathscr{X} \to \underline{\text{Set}}$ is given.

8.8.1 Definition. (i) An $\mathscr{X}$-morphism f is called an $\mathscr{X}$ *embedding* (with respect to φ) iff f is initial with respect to φ and φf is injective.

An object X is called an $\mathscr{X}$*-subspace* (with respect to φ) of an object Y iff φX is a subset of φY and there exists a (unique) $\mathscr{X}$-embedding $f: X \to Y$ such that φf is the inclusion.

The dual notions are: a morphism f in $\mathscr{X}$ is called an $\mathscr{X}$*-quotient-map* (with respect to φ) iff f is final with respect to φ and φf is surjective.

An object X is called an $\mathscr{X}$*-quotient-space* (with respect to φ) of an object Y iff φX is the quotient of φY with respect to some equivalence relation and there exists a (unique) $\mathscr{X}$-quotient-map $f: Y \to X$ such that φf is the projection map associated to the equivalence relation.

Remark. Suppose that $\varphi = \varphi_0 \circ \varphi_1$ for some faithful functor $\varphi_1: \mathscr{X} \to \underline{\text{VS}}$, where φ_0 denotes the obvious functor $\varphi_0: \underline{\text{VS}} \to \underline{\text{Set}}$. Then the notions of $\mathscr{X}$-embedding and $\mathscr{X}$-quotient-map remain the same iff one replaces φ by φ_1.

8.8.2 Proposition. *For any morphism f the following statements are equivalent:*

(1) *f is an isomorphism;*
(2) *f is a surjective (i.e. φf is surjective) embedding;*
(3) *f is an injective (i.e. φf is injective) quotient-map.*

Remark. If initial sources exist then the embeddings are exactly the composites $f \circ g$ with g an isomorphism and f an embedding of a subspace.

And if final sinks exist then the quotient-maps are exactly the composites $g \circ f$ with g an isomorphism and f a quotient map onto a quotient space.

Proposition (8.7.2) yields immediately:

8.8.3 Proposition. (*i*) *The composition of embeddings* (*resp. quotient maps*) *is an embedding* (*resp. a quotient map*).

(*ii*) *If the composition* $f \circ g$ *of two morphisms is an embedding* (*resp. a quotient map*) *then* g *is an embedding* (*resp.* f *is a quotient map*).

BIBLIOGRAPHY

Averbukh, V. I., Smolyanov, O. G. (1968) The various definitions of the derivative in linear topological spaces. *Russian Math. Surveys*, 23, 67–113.

Balanzat, M. (1960) La différentielle d'Hadamard-Fréchet dans les espaces vectoriels topologiques. *Comp. Rend. Acad. Sci. Paris*, 251, 2459–61.

Binz, E., Fischer, H. R. (1981) The manifold of embeddings of a closed manifold. *Lecture Notes in Physics*, 139, 310–29.

Boman, J. (1967) Differentiability of a function and of its compositions with functions of one variable. *Math. Scand.*, 20, 249–68.

Bonic, R., Frampton, J. (1966) Smooth functions on Banach manifolds. *J. Math. Mech.*, 15, 877–98.

Bröcker, T., Jänich, K. (1973) *Einführung in die Differentialtopologie.* Springer, Heidelberg.

Colombeau, J. F. (1979) Infinite dimensional C^∞-mappings with a given sequence of derivatives at a given point. *J. Math. Anal. Appl.*, 71, 95–104.

Colombeau, J. F., Meise, R. (1981) C^∞ functions on locally convex and on bornological vector spaces, in Functional Analysis, Holomorphy, and Approximation Theory, *Lect. Notes Math.*, 843, 195–216.

Dieudonné, J. (1960) *Foundations of Modern Analysis.* Academic Press, New York.

Engelking, R. (1968) *Outline of General Topology.* North-Holland, Amsterdam.

Floret, K. (1973) Folgenretraktive Sequenzen lokalkonvexer Räume. *J. Reine Angew. Math.*, 259, 65–85.

Frölicher, A. (1979) Durch Monoide erzeugte kartesisch abgeschlossene Kategorien. *Seminarberichte Fachbereich Math., Fernuniversität Hagen*, 7–48.

Frölicher, A. (1980) Catégories cartésiennement fermées engendrées par des monoides. *Cahiers Top. Géom. Diff.*, 21, 367–75.

Frölicher, A. (1981) Applications lisses entre espaces et variétés de Fréchet. *Comp. Rend. Acad. Sci. Paris*, 293, 125–7.

Frölicher, A. (1982) Smooth structures, in *Category Theory 1981*, *Lect. Notes Math.*, 962, 69–81.

Frölicher, A. (1986) Cartesian closed categories and analysis of smooth maps, in *Categories in Continuum Physics 1982*, *Lect. Notes Math.*, 1174, 43–51.

Frölicher, A., Kriegl, A. (1985) Convergence vector spaces for analysis, in *Convergence Structures 1984*, Akademie Verlag, Berlin, 115–126.

Frölicher, A., Kriegl, A. (1986) Convenient vector spaces for differentiation theory. *Seminarberichte aus dem Fachbereich Math. und Inform. der Fernuniversität Hagen*, 24, 1–35.

Frölicher, A., Gisin, B., Kriegl, A. (1983) General differentiation theory. *Aarhus Universitet, Various Publication Series*, 35, 125–53.

Gähler, W. (1977) *Grundstrukturen der Analysis I*, Birkhäuser. Basel.

Hamilton, R. S. (1982) The inverse function theorem of Nash and Moser. *Bull. Amer. Math. Soc.*, 7, 65–222.

Hirsch, M. (1976) *Differential Topology*. Springer, New York.

Hogbe-Nlend, H. (1977) *Bornologies and Functional analysis*. North-Holland, Amsterdam.

Hogbe-Nlend, H., Moscatelli, V. B. (1981) *Nuclear and Conuclear Spaces*. North-Holland, Amsterdam.

Horváth, J. (1966) *Topological Vector Spaces and Distributions I*. Addison-Wesley, Massachusetts.

Jarchow, H. (1981) *Locally Convex Spaces*. Teubner, Stuttgart.

Joris, H. (1982) Une C^∞-application non-immersive qui possède la propriété universelle des immersions. *Arch. Math.*, 39, 269–77.

Keller, H. H. (1974) Differential calculus in locally convex spaces. *Lect. Notes Math.*, *417*, Heidelberg.

Kock, A. (1984) Calculus of smooth functions between convenient vector spaces, Preprint, Aarhus.

Kriegl, A. (1980) Eine Theorie glatter Mannigfaltigkeiten und Vektorbündel; thesis, Vienna.

Kriegl, A. (1982) Die richtigen Räume für Analysis im Unendlich-Dimensionalen. *Monatshefte Math.*, 94, 109–124.

Kriegl, A. (1983) Eine kartesisch abgeschlossene Kategorie glatter Abbildungen zwischen beliebigen lokalkonvexen Vektorräumen. *Monatshefte Math.*, 95, 287–309.

Kriegl, A. (1984a) A convenient setting of differential calculus in locally convex spaces. *Abh. Akad. Wiss. DDR*, *2N*, 135–40.

Kriegl, A. (1984b) A cartesian closed extension of the category of smooth Banach manifolds, in *Categorical Topology*, Heldermann, Berlin 323–36.

Kriegl, A., Nel, L. D. (1985) A convenient setting for holomorphy. *Cahiers Top. Géom. Diff.*, 26, 273–309.

Kriegl, A., Michor, P., Schachermayer, W. (1985) Characters on algebras of smooth functions, preprint, Vienna.

Lawvere, L. W., Schanuel, S. H., Zame, W. R. (1981) On C^∞-function spaces, preprint.

Mazur, S. (1952) On continuous mappings on cartesian products. *Fund. Math.*, 39, 229–38.

Meise, R. (1980) Nicht-Nuklearität von Räumen beliebig oft differenzierbarer Funktionen. *Archiv. Math.*, 34, 143–8.

Michor, P. (1983) Manifolds of smooth maps IV. *Cahiers Top. Géom. Diff.*, 24, 57–86.

Mitiagin, B. S. (1961) Approximative dimension and basis in nuclear spaces. *Russian Math. Surveys*, 16, 59–127.

Nel, L. D. (1965) Note on completeness in a pseudo-topological linear spaces. *J. Lond. Math. Soc.*, 40, 497–498.

Nel, L. D. (1986) Enriched locally convex structures, differential calculus and Riesz representation; to appear in *J. Pure Appl. Algebra*, preprint.

Rivlin, T. J. (1974) *The Chebyshev Polynomials*. John Wiley & Sons, New York.

Seeley, R. T. (1964) Extension of C^∞-functions defined in a half space. *Proc. Amer. Math. Soc.*, 15, 625–6.

Seip, U. (1979) A convenient setting for differential calculus. *J. Pure Appl. Alg.*, 14, 73–100.

Shigeta, T. (1986) A characterization of cyclical monotonicity by the Gâteaux derivative. *Tokyo J. Math.*, 9, 247–58.

Torunzcyk, H. (1973) Smooth partitions of unity on some non-separable Banach spaces. *Stud. Math.*, 46, 43–51.

Valdivia, M. (1971) On final topologies. *J. Reine Angew. Math*, 251, 193–9.

Wegenkittl, K. (1987) Topologien auf Räumen differenzierbarer Funktionen; Master thesis, Vienna.

LIST OF SYMBOLS

$\varnothing$	empty set	
$\{*\}$	one-pointed set, terminal object	(1.1.6)
$\mathbb{N}$	the natural numbers	(1.2.8)
$\mathbb{N}_0$	$\mathbb{N}\cup\{0\}$	(1.3.1)
$\mathbb{N}_\infty$	$\mathbb{N}\cup\{\infty\}$	(1.3.1)
$\mathbb{N}_{0,\infty}$	$\mathbb{N}\cup\{0, \infty\}$	(1.3.1)
$\mathbb{Q}$	the rational numbers	(1.3.12)
$\mathbb{R}$	the real numbers	(1.2.8)
$\mathbb{U}$	0-neighborhood filter of $\mathbb{R}$	(2.2.13)
$\aleph_0$	cardinality of $\mathbb{N}$	(1.2.4)
Ω	first non-countable ordinal number	(6.3.4)
$\mathscr{C}$	set of structure curves	(1.1.1)
$\mathscr{F}$	set of structure functions	(1.1.1)
$\mathscr{M}$	set of maps generating the $\mathscr{M}$-spaces	(1.1.1)
R	range of the maps of $\mathscr{M}$	(1.1.1)
S	source of the maps of $\mathscr{M}$	(1.1.1)
$\mathscr{S}$	point separating subset of a dual space	(4.1.9)
Γ	associated set of $\mathscr{M}$-structure curves	(1.1.1)
Φ	associated set of $\mathscr{M}$-structure functions	(1.1.1)
ℓ^∞	set of bounded sequences	(1.2.3)
ℓ^1	set of absolutely summable sequences	(1.2.8)
ℓ^p	set of p-summable sequences	(4.1.23)
$\mathscr{L}\!ip^{-1}$-map	map bounded on M-converging sequences	(4.3.6)
$\mathscr{L}\!ip$	set of locally Lipschitzian functions	(1.4.1)
$\mathscr{L}\!ip^k$	set of k-times Lipschitz differentiable functions	(1.4.1)
C^∞	set of smooth functions	(1.4.1)
$\prod$	product	(8.3.2)
$\coprod$	coproduct (direct sum)	(8.3.2)
$\oplus$	direct sum	(3.4.3)
$\leq_\oplus$	direct summand	(5.3.5)
$\tilde{\otimes}$	tensor product in <u>Con</u>	(3.8.4)
$\otimes$	tensor product in <u>Pre</u>	(3.8.1)
$E^{(J)}$	coproduct of J many summands E	(3.4.1)
E^J	product of J many factors E	(3.3.1)
E'	dual space of E	(3.9.1)
$\tilde{E}$	completion of preconvenient vector space E	(2.6.5)
$X \prod_z Y$	pullback	(4.6.1), (8.3.5)

$\Gamma^k(\pi)$	space of $\mathscr{L}ip^k$-sections of a vector bundle π	(4.6.12)
$\ell^\pi X$	absolutely summable functions with bounded support	(5.1.11)
$c_0 X$	functions falling to zero on bounded subsets	(5.1.13)
$\ell^\infty_c X$	functions of $\ell^\infty(X, \mathbb{R})$ with finite support	(5.1.13)
$\ell^1_c X$	functions of $\ell^1 X$ with finite support	(5.1.11)
$\mathscr{L}ip^k_{j\text{-flat}}(U, F)$	space of j-flat $\mathscr{L}ip^k$-maps	(4.4.30)
$\mathrm{Poly}_j(E, F)$	space of polynomial maps of degree at most j	(4.4.27)
$\mathrm{Homog}_j(E, F)$	space of j-homogeneous maps	(4.4.23)
$C^\infty_{co}(U, F)$	space of smooth functions with usual topology	(5.4.15)
δ^k	difference quotient of order k	(1.3.1)
$\bar{\delta}^k$	extension of difference quotient δ^k	(1.3.22), (4.1.12)
δ^k_{eq}	equidistant difference quotient of order k	(1.3.11)
δ^κ	difference quotient of order κ	(1.3.4)
δ^1_B	difference quotient with respect to $\Vert - \Vert_B$	(4.4.36)
ϑ	directional difference quotient	(4.3.6)
δ^k_i	ith partial difference quotient of order k	(1.3.4)
$\mathrm{D}^{\langle k \rangle}$	domain for a difference quotient of order k	(1.3.1)
D^κ	domain for an extended difference quotient of order κ	(1.3.4)
$D^{\langle \kappa \rangle}$	domain for a difference quotient of order κ	(1.3.4)
U_ϑ	domain of a directional difference quotient	(4.3.6)
$c^\cdot$	derivative of curve c	(2.5.1), (4.1.9)
$c^{(k)}$	kth derivative of curve c	(4.1.9)
f'	derivative of f	(4.3.9)
$f^{(k)}$	derivative of order k of f	(4.3.26)
$\mathscr{D}^k$	derivative of order k for curves	(4.2)
∂_i	ith partial derivative	(4.5.3)
∂^κ	partial derivative of order κ	(1.3.27)
d	differential	(4.3.9)
d^k	differential of order k	(4.3.26)
d_i	ith partial differential	(4.5.3)
T	tangent functor	(4.7.2)
$\langle \_ \rangle$	linear hull or (absolutely) convex hull	(2.1.4)
$\Vert \_ \Vert_B$	Minkowski seminorm generated by a set B	(2.1.15)
E_B	normed subspace generated by $B \subseteq E$	(2.1.15)
$I(E)$	subspace of co-idempotent elements in a co-algebra E	(5.2.6)
$GL(E)$	group of <u>Con</u>-isomorphisms of E	(4.8.1)
$\mathrm{Diff}(X)$	group of smooth diffeomorphisms of X	(1.4.8)
$\mathrm{Emb}(X, Y)$	space of smooth embeddings	(4.7.2)
$\mathrm{Submf}(X, Y)$	space of submanifolds of Y isomorphic to X	(4.7.2)
M-adh^α	αth Mackey adherence	(6.3.4)
f^*p	pullback of bundle projection p along f	(4.6.1)
f^*E	total space of pullback of bundle E along f	(4.6.1)
$\langle \_, \_ \rangle$	dual pairing $\ell^1 X \sqcap \ell^\infty(X, \mathbb{R}) \to \mathbb{R}$	(5.1.12)
$m_{X, Y}$	natural bilinear map $\lambda X \sqcap \lambda Y \to \lambda(X \sqcap Y)$	(5.2.3)
$\sigma(E, \mathscr{S})$	weak topology on E induced by $\mathscr{S} \subseteq E'$	(4.1.9)
$\mathscr{F} \underset{X}{\to} x$	convergence of filter $\mathscr{F}$ towards x in X	(2.2.3)
$f\colon E \supseteq U \to F$	map f defined on M-open subset U of E	(4.1.1)

id_X	identity map of a set X	
pr_j	projection map onto the jth factor	(8.3.2)
in_j	injection map from jth summand	(8.3.2)
ev	evaluation map, $\mathrm{ev}(f)(x) := f(x)$	(8.6.2)
ev_x	point evaluation, $\mathrm{ev}_x(f) := f(x)$	
ins	insertion map, $\mathrm{ins}(x)(y) := (x, y)$	(8.6.2)
comp	composition map, $\mathrm{comp}(f, g) := f \circ g$	(4.4.16)
Δ	diagonal map, $\Delta(x) := (x, \ldots, x)$	
ι	natural inclusion map	(3.9.3)
χ_X	characteristic function of a set X	
sign	signature function	
$(\_)^\vee$	$f^\vee(x)(y) := f(x, y)$	(8.6.2)
$(\_)^\wedge$	$f^\wedge(x, y) := f(x)(y)$	(1.1.7), (8.6.2)
$(\_)^*$	$f^*(g) := g \circ f$	(8.2.1)
$(\_)_*$	$f_*(g) := f \circ g$	(8.2.1)
∇	diagram in category	(8.3.1)
$\mathrm{Id}_{\mathscr{X}}$	identity functor of a category $\mathscr{X}$	(8.2.1)
$\lvert\mathscr{X}\rvert$	class of objects of category $\mathscr{X}$	(8.1.1)
$\mathscr{X}^{\mathrm{op}}$	opposite category of $\mathscr{X}$	(8.2.1)
$\succ$	relation on directed set	(2.2.1)
$\square$	end of proof	

LIST OF CATEGORIES

Arc	arc-generated topological spaces	(2.3.8)
ArcVS	arc-generated vector spaces	(2.3.9)
Born	bornological spaces	(1.2.1)
BornVS	bornological vector spaces	(2.1.2)
bLCS	bornological locally convex spaces	(2.1.12)
bLimVS	bornological convergence vector spaces	(2.2.19)
C^∞	smooth spaces	(1.4.1)
CBS	convex bornological spaces	(2.1.2)
Con	convenient vector spaces	(2.6.3)
Con^∞	convenient vector spaces with smooth maps	(5.3.1)
ConAlg	convenient algebras	(5.2.1)
ConCoAlg	convenient co-algebras	(5.2.1)
DVS	dualized vector spaces	(2.1.1)
LCS	locally convex spaces	(2.1.8)
Lim	convergence spaces	(2.2.3)
LimVS	convergence vector spaces	(2.2.10)
$\mathscr{Lip}$	category generated by $\mathscr{Lip}$	(1.4.1)
$\mathscr{Lip}^k$	category generated by $\mathscr{Lip}^k$	(1.4.1)
ℓ^∞	category generated by ℓ^∞	(1.2.3)
$\mathscr{M}$	category generated by set $\mathscr{M}$	(1.1.1)
$\mathscr{M}$-VS	$\mathscr{M}$-vector spaces	(2.3.1)
Pre	preconvenient vector spaces	(2.4.2)
Set	sets	(1.1.3)
sPre	separated preconvenient vector spaces	(2.5.3)
tCBS	topological convex bornological spaces	(2.1.12)

LIST OF FUNCTORS

$\varphi: \mathscr{X} \to \underline{\mathrm{Set}}$	faithful functor	(8.8.1)
$\mathscr{X}: \underline{\mathscr{X}} \times \underline{\mathscr{X}} \to \underline{\mathrm{Set}}$	hom-functor	(8.2.1)
$\delta: \underline{\mathscr{X}} \to \underline{\mathrm{DVS}}$	duality functor	(2.1.6)
$\sigma: \underline{\mathrm{DVS}} \to \mathscr{X}$	weak structure functor	(2.1.6)
$\iota: \underline{\ell^\infty} \to \underline{\mathrm{Born}}$	embedding functor	(1.2.4)
$\eta: \underline{\mathrm{Born}} \to \underline{\ell^\infty}$	left adjoint to ι	(1.2.4)
$\mathscr{M}: \underline{\mathscr{M}}^{\mathrm{op}} \times \underline{\mathscr{M}} \to \underline{\mathscr{M}}$	internal hom-functor	(1.1.7)
$\mathrm{Born}: \underline{\mathrm{Born}}^{\mathrm{op}} \times \underline{\mathrm{Born}} \to \underline{\mathrm{Born}}$	internal hom-functor	(1.2.10)
$\delta: \underline{\mathrm{CBS}} \to \underline{\mathrm{DVS}}$	duality functor	(2.1.7)
$\sigma_b: \underline{\mathrm{DVS}} \to \underline{\mathrm{CBS}}$	bornology of scalarly bounded sets	(2.1.7)
$\delta: \underline{\mathrm{LCS}} \to \underline{\mathrm{DVS}}$	duality functor	(2.1.9)
$\mu: \mathrm{DVS} \to \mathrm{LCS}$	Mackey topology, left adjoint to δ	(2.1.9)
$\sigma_t: \underline{\mathrm{DVS}} \to \underline{\mathrm{LCS}}$	weak topology, right adjoint to δ	(2.1.9)
$\beta: \underline{\mathrm{LCS}} \to \underline{\mathrm{CBS}}$	von Neumann bornology	(2.1.10)
$\gamma: \underline{\mathrm{CBS}} \to \underline{\mathrm{LCS}}$	left adjoint to β	(2.1.10)
$\iota: \underline{\mathrm{Top}} \to \underline{\mathrm{Lim}}$	embedding functor	(2.2.5)
$\tau: \underline{\mathrm{Lim}} \to \underline{\mathrm{Top}}$	left adjoint to ι	(2.2.6)
$\xi: \underline{\mathrm{BornVS}} \to \underline{\mathrm{LimVS}}$	Mackey convergence	(2.2.15)
$\zeta: \underline{\mathrm{LimVS}} \to \underline{\mathrm{BornVS}}$	right adjoint to ξ	(2.2.15)
$\delta: \underline{\mathscr{M}\text{-VS}} \to \underline{\mathrm{DVS}}$	duality functor	(2.3.2)
$\sigma: \underline{\mathrm{DVS}} \to \underline{\mathscr{M}\text{-VS}}$	weak $\mathscr{M}$-structure, right adjoint to δ	(2.3.2)
$\sigma_l: \underline{\mathrm{DVS}} \to \underline{\ell^\infty\text{-VS}}$	weak ℓ^∞-structure, right adjoint to δ	(2.3.3)
$\sigma_k: \underline{\mathrm{DVS}} \to \underline{\mathscr{Lip}^k\text{-VS}}$	weak $\mathscr{Lip}^k$-structure, right adjoint to δ	(2.3.5)
$\tau: \underline{\mathrm{LimVS}} \to \underline{\mathrm{Top}}$	associated topology	(2.3.7)
$\tau_f: \underline{\mathscr{Lip}^k\text{-VS}} \to \underline{\mathrm{Top}}$	final topology induced by $\mathscr{Lip}^k$-curves	(2.3.7)
$\alpha: \underline{\mathrm{Top}} \to \underline{\mathrm{Arc}}$	arc-generated topology	(2.3.8)
$\tau_M: \underline{\mathrm{DVS}} \to \underline{\mathrm{Arc\text{-}VS}}$	Mackey closure topology	(2.3.14)
$\tau_k: \underline{\mathrm{Pre}} \to \underline{\mathrm{Top}}$	compactly generated topology	(6.1.1)
$\tau_s: \underline{\mathrm{Pre}} \to \underline{\mathrm{Top}}$	final topology induced by sequences	(6.1.1)
$\omega: \underline{\mathrm{Pre}} \to \underline{\mathrm{sPre}}$	separation functor	(2.5.6)
$\bar{\omega}: \underline{\mathrm{Pre}} \to \underline{\mathrm{Con}}$	completion functor	(2.6.5)

$L: \underline{\text{Pre}}^{op} \times \underline{\text{Pre}} \to \underline{\text{Pre}}$	space of linear maps	(3.6.2)
$L: \underline{\text{Con}}^{op} \times \underline{\text{Con}} \to \underline{\text{Con}}$	space of linear maps	(3.6.3)
$\otimes: \underline{\text{Pre}} \times \underline{\text{Pre}} \to \underline{\text{Pre}}$	tensor product in $\underline{\text{Pre}}$	(3.8.1)
$\tilde{\otimes}: \underline{\text{Con}} \times \underline{\text{Con}} \to \underline{\text{Con}}$	tensor product in $\underline{\text{Con}}$	(3.8.4)
$(\_)': \underline{\text{Con}}^{op} \to \underline{\text{Con}}$	internal duality functor	(3.9.1)
$(\_)^t: \underline{\text{CBS}} \to \underline{\text{LCS}}$	space of bornological linear functions	(5.4.1)
$(\_)^b: \underline{\text{LCS}} \to \underline{\text{CBS}}$	space of continuous linear functions	(5.4.1)
$(\_)^s: \underline{\text{LCS}} \to \underline{\text{LCS}}$	strong dual	(5.4.1)
$\ell^1: \underline{\ell^\infty} \to \underline{\text{Con}}$	left adjoint to forgetful functor	(5.1.23)
$\ell^1: \underline{\ell^\infty} \to \underline{\text{ConCoAlg}}$	left adjoint to forgetful functor	(5.2.5)
$\mathscr{M}(\_, \mathbb{R}): \underline{\mathscr{M}} \to \underline{\text{ConAlg}}$	algebra of $\mathscr{M}$-functions	(5.2.16)
$(\_)': \underline{\text{ConCoAlg}}^{op} \to \underline{\text{ConAlg}}$	duality functor	(5.2.15)
$\lambda: \underline{\mathscr{M}} \to \underline{\text{ConCoAlg}}$	left adjoint to forgetful functor	(5.1.1)
$\Gamma^k: \underline{\mathscr{Lip}^k\text{-VB}} \to \underline{\text{Con}}$	space of sections for vector bundles	(4.6.18)
$\mathscr{Lip}^k: (\underline{\mathscr{Lip}^k})^{op} \times \underline{\text{Con}} \to \underline{\text{Con}}$	space of $\mathscr{Lip}^k$-maps	(4.4.3)
$\ell^\infty: (\underline{\ell^\infty})^{op} \times \underline{\ell^\infty} \to \underline{\ell^\infty}$	internal hom-functor	(1.2.8)
$\ell^\infty: (\underline{\ell^\infty})^{op} \times \underline{\text{Con}} \to \underline{\text{Con}}$	space of ℓ^∞-maps	(3.6.1)
$C^\infty: (\underline{C^\infty})^{op} \times \underline{C^\infty} \to \underline{C^\infty}$	internal hom-functor	(1.4.3)
$C^\infty: (\underline{C^\infty})^{op} \times \underline{\text{Con}} \to \underline{\text{Con}}$	space of smooth maps	(4.4.3)

INDEX